C. Clément E. Pacini J.-C. Audran (Eds.)

Anther and Pollen

Springer
Berlin
Heidelberg
New York
Barcelona
Hong Kong
London
Milan
Paris
Singapore
Tokyo

Christophe Clément
Ettore Pacini
Jean-Claude Audran (Eds.)

Anther and Pollen

From Biology to Biotechnology

With 92 Figures

Springer

Dr. CHRISTOPHE CLÉMENT
Professor Dr. JEAN-CLAUDE AUDRAN
Université de Reims Champagne Ardenne
Biologie et Physiologie Végétales
B.P. 1039
F-51687 Reims Cedex 2, France

Professor Dr. ETTORE PACINI
Università di Siena
Dipartimento di Biologia Ambientale
Via P.A. Mattioli 4
53100 Siena, Italia

Library of Congress Cataloging-in-Publication Data

Anther and pollen: from biology to biotechnology / Christophe Clément, Ettore Pacini, Jean-Claude Audran (eds.).
p. cm. Includes bibliographical references and index.

ISBN-13: 978-3-642-64209-8 e-ISBN-13: 978-3-642-59985-9
DOI: 10.1007/978-3-642-59985-9

1. Pollen - Biotechnology. 2. Anther - Biotechnology. 3. Pollen. 4. Anther.
I. Clément, Chrsitophe, 1964- . II. Pacini, E. (Ettore), 1942- . III. Audran, Jean-Claude, 1939-
SB106.B56A78 1999 575.6-dc21 99-17766 CIP

Softcover reprint of the hardcover 1st edition 1999

Typesetting: Camera-ready by the editors
Cover-design: *design & production* GmbH, Heidelberg

SPIN 10679770 31/3137 - 5 4 3 2 1 0 - Printed on acid-free paper

Preface

Professor
M Favre-Duchartre

A passionate researcher, a fascinating teacher...

Professeur Michel Favre-Duchartre left us on April 5, 1997.

He was an exacting and rigorous master in his job at the University, an understanding and cordial person with everyone. He had a huge knowledge in Botany and an innovative approach of the Plant world so that he could simply keep up a rare magnetism with his many speakers.

I am proud and honoured to have been one of his students; I hope to still be, as he used to be permanently, according to his favorite phrase "an adult with newborn eyes, unceasingly discovering and rediscovering the surrounding world".

Professor Jean-Claude Audran

Contents

1 Types of pollen dispersal units and pollen competition

Pacini*[1] E and Franchi[2] GG

1.1 Introduction

The reproductive structures of algae are dispersed actively by means of flagella, or passively, by means of water currents. Likewise in the most primitive terrestrial plants, dispersal may be active or passive, but as well as water currents, air currents are also exploited (Faegri and van der Pijl 1979). Originally in gymnosperms, but now mainly in angiosperms, animals attracted to the reproductive structures by seduction or deception may be agents of dispersal (Dafni 1984, Pacini 1992).

The spores of land plants and pollen grains of gymnosperms, with a few exceptions, are always dispersed singly; not until the angiosperms did plants evolve "pollen dispersal units" (PDU) consisting of a number of pollen grains held together by various means. Such units are found in at least 49 families (Walker and Doyle 1975; Pacini and Franchi 1998). One of the reasons why these dispersal units are only found in angiosperms is the fact that this group has ovaries that may contain from one to several hundred thousand ovules (Pacini and Franchi 1998). Also Wilson (1979) considered the evolution of PDUs.

Competition for limited resources is a biological fact and exists at different levels, for example between species, within species and within individuals. Irrespective of the type of PDU, competition may occur throughout development of the organism but as far as the male gametophyte is concerned, mainly in the progamic phase, namely, between pollination and fertilization (Ottaviano and Mulcahy 1989).

In this paper various types of PDUs and types/modes of competition occurring during pollen development and germination are considered.

---

*[1] for correspondence: Department of Environmental Biology; Botany Section; Siena University; 4, via PA Mattioli; 53100 Siena; Italy

[2] Institute of Pharmacology; Siena University; 6, via delle Scotte 53100 Siena; Italy

1.2
Types of pollen dispersal units

Some types of PDUs have a constant number of pollen grains. These include monads and pollen held together by common walls in tetrads, polyads and compact pollinia. These land on the stigma as they were when they left the anther. Other types, in which the pollen is held together by pollenkitt, tryphine, viscin threads or elastoviscin or as mealy pollinia, may lose or gain pollen grains between the anther and the stigma. They lose pollen in anemophilous species in which the pollen forms clumps with the aid of pollenkitt (Lisci et al. 1996), because the PDUs may split up. They gain or lose pollen grains when transported by insects, in which case pollen is lost and acquired every time a hermaphrodite flower is visited. For example it may occur during successive visits to flowers of the same species when not all of the pollen is unloaded each time. If the flower is unisexual, loading and unloading of pollen mainly occur when the insect visits a flower of opposite sex to the previous one (Nepi and Pacini 1993). When two or more male flowers are sequentially visited, loading and loss of pollen occur, and not only total pollen load varies, but some grains coming from different male flowers may join the same dispersal unit. In these cases, pollen from different plants may unite in the same PDU, favouring genetic assortment and competition (Levin and Waller 1989). In anemophiles, this does not happen; at most, during wind transport, a PDU, held together by viscous fluids, may conceivably break up, with a decrease in pollen number (Lisci et al. 1996); but this does not mean that wind-dispersed pollen of different origins cannot arrive on the same stigma and that pollen can be subject to male-male competition.

Pollen grains may be dispersed singly or grouped in various ways: 1/ by becoming tangled in the case of extremely long pollen which untangles when dispersed by water; 2/ by means of viscous substances such as pollenkitt, tryphine and elastoviscin which also favour adhesion to pollinator bodies; 3/ by means of sporopollenin filaments or viscin threads which are extensions of the exine; 4/ by means of common walls of exine and/or intine; 5/ by means of filaments or other devices derived from the anther walls (Pacini and Franchi 1998). Table 1 shows the various ways in which pollen is held together, with examples.

1.3
Competition during development of the male gametophyte

Table 2 shows the ways in which competition is reduced during pollen development. Monad pollen is always dispersed by air currents and normally these plants have only one ovule per ovary (Table 3). The kiwi is an exception because the female flower has many hundreds of ovules. The pollen tubes emitted by compound pollen normally all penetrate the stigma and adhere to it, but there is at least one exception to this apparently logical rule. In *Typha latifolia* the pollen tubes emitted by the tetrad do not all penetrate the stigma to which the tetrad adheres, but other nearby stigmas (Nicholls and Cook 1986). This is because the inflorescence is monoecious with male flowers separate from female flowers; the latter are very small and crowded together.

Table 1. Modes of pollen adhesion to form different types of pollen dispersal units and examples with number of ovules/ovary

1 Single grains (monad), 1 ovule/ovary (Betulaceae, Urticaceae, Poaceae)
2 Extremely elongated grains forming a tangle: 1 ovule/ovary (Zosteraceae)
3 Monad pollen grouped by viscous fluids derived from tapetum
- **3.1** pollenkitt, highly variable ovule/ovary ratio: from 1 (cherry) to hundreds (Cistaceae)
- **3.2** tryphine, 1-20 ovules/ovary (Cruciferae)
- **3.3** elastoviscin, several hundred thousands ovules/ovary (some Orchidaceae)
- **3.4** viscin threads, several hundreds ovules/ovary (*Oenothera*, *Fuchsia*)

4. Monads grouped by devices derived from other anther parts
- **4.1** mucilage derived from connective tissue, e.g. *Tylosema esculentum* (Caesalpiniaceae): about 10 ovules/ovary (Frey et al. 1992)
- **4.2** threads, e.g. *Gymnocalicium* sp. (Cactaceae): a few hundred ovules/ovary (Halbritter et al. 1997)

5. Tetrad pollen
- **5.1** tetrads of pollen, 1 ovule/ovary (Typhaceae)
- **5.2** tetrads grouped by
  - **5.2. 1** pollenkitt, 50 - 100 ovules/ovary (some Ericaceae)
  - **5.2.2** viscin threads, 50 - 100 ovules/ovary (some Ericaceae)
  - **5.2. 3** common walls
    - **5.2.3.1** 2-8 tetrads, each grain has intine and exine, less than 50 ovules/ovary (Mimosaceae)
    - **5.2.3.2** many tetrads grouped to form a pollinium
      - **5.2.3.2.1** pollinium composed of massulae
        - **5.2.3.2.1.1** inner grains of massula with exine and intine; several hundred (some Asclepiadaceae) to several hundred thousand (some Orchidaceae) ovules/ovary
        - **5.2.3.2.1.2** inner grains of massula with only intine: several hundred thousand ovules/ovary (some Orchidaceae)
      - **5.2.3.2.2** compact pollinium: one thousand to several hundred thousand ovules/ovary (some Orchidaceae)

In the reproductive structures such as pollen and ovules, there is competition for nutrients and to reach the partner. Competition occurs between male gametophytes of the same and different individuals, all deposited on the same stigma. Female gametophytes instead compete mainly for the male gametes.

Competition has a genetic basis and is influenced by chance occurrences; a pollen grain that falls on a site of the stigma far from the ovary is at a disadvantage with respect to one that falls closer to the ovary (Ottaviano and Mulcahy 1989). Hence stigma form may increase competition (Heslop-Harrison and Shivanna 1977). The number of pollen grains in the PDU, and which therefore reach the stigma together, may trigger male and female competition (Table 3).

1.3.1 Pollen development

Competition between pollen grains during development may depend on genetic and structural factors. There are essentially three genetic factors: 1/ asynchrony of the second meiotic division in the meiocyte; 2/ asynchrony of the first haploid mitosis; 3/ segregation of certain genes (Pacini 1994). The cells of the tapetum develop synchronously. The first meitoic division is synchronous because there are cytomictic channels. Dehydration of the anther and/or pollen are also synchronous phenomena. However, in the interval between these two stages, the pollen develops asynchronously (Pacini 1994), as explained above: the more asynchronous the development, the greater the competition. When dehydration begins, some pollen grains may not have completed development and are at a disadvantage; they may even die when the anther opens.

Structural factors, on the other hand, may depend on anther structure and the type of PDU. If the loculus contains many pollen grains but nutrition is not uniform, competition will be high and the pollen in the centre of the loculus will be penalized and may die. This is why ways of ensuring nutrition to all pollen grains have evolved, especially in the case of compound pollen (Table 2). The ways may be different according to whether the pollen is in monads or compound form.

1.3.2. Type of pollen dispersal unit, pollination and paternity

The type of PDU is related to the type of pollination. Anemophilous pollination occurs with monad and isolated or pollenkitt-united tetrad pollen. Hydrophilous pollination occurs with single monads and extremely long tangled pollen. Zoophilous pollination occurs with all types of dispersal units containing more than one pollen grain (Pacini and Franchi 1998). When animals visit flowers, the number of pollen grains forming units held together by pollenkitt, tryphine, viscin or elastoviscin may vary: during transport some pollen grains are deposited, others adhere and others fall (Nepi and Pacini 1993). It follows that the greater the number of visits, the more types of genomes in the pollen, especially in the case of herbaceous plants.

Some PDUs contain enough pollen to fertilize all the ovules in the ovary. This is possible in the Mimosaceae (Kenrich and Knox 1989, Muona et al. 1991) and

Asclepiadaceae (Gold and Shore 1995). In the Orchidaceae, there may be several million ovules per ovary and the pollinia contain the same order of magnitude of pollen grains; however it has been observed experimentally that an increase in the number of pollinia deposited on the stigma does not always lead to an increase in the weight of the fruits and hence in the number of seeds produced (Proctor and Harder 1994). In *Raphanus sativus*, which rarely produces more than ten mature seeds and the pollen of which is dispersed by bees in clumps held together by tryphine, 85% of the fruits have multiple paternity (Ellstrand 1984). Similar values are reported also for *Ipomopsis aggregata* (Polemoniaceae), having 2-14 seeds per fruit (Campbell 1988).

Table 2. Different ways to reduce competition for nutrition between pollen grains during pollen development

1 The loculus contains many spaced grains and locular fruid; tapetal cells of the parietal type lose their walls and have cycles of hyperactivity responsible for pollen and locular fluid movements; monad or tetrad pollen e.g. *Nymphaea colorata*, *Olea europaea*				
2 Pollen grains do not move inside the loculus; locular fluid present. Tapetal cells, of the parietal type, usually lose their walls and may have cycles of hyperactivity. Monad or tetrad pollen.	2.1 Locular fluid flows between grains which are locked together. e.g. gymnosperms with saccate grains, like *Pinus sylvestris*.	2.2 The loculus cross section contains very few (2-3) rows of closely packed pollen e.g. Cucurbitaceae and Ericaceae	2.3 The anther is divided into several subloculi (septate anther), each containing a few pollen grains (10-20) e.g. Impatiens sultani	2.4 The loculus cross section contains a few pollen grais (6-12), all in contact with the tapetum through their pore; anthers are very long. Locular fluid movement is not necessary. No cycles of hyperactivity in tapetal cells e.g. Graminaceae
3 Pollen does not move inside the loculus, locular fluid extremely reduced or absent, compound pollen. Tapetal cells keep their walls until degeneration. e.g. Orchidaceae.	3.1 Anthers are septate, compoud pollen as poliads, with 16-32 pollen grains, each in contact with the tapetum. e.g. *Acacia*.	3.2 Compound pollen in soft or mealy pollinia consisting of many (15-30) massulae. Each massula contains a few hundred of pollen grains; only the external ones are in contact with the tapetum; the inner ones are often devoid of exine and cytomictic channels may persist up to just after the first haploid mitosis e.g. massulate orchids.	3.3 Compound pollen in a compact pollinium (thousands or hundreds of thousands of pollen grains). Only external pollen is in contact with the tapetum, inner pollen is devoid of exine; cytomictic channels persist at least up to just after the first haploid mitosis e.g. *Cleisostoma racemifera*.	
4 Pollen grain is completely enveloped by tapetum of the periplasmodial type (also known as amoeboid). Anthers contain few pollen grains per loculus e.g. Berberidaceae, Compositae, Araceae, Commelinaceae.				

1.3.3
Pollen grain dehydration and competition during germination

The pollen of gymnosperms is always dispersed in partially dehydrated form, that is, with a water content of less than 30%. In angiosperms, the pollen of many

Table 3. Relationship between pollen dispersal unit, number of ovules/ovary, pollen/ovuley ratio, paternity and competition. Paternity can be single in the case of a single ovule/ovary or many thousands of ovules/ovary as in orchids with compact pollinia

types of pollen dispersal unit	nb of ovules/ ovary (1)	pollen/ovule ratio	male competition	female competition	paternity	examples
monad pollen	1	high	strong (2)	absent	single	Betulaceae, Urticaceae, Poaceae
	few - several hundred	low	reduced	strong	multiple	tomato, kiwi
a tangle of elongated monad pollen	1	high	not as strong as above	absent	single	Zoosteraceae
monad pollen grouped by pollenkitt or tryphine	1	high	strong	absent	single	cherry
	tens to several hundreds	moderate	reduced	reduced	multiple	Cistaceae, Myrtaceae, *Passiflora*
monad pollen grouped by viscin threads	about ten	moderate	very low	moderate	multiple	some Caesalpinaceae
	tens	moderate	very low	moderate -high	multiple	*Oenothera*
monad pollen grouped by elastoviscin	several hundred	very low	very low	moderate - high	multiple	slippery orchids
tetrad pollen	1	low	moderate	absent	single	Typhaceae
	3-several hundred	moderate	high	moderate	multiple	Juncaceae
tetrads grouped by pollenkitt	several tens	moderate	low	moderate	multiple	some Ericaceae
	1-9	moderate	low	moderate	single or multiple	some Araceae
tetrads grouped by viscin threads	several tens	moderate	low	moderate	multiple	some Ericaceae
polyads	few tens	very low	very low	very low	often single	Mimosaceae
groups of tetrads united by different devices	thousands	very low	very low	reduced	multiple	some Orchidaceae (*Polystachia pubescens* and members of tribes of Orchideae and Neottieae)
massulate	thousands	very low	very low	very low	usually multiple	massulate orchids
compact pollinium	hundreds of thousands	very low	very low	very low	single	some Orchidaceae and Asclepiadaceae

(1) data from Chadefaud and Emberger 1960; (2) depends on stigma morphology

species is dehydrated as in gymnosperms, but some is only slightly dehydrated, that is, it has a water content greater than 30% (our unpublished data). The fact that the latter is dispersed in an active metabolic state (Heslop-Harrison et al. 1997) has advantages and disadvantages. The main advantage is that it germinates in minutes on the stigma; the main disadvantage is vulnerability to water loss, in the absence of some form of protection, so that transport needs to be quick to maintain viability. It follows that because partially dehydrated pollen resists environmental conditions well, it reaches the stigma with practically the same water content as at the outset. The condition of slightly dehydrated pollen varies to a greater extent. Grains which lose much water can die, or they can take longer to rehydrate and emit the pollen tube. The percentage of water at dispersal and the ease with which it is maintained enhances competition during dispersal.

1.3.4 Competition during pollination

During pollination, especially the entomophilous type, competition between co-occurring plants for pollinators can be intense. Species growing in the same environment may compete for pollinators and so do plants of the same species (Waser 1983). The latter type of competition occurs, for instance, when the sugar concentrations of the nectar differ between plants of the same species growing in the same environment (see Freeman and Head 1990; Wyatt et al. 1992). The quantity of nectar produced by a flower is an index of the number of insect visits (Nepi and Pacini 1993); the more a flower is visited, the more likely it is for pollen of different genotypes to fall on the stigma, enhancing competition. In many Geraniaceae, Labiatae and Compositae, the volume of nectar produced is so small (Fahn 1979) that a single insect visit is presumably enough to bring sufficient pollen to fertilize the ovules. In this case pollen competition is very reduced, especially if the pollen load origins from a single individual.

1.3.5. Competition during the progamic phase

This type of competition depends on chance and/or the structure of certain floral parts. Pollen with pollenkitt dispersed singly is advantageous because it sticks to the stigma, hydrates more quickly and immediately emits the pollen tube (Nepi and Pacini 1993). The grains in contact with the stigma surface of pollen that is dispersed in clumps hydrates sooner and germinates sooner than the rest of the grains. The position of the pollen pore/s with respect to the stigma is also important; pollen adhering with a pore in contact germinates more quickly than pollen adhering in an interporal region (Heslop-Harrison et al. 1975). Stigma structure, too, is important. If the stigma has a small flat landing surface, all pollen adhering at the same time will have the same advantages. If the stigma is long, like the silk of maize, the site where the pollen lands determines the degree of competition. Pollen landing near the ovary is favoured (Ottaviano and Mulcahy 1989). Competition may also depend on the number of ovules per ovary. Male competition is at a maximum with only one ovule per ovary; when there are many ovules, male competition decreases and female competition increases (Table 3). The duration of female, or rather stigma, receptivity is important. The percentage germination of pollen on the stigma also varies in relation to pistil age. In *Cucurbita pepo*, 53% of pollen germinates the day before anthesis, 71% at anthesis, 27% the day after and 16% two days after (Nepi and Pacini unpublished data). The longer the pollen awaits the vector in the anther, and the longer the duration of dispersal, the greater the water loss, and if the pollen is still viable, the longer it takes to hydrate and germinate (Pacini et al. 1997).

The number of pollen grains falling on the stigma is another factor influencing competition (Hormanza and Herrero 1996). The percentage of germinated pollen grains penetrating the stigma and traveling down the style has been found to decrease with pollen load size (Lisci and Pacini 1994, Niesenbaum and Schueller 1997). It is necessary to remind also the influence of a mass effect, the "pollen population effect" (Brewbaker and Majamder 1961, Stanley and Linskens 1974).

Competition during pollen tube growth is not always constant at different pistil heights; it is determined by morphological and anatomical constraints that stress competition at certain points and decrease it at others (Herrero and Hormanza 1996). These points are mainly the base of the style, the upper part of the ovary, and the funiculus (Herrero and Arbeloa 1989). For example in maize and *Cucurbita pepo*, the transmitting tissue decreases in thickness so that there is less space for the pollen tubes to grow (Heslop-Harrison et al. 1985; Nepi and Pacini unpublished data). The term pollen attrition indicates this decrease in tube number during their growth (Smith-Huberta 1997); pollen attrition is obviously proportional to the number of pollen grains that fall on the stigma (Smith-Huberta 1997; Winsor et al. 1987).

1.4 Conclusions

The factors that contribute to competition are listed above. Now let us consider how they combine with the number of ovules per ovary and paternity. First of all, the number of ovules per ovary increases with the number of pollen grains per dispersal unit (Cruden 1977; Cruden and Jensen 1979; Pacini and Franchi 1996). When pollen is in monads, many grains from different individuals can fall on the stigma and compete; when pollen is in compact pollinia, this may fertilize all the ovules in the ovary and all the seeds will have the same mother and father. In this case, male and female competition are reduced and postponed to when the seeds germinate. Table 3 shows some case histories in which type of PDU, number of ovules per ovary, male and female competition and paternity are considered.

The type of PDU affects the timing of competition. In some cases this begins in the anther, before anthesis, as in the case of massulate orchids. In other cases it begins only after the pollen lands on the stigma, e.g. when the PDU consists of monads clumped by viscous fluids. Inside the anther, the competition is for nutrients. Later it is to reach the stigma, ovary and ovules.

Acknowledgements

This research was supported by a grant from MURST - Rome (Italy).

The authors are grateful to Ruth Neiland for stimulating discussion and linguistic improvement of the manuscript.

References

Brewbaker JL, Majamder SK (1961) Cultural studies on the pollen population effect and the self-incompatibility inhibition. Am J Bot 48: 457-464

Campbell DR (1988) Multiple paternity in fruits of *Ipomopsis aggregata* (Polemoniaceae). Am J Bot 85: 1022-1027

Chadefaud M, Emberger L (1960) Traité de Botanique (Systematique). Masson, Paris

Cruden RW (1977) Pollen-ovule ratios: a conservative indication of breeding systems in flowering plants. Evolution 31: 32-46

Cruden RW, Jensen KG (1979) Viscin threads, pollination efficiency and low pollen-ovule ratio. Am J Bot 66: 875-879

Dafni A (1984) Mimicry and deception in pollination. Ann Rev Ecol Syst 15: 259-278

De Frey HM, Coetzer LA, Robbertse PJ (1992) A unique anther-mucilage in the pollination biology of *Tylosema esculentum*. Sex Plant Reprod 5: 303

Ellstrand NC (1984) Multiple paternity within fruits of the wild radish, *Raphanus sativus*. Am Natural 123: 819-828

Faegri K, Pijl L van der (1979) The principles of pollination ecology. Pergamon Press, Oxford

Fahn A (1979) Secretory tissues in plants. Academic Press, London.

Freeman CE, Head KC (1990) Temperature and sucrose composition of floral nectars in *Ipomopsis longiflora* under field conditions. Southwestern Natural 35: 423-426

Gold JJ, Shore JS (1995) Multiple paternity in *Asclepias syriaca* using paired-fruit analysis. Can J Bot 73: 1212-1216

Halbritter H, Hesse M, Buchner R (1997) Pollen-connnecting threads in *Gymnocalycium* (Cactaceae): their origin, function, and systematic relevance. Grana 36: 1-10

Herrero M, Arbeloa A (1989) Influence of the pistil on pollen tube kinetics in peach (*Prunus persica*). Am J Bot 76: 1441-1447

Herrero M, Hormanza JI (1996) Pistil strategies controlling pollen tube growth. Sex Plant Reprod 9: 343-347

Heslop-Harrison J, Knox RB, Heslop-Harrison Y, Mattsson O (1975) Pollen-wall proteins: emission and role in incompatibility responses. In : Ducker JC, Racey PA (eds) Biology of the male gamete. Academic Press, London, pp 188-202

Heslop-Harrison J, Heslop-Harrison Y, Heslop-Harrison JS (1997) Motility in ungerminated grass pollen: association of myosin with polysaccharide-containing wall-precursor bodies (P-particles). Sex Plant Reprod 10: 65-66

Heslop-Harrison Y, Shivanna KR (1977) The receptive surface of the Angiosperm stigma. Ann Bot 41: 1233-1258

Heslop-Harrison Y, Heslop-Harrison J, Reger BJ (1985) The pollen-stigma interaction in the grasses. 7. Pollen-tube guidance and the regulation of the tube number in *Zea mays* L.. Acta Bot Neerl 34: 193-211

Hormaza JI, Herrero M (1996) Dynamics of pollen tube growth under different competition regimes. Sex Plant Reprod 9: 153-160

Kenrick J, Knox RB (1989) Pollen-pistil interactions in Leguminosae (Mimosoideae). In: Stirton CH, Zarucchi JL (eds) Advances in legume biology. Monogr Syst Bot Missouri Bot Gard 29: 127-156

Levin MD, Waller GD (1989) The role of pollinating insects in future world food production. Apiacta 24: 18-21

Lisci M, Tanda C, Pacini E (1994) Pollen ecophysiology of *Mercurialis annua* L. (Euphorbiaceae), an anemophilous species flowering all year round. Ann Bot 74: 125-135

Lisci M, Cardinali G, Pacini E (1996) Pollen dispersal and role of pollenkitt in *Mercurialis annua* L. (Euphorbiaceae). Flora 191: 385-391

Muona O, Moran GF, Bell JC (1991) Hierarchical patterns of correlated mating in *Acacia melanoxilon*. Genetics 127: 619-626

Nepi M, Pacini E (1993) Pollination, pollen viability and pistil receptivity in *Cucurbita pepo*. Ann Bot 72: 527-536

Nicholls MS, Cook DK (1986) The function of pollen tetrads in *Typha* (Typhaceae). Veröff Geobot Inst ETH 87: 112-119

Niesenbaum RA, Schueller SK (1997) Effects of pollen competitive environment on pollen performance in *Mirabilis jalapa* (Nyctaginaceae). Sex Plant Reprod 10: 101-106

Ottaviano E, Mulcahy DL (1989) Genetics of angiosperm pollen. Adv Gen 26: 1-64

Pacini E (1992) Seduction and deception in pollen and seed dispersal. Giorn Bot Ital 126: 161-168

Pacini E (1994) Cell biology of anthers and pollen development. In: Williams EG, Clarke AE, Knox RB (eds) Gamete control of self-incompatibility and reproductive development in flowering plants. Kluwer Academic Publishers, Dordrecht, pp. 289-308

Pacini E, Franchi GG (1996) Some cytological, ecological and evolutionary aspects of pollination. Acta Soc Bot Pol 65: 11-16

Pacini E, Franchi GG (1998) Pollen dispersal units, gynoecium and pollination. In: SJ Owens, PJ Rudall (eds) Reproductive Biology. Royal Botanic Gardens, Kew, in press

Pacini E, Franchi GG, Lisci M, Nepi M (1997) Pollen viability related to type of pollination in six angiosperm species. Ann Bot 80: 83-87

Pandolfi T, Pacini E (1995) The pollinium of *Loroglossum hircinum* (Orchidaceae) between pollination and pollen tube emission. Pl Syst Evol 196: 141-151

Pandolfi T, Pacini E, Calder DM (1993) Ontogenesis of monad pollen in *Pterostylis plumosa* (Orchidaceae, Neottioideae). Pl Syst Evol 186: 175-185

Proctor HC, Harder LD (1994) Pollen load, capsule weight, and seed production in three orchid species. Can J Bot 72: 249-255

Smith-Huberta NL (1997) Pollen tube attrition in *Clarkia tembloriensis* (Onagraceae). Int J Plant Sci 158: 519-524

Stanley RG, Linskens HF (1974) Pollen: Biology Biochemistry Management. Springer, Berlin

Walker JW, Doyle JA (1975) The bases of angiosperm phylogeny: Palynology. Ann Miss Bot Gard 62: 664-723

Waser NM (1983) Competition for pollination and floral character differences among sympatric plant species: a review of evidence. In: Jones CE, Little RJ (eds) Handbook of Experimental Pollination Biology. Van Nostrand Theinhold, New York, pp. 277-293

Wilson M (1979) Sexual selection in plants. Amer Natural 113: 777-790

Wyatt R, Broyles SB (1994) Ecology and evolution of reproduction in milkweeds. Ann Rev Ecol Syst 25: 423-441

Wyatt R, Broyles SB, Derda GS (1992) Environmental influences on nectar production in milkweeds (*Asclepias syriaca* and *A. exaltata*). Am J Bot 79: 636-642

Winsor JA, Davis LE, Stephenson AG (1987) The relation between pollen load and fruit maturation and the effect of pollen load on offspring vigor in *Cucurbita pepo*. Amer Natural 129: 643-656

2 What may be the significance of polysiphony in *Lavatera arborea* ?

Nepi* M and Pacini E

2.1 Introduction

In lower gymnosperms (e.g. *Cycas*, *Zamia* and *Ginkgo*) pollen nuclei do not move inside pollen tubes and these and are not related to fertilisation but play a haustorial role (Johri 1992). Like the most evolved gymnosperms, angiosperms are siphonogamous. This means that the male gametes are conveyed by the pollen tube to the embryo sac for double fertilisation (Heslop-Harrison 1978). Beside this principal function, the pollen tubes of certain angiosperms also have a haustorial function during the development of the ovule after fertilisation (Johri 1992). Almost all pollen grains are monosiphonous, i.e. each pollen grain produces only one pollen tube. Other pollen, such as that of some species of the families Malvaceae, Cucurbitaceae and Campanulaceae are polysiphonous (Steffen 1963), each grain emitting more than one pollen tube.

The phenomenon of polysiphony has been known for many years and is mentioned but not explained in many plant embryology text books (Maheshwari 1950, Davis 1966). The aim of the present study was to determine the function of polysiphony in *Lavatera arborea* L.

2.2 Material and methods

2.2.1 Light microscopy

Mature anthers of *Lavatera arborea* were collected, fixed in 5% glutaraldehyde in phosphate buffer at pH 6.9, dehydrated in an ethanol series and embedded in Technovit 7100 (Heraeus Kulzer GmbH). Semi-thin sections were stained with: 1/

* for correspondence: Department of Environmental Biology; Botany Section; University of Siena; Via P.A. Mattioli 4; 53100 Siena; Italy

PAS (periodic acid Shiff reaction) for total insoluble polysaccharides (O'Brien and McCully 1981); 2/ Auramine O (Heslop-Harrison 1977); 3/ Aniline blue for callose (O'Brien and McCully 1981); 4/ Toluidine blue (O'Brien and McCully 1981) as general stain; 5/ FITC (fluorescein isothiocyanate) for total proteins (Pearse 1968).

2.2.2
Stigma contact and pollen germination

In order to follow tube growth in vivo, branches of hand pollinated stigma were collected at different times after pollination, softened in 8N NaOH for 1 h at 60°C, stained with aniline blue and examined by epifluorescence microscopy (Martin 1959).

In order to verify the influence of stigma contact on pollen germination the following experiment was carried out. A single pollen grain was collected from recently opened anther with an eyelash and placed on a slide. Then two branches of unpollinated stigma were brought simultaneously into contact with different sides of the grain and kept thus for some minutes to allow germination. The slide was than stained with aniline blue to detect pollen tube emission.

In vitro germination tests were effected up to 90 min after sawing in Brewbaker-Kwack medium (Kearns and Inouye 1993) modified in calcium nitrate and sucrose concentrations (0.6 g/l and 20% respectively). Experiments were performed in triplicate in Petri-dishes (35 mm in diameter), at room temperature (18-20°C) and under the same indirect light.

2.2.3
Adhesion of pollen

In order to determine the degree of pollen adhesion, stigma branches were saturated with pollen (previously tested for viability by FCR: fluorochromatic reaction) and removed at different times (1, 3, 6, 10, 15, 20, 30 and 40 min after pollination). The number of grains deposited was determined in 20 stigma branches. The number of germinated pollen grains on 3 stigma branches at each time was counted, together with the number remaining after washing in: 1/ water (10-15 sec at room temperature); 2/1N NaOH (60°C for 10 min) according to Ferrari et al. (1985). Washing was performed in a 20 ml baker by pippetting the washing medium on the stigma branches.

2.3 Results

2.3.1 Flower morphology and flower anthesis

The corolla is choripetalous with five petals and radial symmetry. The stamens form a column in the centre of the flower. The column bears 60-80 anthers each containing about a hundred pollen grains. The style is situated inside the central

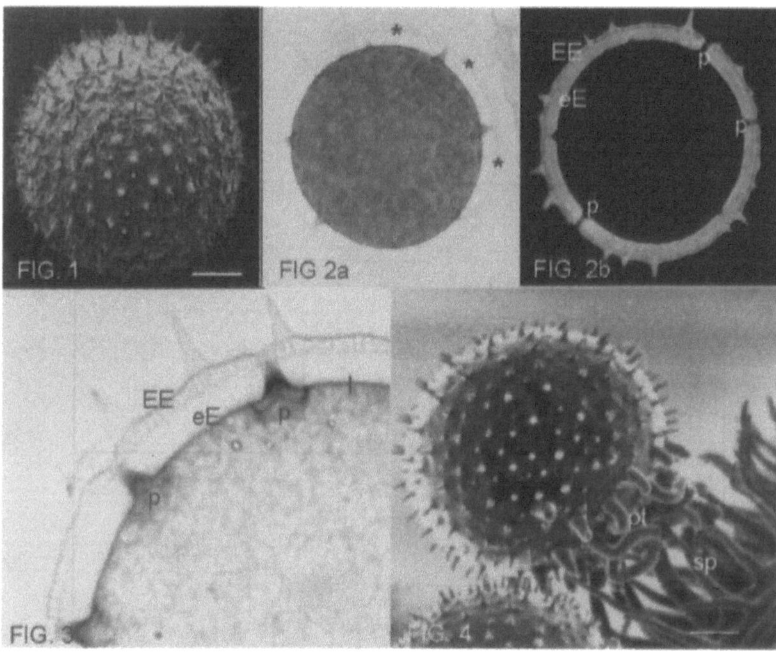

Fig. 1. Scanning electron micrograph of *Lavatera arborea* pollen grain. The pollen is 130 µm in diameter with echinate exine. It is coated with pollenkitt that forms folds on the pollen surface. Bar=20µm. **Fig. 2.** Cross section of mature pollen of *Lavatera arborea* stained with PAS and Auramine O and observed: in a bright field (a) and with UV light (b). a. The cytoplasm of the vegetative cell is PAS-positive and contains many amyloplasts. A weakly positive reaction is also evident in the lumina of ectexine (asterisks). b. The exine has thick endexine (eE) and ectexine (EE) with tectum, bacula and lumina. The pores (p) do not have opercula. X 180. **Fig. 3.** Cross section of mature pollen grains of *Lavatera arborea* stained with toluidine blue O. The intine (I) is thicker in the pore regions (p) where it consists of three layers; eE= endexine; EE=ectexine. X 600. **Fig. 4.** A pollen grain on the stigma of *Lavatera arborea* about 10 min after deposition. Note that the longest pollen tubes arises from the centre of pollen tubes emission region. The peripheral pollen tubes cease to grow at this stage; sp=stigmatic papillae; pt=pollen tubes. Bar=20µm

column and carries 6-8 stigma branches. The stigma is of the dry type, with unicellular papillae (Heslop-Harrison and Shivanna 1977).

The flowers open in the first hours of the morning and are proterandrous. Anther shedding and pollen exposure occur on the first day of anthesis but the stigma branches do not protrude from the central column of the flower until the second day. The corolla closes in the afternoon of the second day of anthesis. To obtain unpollinated stigma it is sufficient to emasculate the flower and keep it bagged on the first day of anthesis.

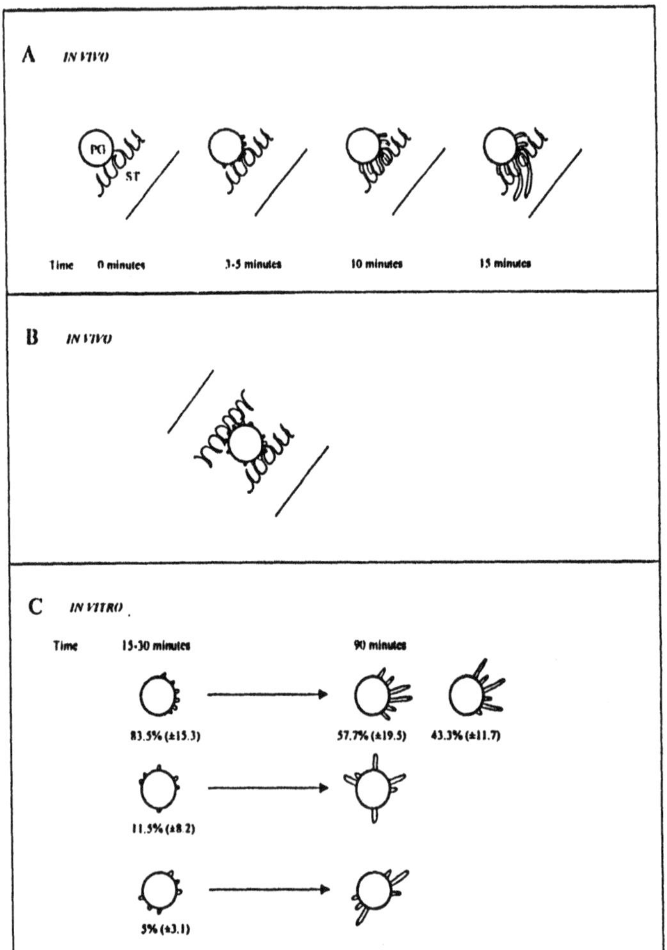

Fig. 5. Diagram showing pollen germination and pollen tube growth of *Lavatera arborea* at different time after pollen deposition. A. In vivo, the pollen tubes are emitted 3-5 min after pollen deposition. Only the central pollen tubes grow to penetrate the stigma, the peripheral ones (arrows) stop growing after about 10 minutes. B. A pollen grain in contact with two stigmas. Pollen tubes are emitted from both the contact regions. C. Pollen germination in vitro takes longer (15-20 min) and is less regular than in vivo. The different patterns of germination and pollen tube growth are indicated with the corresponding mean percentages (± SD). PG=pollen grain; ST=stigma

2.3.2 Pollen morphology and cytology

The pollen of *Lavatera arborea* is bicellular, spherical (130 μm in diameter), with an echinate surface coated with pollenkitt (Fig. 1). It is pantoporate and the pores do not have opercula (Figs. 2a, b). The sporoderm consists of:

1/exine divided into thick compact endexine and ectexine having tectum, bacula and lumina (Figs. 2a, 2b and 3), the latter containing PAS (Fig. 2a) and FITC positive material, derived from the degeneration of the amoeboid or periplasmodial tapetum (Davis 1966);

2/ thin intine, thicker in the pore regions,where it consists of three layers (Fig. 3): an outer one containing gametophytic proteins, a middle one and an inner one made of cellulose;

3/ a very thin layer of callose beneath the intine.

The vegetative cell contains many amyloplasts with a single starch grain and has a PAS-positive cytoplasm (Fig. 2b).

2.3.3 Pollen germination and pollen tube growth

Pollen grains adhere to 8-12 stigmatic papillae and emit 15-20 pollen tubes within 3-5 min (Fig. 5A). The tubes always emerge from an area of the pollen grain surface equal to about one sixth of the total surface area on the side in contact with the stigma surface. In vivo there is a dominance of the central pollen tubes over the peripheral ones. After about 10 min (Figs. 4, 5A), the tubes in the central area are longer (Fig. 4) and the others stop growing, as shown by the presence of callose at their tips. After 15 min, two to eight pollen tubes of the central area, occasionally only one, penetrate the stigma (5A).

Because it is difficult to follow the path of individual pollen tubes, it cannot be established whether more than one pollen tube from a given pollen grain reaches the ovary and fertilises it. However it was noted that the number of pollen tubes reaching the ovary was about half the number of pollen grains on the stigma, so it seems reasonable that one pollen tube takes the lead and the others stop growing between stigma and ovules.

The double contact experiment showed that the pollen germinates at both regions of contact with the two stigma branches (Fig. 5B).

The time taken for in vitro germination was about 15-30 min and the pattern of pollen tube growth was less regular than in vivo. Eighty five percent of the grains emitted pollen tubes from only one side of pollen surface, successively. In a few cases (10% on average) the pollen tubes were randomly emitted or were emitted by two areas of the pollen grain (5%) (Fig. 5C). The peripheral pollen tubes often grew longer than the central ones (45%) (Fig. 5C). The typical dominance of the central pollen tubes observed in vivo was strongly reduced in vitro.

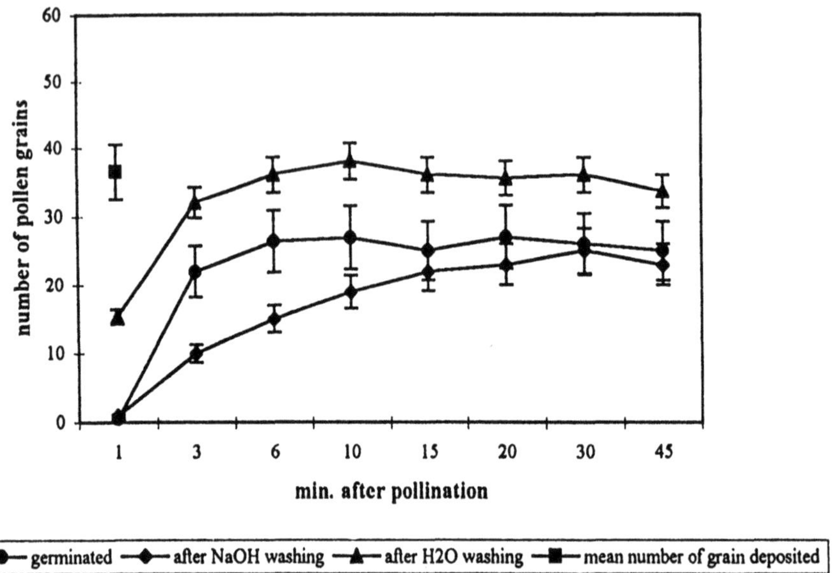

Fig. 6. Graph summarising the pollen adhesion experiment. Water did not remove pollen grains after pollen tubes emission (about 3-5 min after pollen deposition). NaOH removed some germinated pollen grains between tubes emission and stigma penetration (about 15 minutes after pollen deposition). NaOH subsequently failed to remove germinated pollen

2.3.4 Pollen adhesion

In the first minute after pollination, all the grains were removed by NaOH treatment but only about one third by water (Fig. 6). Pollen germination started at 3 min. From 3 min onwards, water removed very few or no pollen grains (Fig. 6). NaOH treatment removed some germinated pollen grains 3 to 15 min after pollination (Fig. 6). The pollen tubes penetrated the stigma 15 min after pollination. After this, NaOH treatment had almost no effect on the germinated pollen grains (Fig. 6).

2.4 Discussion

The time taken to emit pollen tubes varies from a few minutes in *Zea mays*, *Secale cereale*, *Sorghum vulgare*, and *Cucurbita pepo* to several days in *Garrya elliptica* (Maheshwari 1950; Heslop-Harrison 1979; Nepi et al. 1993). The pollen is generally partially dehydrated when it reaches a stigma, where it rehydrates and emits a pollen tube (Speranza et al. 1997). The time required for pollen germination may depend on the dehydration state of the pollen and its rehydration capacity. In the case of *Lavatera arborea*, rapid emission of pollen tubes is possible because the pollen is only slightly dehydrated (Pacini 1996) and there is already a callosic layer in the sporoderm. The same happens in *Cucurbita pepo*, the pollen of which is likewise dispersed in a slightly dehydrated state and germinates in 3-5 min, but emitting a single pollen tube (Nepi and Pacini 1993; Nepi et al. 1995). Polysiphony is not surprising in species of the families Malvaceae, Cucurbitaceae and Campanulaceae (Steffen 1963), because these families have species with only slightly dehydrated pollen and many pores (our unpublished data).

The pollen of *Lavatera arborea* is quite large and the stigma is dry with tiny papillae. Adhesion of the pollen may therefore be precarious, despite the presence of pollenkitt. The present results suggest that pollen tube function is initially (3-15 min after pollination) to anchor the pollen to the stigma; subsequently, only a few pollen tubes continue to grow and only one reaches the ovary. This mechanism presumably functions like a velcro band to attach the pollen to stigma papillae.

The pollen of *Lavatera arborea* is an interesting model of the process of germination and the migration of male germ unit in the pollen tube. In this case two kind of choice must operate. The first choice concerns the region of pollen grain from which pollen tubes are emitted. This process seems to be determined by pollen-stigma interaction. In fact in vitro pollen tube emission was more dispersed on the pollen surface. Our results show that all the regions of the pollen grain are potentially capable of emitting pollen tubes under the stimulus of stigma contact. An interesting aspect is also the dominance of the central pollen tubes over the peripheral ones that stop growing about 10 min after pollination. This feature was not the rule in vitro where the peripheral pollen tubes were often longer than the central ones.

The second choice concerns the pollen tube in which the male germ unit will move. Because of the many pollen tubes for each pollen grain of this species, the sperm cells must be "guided" by biochemical, cytological or physiological target. Detailed studies are needed to clarify all these aspects.

Acknowledgements

The authors are much indebted to Dr. E. S. Pierson for the critical revision of the manuscript. This work was supported by a grant of M.U.R.S.T.

References

Davis GL (1966) Systematic embryology of the Angiosperm. John Wiley and Sons, New York

Erdtman G (1960) The acetolysis method. A revised description. Svensk Botanischen Tidskript 54: 561-564

Ferrari TE, Best V, More TA, Comstock P, Muhammad A, Wallace DH (1985). Intercellular adhesions in the pollen-stigma system: pollen capture, grain binding, and tube attachments. Am J Bot 72(9): 1466-1474

Heslop-Harrison Y (1977) The pollen stigma interaction: pollen tube penetration in *Crocus.* Ann Bot 41: 913-922

Heslop-Harrison J (1979) Aspects of the structure, cytochemistry and germination of pollen of rye (*Secale cereale* L.). Ann Bot 44 (Suppl. 1): 1-47

Heslop-Harrison J (1987) Pollen germination and pollen tube growth. Int Rev Cytol 107: 1-78

Heslop-Harrison Y, Shivanna KR (1977) The receptive surface of the Angiosperm stigma. Ann Bot 41: 1233-1258

Johri BM (1992) Haustorial role of pollen tubes. Ann Bot 70: 471-475

Kearns CA, Inouye DW (1963) Techniques for pollination biologists. University Press of Colorado, Niwot, Colorado

Maheshwari P (1950) An introduction to the embryology of Angiosperms. McGraw-Hill, New York

Martin FW (1959) Staining and observing pollen tubes in the style by means of fluorescence. Stain Technol 34: 125-128

Nepi M, Pacini E (1993) Pollination, pollen viability and pistil receptivity in *Cucurbita pepo*. Ann Bot 72: 527-536

Nepi M, Ciampolini F, Pacini E (1995). Development of *Cucurbita pepo* pollen: ultrastructure and histochemistry of the sporoderm. Can J Bot 73: 1046-1057

Pacini E (1996) Types and meaning of pollen carbohydrate reserves. Sex Plant Reprod 9: 362-366

Pearse AG (1968) Histochemistry - theoretical and applied. Churchill Ltd, London

Speranza A, Calzoni GL, Pacini E (1997) Occurrence of mono- or disaccharides and polysaccharide reserves in mature pollen grains. Sex Plant Reprod 10: 110-115

Steffen K (1963) Male gametophyte. In Maheshwari P (ed) Recent advances in plant embryology. Catholic Press, Ranchi, pp 15-40

3 Sexual dimorphism and stamen development in the dioecious white campion

Lardon A, Georgiev S, Barbacar N, Monéger F, Delichère C and Negrutiu* I

3.1 Introduction

Among the flowering plants, hermaphrodite sexuality (plants bearing bisexual flowers with both stamens and pistil) represents the most common mode of reproduction (near 72% of species). Approximately one-tenth of all angiosperms are strictly dioecious (male plants bearing only staminate flowers and female plants only pistilate flowers) or monoecious (separate male flowers and female flowers on the same plant). Intermediate forms of sexual dimorphism, including gynodioecy and androdioecy, represent 7% of plant species, whereas 10% of plant species possess both unisexual and bisexual flowers (Dellaporta and Calderon-Urrea 1993). Most of the plants used as model have hermaphrodite flowers, which makes the cloning of genes expressed specifically during either male or female reproductive organ development difficult.

In this context, we have chosen the dioecious white campion (*Silene latifolia*) as a working model for sexual development. This species is unusual in there is no difference, at the physiological level, between the presence of male and female plants to environmental stimuli, such as day length, or to ectopic hormone application. This makes it an excellent model for genetic studies, since the sex consistently reflects the genotype. It is also one of the few dioecious species with morphologically distinguishable sex chromosomes. Female have two X chromosomes that could recombine during meiosis, while males have an X and a morphological different Y chromosome which can only recombine through a short pseudo-autosomal region present also on the X chromosome (Grant et al. 1994).

At a very early stage, both male and female flower meristems display stamen and carpel primordia (Fig. 1). Unisexuality is achieved through the arrest of

* for correspondence: Ecole Normale Supérieure de Lyon; RDP - UMR 9938 CNRS/INRA/ENS; 46 Allée d'Italie; 69048 Lyon; France

development of reproductive organs of the opposite sex to that which will be expressed.

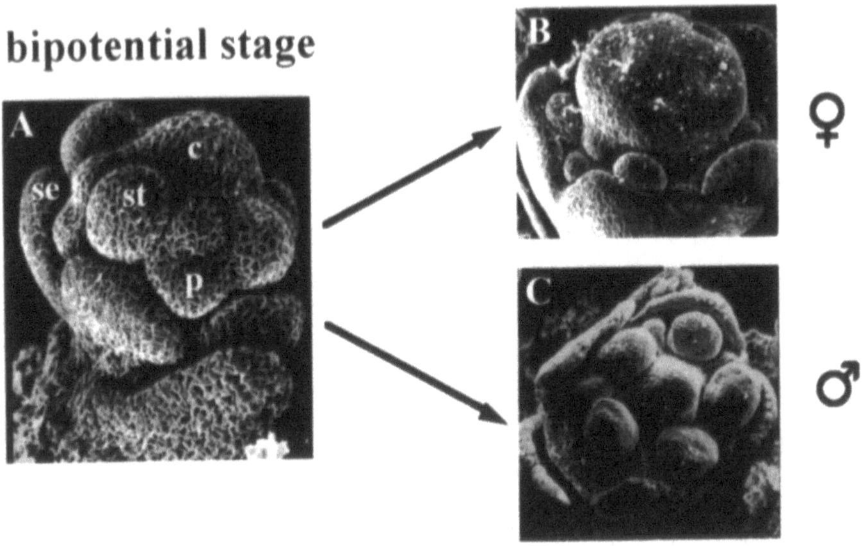

Fig. 1. Early sexual dimorphism in white campion. *c*, carpel; *p*, petal; *se*, sepal; *st*, stamen

The male plant contains all the genetic information to produce a functional hermaphrodite flower. However, the presence of the Y located "Gynoecium Suppressing Function" (GSF) results in the sudden arrest of cell proliferation in whorl 4 of male flowers. This is the earliest event in the establishment of sexual dimorphism in white campion (Farbos et al. 1997; Grant et al. 1994) and corresponds to the morphological partitioning between the third and fourth whorls. As a result, a filamentous structure develops in the centre of the male flower where 5 carpels normally appear in the female flower. This substitution organ has no structural similarity with the wild type pistil and does not result from degenerative processes. Rather, the viable filament is produced by cell elongation from a whorl 4 primordium in which cessation of cell proliferation and a 5-fold deficit in cell number can no longer produce carpellar structures.

In female flowers, the block in stamen development is seen as a lack of parietal initials and the degeneration of sporogenous cell initials during anther differentiation (Farbos et al. 1997). Thus, the female plant contains the genetic information necessary to initiate anther development up to the early sporogenous stage. Subsequently, anther development is arrested because the genes required to proceed beyond that stage are not expressed or are missing. The slow regression of sporogenous cells into parenchymatous cells within the anther, without apparent cell lysis, tends to support the view that a "Stamen Promoting Function" (SPF) present in male plants (XY constitution), could be missing in the female plants (XX constitution).

3.2 Genetic of white campion

Chromosomal aberrations produced by crosses between diploid and polyploid lines (Westergaard 1946), by interspecific crosses between *S. latifolia* and *S. dioica* (Van Nigtevecht 1966), by X-ray mutagenesis (Donnison et al. 1996) or by gamma-ray irradiation of pollen (Lardon et al. 1998) induce sexual mutant phenotypes. Three groups of mutant can be generated by these treatments:

(1) Some of them exhibit asexual phenotypes (*asexua* mutants) with an early arrest in stamen development similar to that observed in aborted stamens of female flowers. Thus, these mutants cumulate two developmental blocks that separately characterise the WT male flower (inhibition of female organs) and the WT female flower (arrest of the male development). The *asexua* phenotypes could correspond to a specific deletion of the SPF locus on the Y chromosome (Fig. 2).

(2) A second set of plants display hermaphrodite phenotypes (*bisexua* mutants), with both stamen and pistil in the same flowers. These mutants could correspond to a deletion of the GSF locus on the Y chromosome which would permit the restoration of a pistil in the male flower (Fig. 2).

(3) Other mutants are affected at different stages of anther formation and become male sterile.

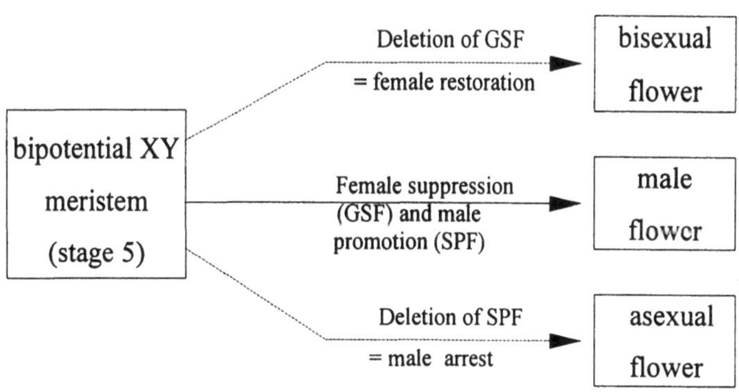

Fig. 2. Sexual mutants in white campion (XY constitution). *GSF*, Gynoecium Suppressing Function; *SPF*, Stamen Promoting Function

Genetic and cytogenetic analysis revealed that most of these mutants are deleted on the Y chromosome. Westergaard (1946) postulated that GSF and SPF are located on the short arm (p-arm) of the Y chromosome whereas long arm (q-arm) bears male fertility genes. In our collection of mutant, the largest deletion found represents more than 50% of the (p)-arm of the Y chromosome in a *bisexua* mutant, and 40% of the same arm in an *asexua* phenotype. No other abnormality than flower phenotypes was observed in these mutants. This suggests the presence of only a few genes (corresponding to probably only SPF and GSF) on the short arm of this chromosome.

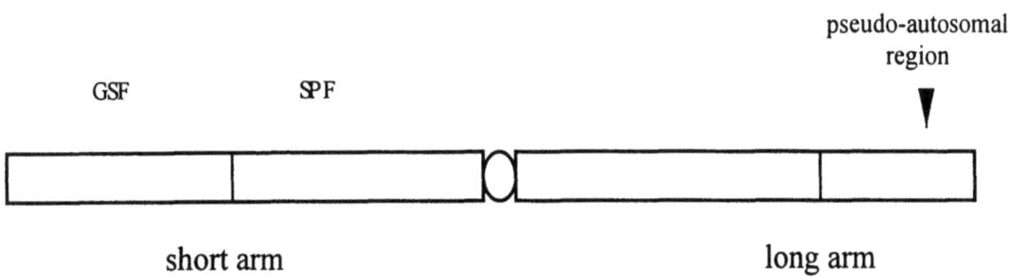

Fig. 3. Relative location of GSF ans SPF on the white campion Y chromosome

In order to determine the relative locations of GSF and SPF genes on the short arm, in situ hybridisation on chromosomes was performed using the subtelomeric probe X.43 in a *bisexua* mutant and an *asexua* mutant. The results show that the X.43 signal is present at both ends of the Y chromosome in wild-type and in the *asexua* mutant, but that the signal is missing in the differential (p)-arm of the *bisexua* mutant. This indicates that the deletion in the *asexua* mutant was interstitial, whereas the deletion in the *bisexua* mutant in distal. We propose the model reported in Fig. 3 concerning the linkage configuration between the two loci.

3.3 Subtraction cloning

With the aim to identifying the SPF gene(s), and other genes linked with stamen development, we have used an *asexua* mutant from our collection to develop a molecular subtraction approach (Fig. 4) using male and asexual flower buds (Barbacar et al. 1997). This resulted in the cloning of 22 cDNA clones expressed in stamen at distinct stages of development. Fourteen of these clones corresponded to genes whose expression was detected in pre-meiotic stamens, a stage of development for which very little information is presently available. Furthermore, the absence of similarities with sequences from the data base for 10 clones suggests that they represent novel genes (Table 1). In parallel, these clones are being exploited as developmental markers of early differentiation within the flower and their expression patterns are being analysed by in situ hybridisation.

One of the late clones, CCLS6, was specifically expressed after meiosis in the tapetum of stamen (Fig. 5A and 5B). It shows high homology with an anther-specific gene (a chalcone-synthase-like) from *Brassica* which could be involved in the flavonoid synthesis pathway (Shen and Hsu, 1992). The expression of CCLS6 in the tapetum is consistent with the role of flavonoids in reproduction (pollen fertility, pollen tube growth). Indeed, products of flavonoid biosynthesis are known to be involved in tapetal breakdown of the anther and flavonoids are deposited in the pores of the pollen exine. We can hypothesise that CCLS6 could be involved in these processes.

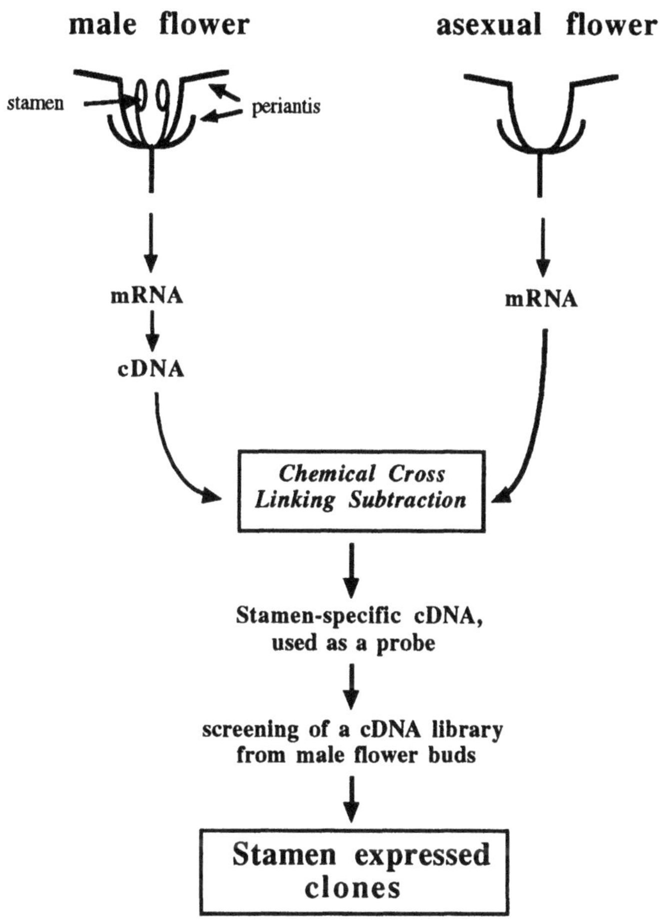

Fig. 4. Subtraction cloning

The early clone CCLS4 is a gene specifically expressed in anthers from pre- to post-meiotic stages in white campion. The sequence contains an open reading frame of 678bp coding for a putative protein of 217 amino acids. The predicted protein has a hydrophobic amino terminus with features characteristic of a secretory signal peptide and a potential cleavage site between aminoacid 24 and

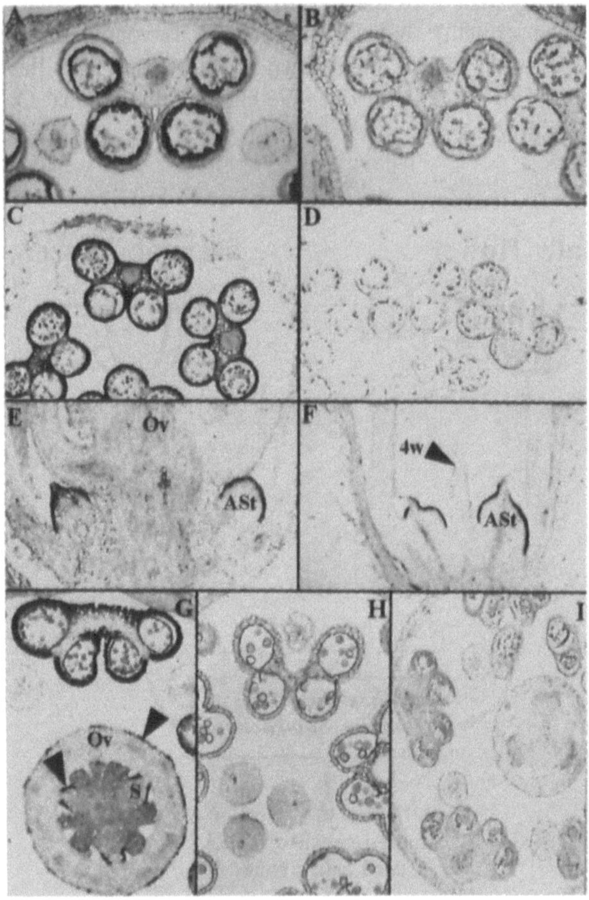

Fig. 5 In situ hybridisation in *Silene*. A: cross section of male flower bud of *S.latifolia*, hybridised with an CCLS6 antisense probe; B: cross section of male flower bud of *S.latifolia*, hybridised with an CCLS6 sense probe; C: cross section of male flower bud of *S.latifolia*, hybridised with an CCLS4 antisense probe; D: cross section of male flower bud of *S.latifolia*, hybridised with an CCLS4 sense probe; E: longitudinal section of female flower bud of *S.latifolia*, hybridised with an CCLS4 antisense probe (*Ov*, ovary; *Ast*, arrested stamen); F: longitudinal section of an asexual mutant flower bud, hybridised with an CCLS4 antisense probe (*4w*, fourth whorl; *Ast*, arrested stamen); G: cross section of a flower bud of *S.conica*, hybridised with an CCLS4 antisense probe; H: cross section of a flower bud of *S.noctiflora*, hybridised with an CCLS4 antisense probe; I: cross section of a flower bud of *S.vulgaris*, hybridised with an CCLS4 antisense probe

25. No known domains or major signature was detected using PROSITE program. In situ hybridisation using this gene shows that it is a marker of the floral 3rd whorl (Hinnisdaels et al. 1997). Namely, in male buds the gene is expressed in normally developing stamens (Fig. 5C and 5D). During meiosis (flower buds of 2 - 2.8mm), the gene was strongly expressed in the epidermal and the differentiating endothecial cell layers, but also, albeit weakly, in the connective tissue. No signal was detected in the stomium, tapetum, pollen mother cells, filament or vascular bundle. After meiosis (flower buds larger than 2.8 mm), the signal decreased gradually. The maximum expression of the clone during meiosis appears to overlap with the deposit of thickenings in the endothecium. At the same time, the cells of the stomium, while in the continuity of epidermis and endothecium, exhibit no cell wall deposition. This ensures the opening of the mature anther through the localised breakdown of the stomium. So, we hypothesise that CCLS4 could be involved in the anther dehiscence process.

Furthermore, as a negative control, we performed in situ hybridisation with CCLS4 in female flowers. Surprisingly we revealed transcripts in the epidermis at the base of the arrested stamen (Fig. 5E). The marked epidermis corresponds thus to the L1 layer of the 3rd whorl domain. When considered together, these results show that CCLS4 expression is tightly restricted to the 3rd whorl. We find the same signal in the epidermis of the third whorl of an *asexua* mutant (Fig. 5F). Our results suggest that CCLS4 may fulfil different functions during pre- and post-meiotic anther development and reveal the complex role parietal cells may play in cellular signalling during early stamen formation.

To verify the specificity of the CCLS4, we performed hybridisation with flower buds from other species in the genus *Silene*. The results reveal a signal within flowers of the dioecious species *S. dioica* and within flowers of the hermaphrodite species *S. conica* (Fig. 5G), but not in *S. noctiflora* (Fig. 5H) nor in *S. vulgaris* (Fig 5I). In *S. conica*, the same hybridisation pattern as in *S. latifolia* is obtained in stamens (epidermis, endothecium, connective tissues), and also, surprisingly, at the level of the pistil (the external epidermis of the ovary and the epidermis of carpel septa) (Fig. 5G; arrows). The role of this gene in female organs is still not known, but interestingly a phylogenetic analysis of species in the genus *Silene*, by comparison of sequences of the internal transcribed spacer (ITS) region of the nuclear ribosomal DNA, revealed that *S. latifolia* , *S. dioica* and *S. conica* were members of the same clade (Desfeux and Lejeune 1996). On the other hand, this analysis indicated that *S. noctiflora* and *S. vulgaris* present a weaker relationship with *S. latifolia*.

Southern blot analysis on genomic DNA from male and female plants did not seem to indicate that the CCLS genes are located on the Y chromosome. It is therefore unlikely that they correspond to the SPF. However, we cannot exclude the possibility that these genes are the targets of the SPF. More work needs to be done in order to position these genes in the regulatory cascades operating in the developing anther.

Table 1. CCLS sequence similarities. Expression of each of the 22 clones was analysed by Northern blot. According to the stage at which the transcript was firstly detected, each clone was classified in "very early" (pre-sporogenous), "early" (pre-meiotic) and "late" (post-meiotic). For each clone, the insert has been partially or completely sequenced and similarities with database sequences reported (Barbacar et al. 1997).

CCLS n°	Accession number	Homologies found
Very early clones		
37.1	Z93048	none
52	Y12529	Leucine-rich repeats in the receptor-like protein kinase precursor AtLRK5 in *Arabidopsis thaliana*
54	Z93053	multicystatin in potato
57.05	Z93057	Leu-rich protein in yeast
62	Z 93055	portion of a pectin methyl esterase (AtPME2) in *Arabidopsis thaliana*
65	Y12603	central domain of karyopherin alpha (importin) SRP1 in yeast
70	Z93050	none
79.1	Z93054	none
86.1	Z93059	none
Early clones		
4	X94358	MROS3 (Male Reproductive Organ Specific) in white companion
17	Z93058	EST (expressed sequence tag) in *Brassica*
57.1	Z93063	portion of DNA binding protein ENPB1 binding to the ENOD12b gene promoter in *Vicia sativa*
96.1	Z93056	none
120.2	Z93047	none
Late clones		
1	Z93051	phospholipid transfer protein (LTP) in castor bean
6	Z93049	anther_specific in *Brassica*
8	Z93046	none
29	Z93060	none
30.1	Z93061	portion of a DNA_binding Zn-finger type gene (GLO3) in yeast
30.2	Z93043	coumarate CoA ligase in *Pinus*
30.3	Z93044	microtubule-associated protein (MAP) in yeast
37.2	Z93045	none

3.4 Genes located on the Y chromosome

In order to have a better knowledge of the Y chromosome, both at the structural and functional levels, we attempted to identify coding sequences located on this chromosome. For this purpose, Y chromosomes microdissected from metaphase plates were used in DOP-PCR amplification. The Y-specific DNA obtained was used as a probe against a cDNA library from male flower buds at a stage when the genes coding for the SPF and GSF are thought to be expressed. We purified 115 positive clones and hybridised them one by one to genomic DNA from male and female plants. This allowed us to identify 5 partial cDNAs located on the Y chromosome. The location of these genes was confirmed by segregation analysis using Southern blot analysis: each of the cDNA hybridised to a genomic DNA fragment which strictly segregates with the male plants. Interestingly, all 5 clones seem to have a homologue on the X chromosome. The sequence analysis of these clones revealed that they are all similar to EST sequences (Expressed Sequence Tags) of *Arabidopsis* or rice. One of them has 2 calcium binding domains similar to those found in calmodulin. Another one has strong similarity to Rb-binding proteins found in mammals and we thereofre focused our work on this clone. We first tried to identify specific sequences for each gene in order to have probes specific for the allele on the Y chromosome and for the allele on the X chromosome. The analysis is underway and should allow investigation of the expression profile of each gene specifically by Northern blot and in situ hybridisation in order to determine if these genes could be involved in sexual dimorphism in white campion. FISH experiments should allow us to position these different genes on the Y chromosome. Functional analysis will be done by using *Arabidopsis* as a transformation system. Ultimately, the goal is to understand the function of these genes and to determine whether they have acquired specialised functions during evolution. The investigation of our deletion mutants should give interesting information regarding the role of these genes in sexual dimorphism. From an evolutionary point of view, the presence of a homologue of the Y genes on the X chromosomes supports the idea that the XY pair has evolved from a pair of autosomes.

References

Barbacar N, Hinnisdaels S, Farbos I, Monéger F, Lardon A, Delichère C, Mouras A, Negrutiu I (1997) Isolation of early genes expressed in reproductive organs of the dioecious white campion (*Silene latifolia*) by subtraction cloning using an asexual mutant. Plant J 12: 805-817

Dellaporta S L, Calderon-Urrea A (1993) Sex determination in flowering plants. Plant Cell 5: 1241-1251

Desfeux C, Lejeune B (1996) Systematics of euromediterranean *Silene* (Caryophyllaceae): evidence from a phylogenetic analysis using ITS sequences. C R Acad Sci Paris Life Sci 319: 351-358

Donnison I S, Siroky J, Vyskot B, Saedler H, Grant S R (1996) Isolation of Y chromosome-specific sequences from *Silene latifolia* and mapping of male sex-determining genes using representational difference analysis. Genetics 144: 1893-1901

Farbos I, Oliveira M, Negrutiu I, Mouras A (1997) Sex organ determination and differentiation in the dioecious plant *Melandrium album* (*Silene latifolia*): a cytological and histological analysis. Sex Plant Reprod 10: 155-167

Grant S, Hunkirchen B, Saedler H (1994) Developmental differences between male and female flowers in the dioecious plant *Silene latifolia*. Plant J 6: 471-480

Hinnisdaels S, Lardon A, Barbacar N, Negrutiu I (1997) A floral third whorl-specific marker gene in the dioecious species white campion is differentially expressed in mutants defective in stamen development. Plant Mol Biol 35: 1009-1014

Lardon A, Georgiev S, Aghmir A, Le Merrer G, and Negrutiu I (1998) Compex control of carpel number is revealed by Y chromosome deletions in dioecious white campion. Genet (*submitted*)

Shen JB, Hsu F C (1992) *Brassica* anther-specific genes: characterization and *in situ* localization of expression. Mol Gen Genet 234: 379-389

Van Nigtevecht G (1966) Genetic studies in dioecious *Melandrium*. I. Sex-linked and sex-influenced inheritance in *Melandrium album* and *Melandrium dioicum*. Genetica 37: 281-306

Westergaard M (1946) Aberrant Y chromosomes and sex expression in *Melandrium album*. Hereditas 32: 419-443

4 Polyamines as markers in sexual reproduction of kiwifruit

Biasi*[1] R, Franceschetti[2] MM, Falasca[3] G, Altamura[3] MM, Bagni[2] N

4.1 Introduction

Aliphatic polyamines are plant growth regulators that affect mainly the induction and sustenance of cell division (Bagni 1989). They are involved in all the steps of protein synthesis (replication, transcription and translation) through specific and aspecific interactions with DNA and different RNAs (Feuerstein and Marton 1989). In particular, the specific interactions of two molecules of spermine or spermidine, one located in the major groove at one end of the anticodon stem, the other near the variable loop of tRNA, are able to switch tRNA conformation from the inactive to the active form (which is then able to accept the specific aminoacid) (Quigley et al. 1978). This observation provides one of the most plausible explanations for the mode of action of these compounds in the cell.

Normal and/or tumor growth processes have always been related to high polyamine levels (Bagni and Serafini-Fracassini 1979).This is particularly true for different organs involved in sexual plant reproduction: for example the anthers and ovaries of wheat and tomato (Egea-Cortines and Mizrahi 1991; Bagni et al. 1967) show a high polyamine content, and the mature male gametophyte of apple and tobacco contains huge amount of polyamines, especially spermidine (Bagni 1986).

Until now, the gynoecium has been less investigated than the androecium (Torrigiani et al. 1998).

It has also been proposed that the change in ratio between spermidine and putrescine is related to some extent to sex determination (Biasi et al. 1997). In

[1]* for correspondence: Dipartimento di Produzione Vegetale; Università della Basilicata; Potenza; Italy
[2] Dipartimento di Biologia e Centro Interdipartimentale per le Biotechnologie; Università di Bologna; Bologna; Italy
[3] Dipartimento di Biologia Vegetale; Università di Roma 'La Sapienza'; Roma; Italy

addition, the switch between free polyamines, the active form responsible for cell division (Bagni 1989), and polyamines conjugated to hydroxycinnamic acid could be responsible for the fate of a specific organ. In fact polyamines conjugated to monomers, dimers and more complex structures (Bokern et al. 1995) are not only important for the regulation of the polyamine pool inside the cell (Bagni and Pistocchi 1990), but also for modifying the interactions between cell wall components. These alterations could influence cell differentiation and/or be involved in programmed cell death.

Moreover, in plants there is a biosynthetic link between polyamine and ethylene synthesis through the pivotal role of S-adenosylmethionine, whose regulation is able to affect growth and senescence (Altman 1989). In comparison with other plant growth regulators, spermidine and spermine are abundant in tissues related with sexual reproduction (Bagni 1989), as in animals. Genetic evidence for a role of spermidine and spermine in plant sexual reproduction comes from the isolation of tobacco mutants that are partially blocked in polyamine synthesis or resistant to methylglyoxal bis (guanylhydrazone), a competitive inhibitor of S-adenosylmethionine decarboxylase (SAMDC). These mutants show alterations in floral morphology such as anthers partially turned into petals, ovules transformed into stamens and sexual reversion (Malmberg and McIndoo 1983). Viceversa petunia mutants with irregular floral development display abnormal polyamine levels (Gerats et al. 1988).

In dioecious species, such as *Asparagus officinalis*, the transition from hermaphroditic to unisexual stages is accompanied by some biochemical changes such as a peak of auxin in male flowers which correlate negatively with carpel growth and positively with anther growth (Marziani Longo et al. 1990). At the same time, a peak of RNase activity occurs during anther degeneration in female flowers. Furthermore, the polypeptide pattern of male and female flowers, which is very similar during hermaphroditic developmental stages, become different at the transition to unisexual flower (Caporali et al. 1994).

In the same plant, cytokinins/gibberellins ratio has also been proposed as a parameter controlling sexual differentiation (Ombrello and Garrison 1987).

Kiwifruit, a dioecious species exibiting morphologically hermaphrodite but physiologically unisexual flowers (Hopping 1990), can represent a good model for studies on the control of sexual expression.

The aim of this work is the anatomical characterisation of female and male ovaries of kiwifruit at different stages and the utilization of polyamines as biochemical markers of specific events in reproductive development. To this aim, free and conjugated polyamine levels and biosynthetic enzyme activities (arginine-, ornithine-,S-adenosylmethionine-decarboxylase) were determined.

4.2 Material and methods

The biochemical and histological investigations were performed on developing pistillate (cv. Hayward) and staminate (cv. Tomuri) flower buds, sampled from kiwifruit (*Actinidia deliciosa* var. deliciosa (A. Chev.), C.F. Liang et A.R. Ferguson) vines, starting from 18 days after bud burst (36 days before flowering) until full bloom.

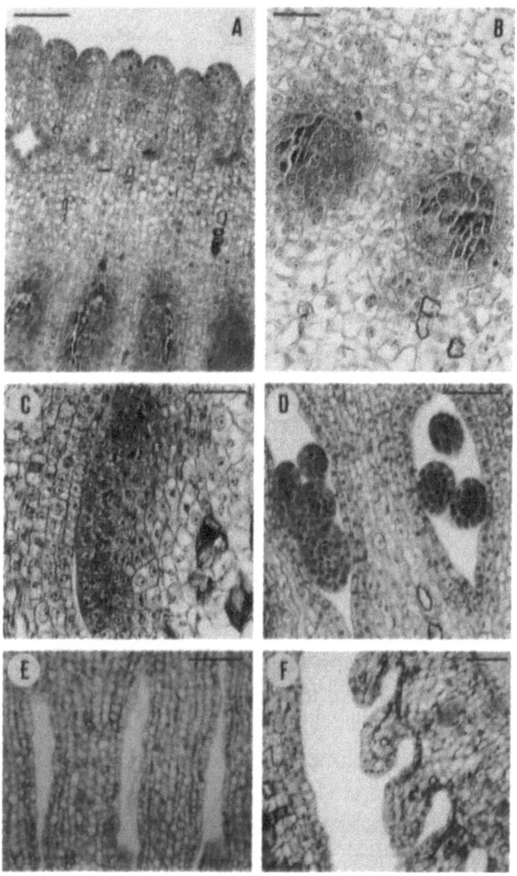

Fig. 1. Carpel development in female and male flowers from 18 days after bud burst up to one month before anthesis, i.e. 28 days after bud burst. A-C. Meristematic regions below the stigmatic surface in female (A and C, bars= 100 μm and 38 μm, respectively) and male (B, bar= 50 μm) ovaries. D. Ovule differentiation in female flowers at day 28 after bud burst (one month before anthesis) (bar= 100 μm). E-F. Lack of ovule formation in the male flowers at the same day: E, empty carpel loculi (bar= 100 μm). F, pseudo-parenchymatic ovule-like structures occasionally present in the carpel loculi (bar= 100 μm)

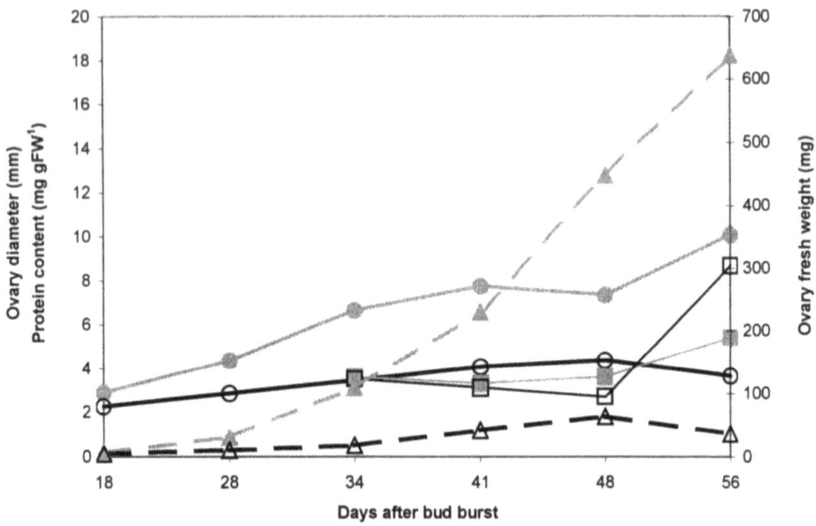

Fig. 2. *In planta* ovary growth for female and male kiwifruit flowers from the initiation of the gynoecium (18 days after bud burst) up to flowering (56 days after bud burst): mean ovary diameter (female —●— ; male —⊖—), mean protein content (female ···■··· ; male .—⊟—) and mean fresh weight (on the secondary ax , female —▲— ; male —▽—)

4.2.1
Histochemical preparations

Flower buds from both female and male plants were periodically collected during floral morphogenesis and fixed in FAA (90% ethanol, 5% acetic acid, 5% formaldehyde). At 18, 28, 34, 41, 48 days after bud burst and at anthesis (56 days after bud burst), six central flowers of the inflorescence of the pistillate and staminate plants were dehydrated in ethanol series and embedded in paraffin. Ovary development and megagametogenesis were monitored on paraffin-embedded longitudinal sections (10 μm thick), stained with safranin and fast-green (Jensen, 1962) and observed under a light microscope (Axiophot-Zeiss). Dissected ovaries from both pistillate and staminate flowers were also weighted and directly observed with a confocal laser scanning microscope (Zeiss LSM3, laser 488 nm, filter BP 515-565) for monitoring carpel growth and stigmatic surface development.

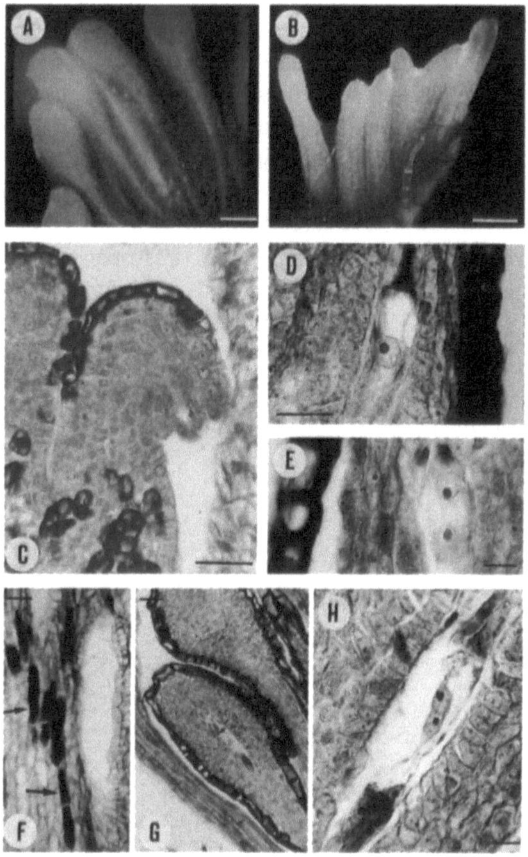

Fig. 3. Carpel development in female and male flowers from 18 days after bud burst to anthesis (56 days after bud burst). A-B. Stigmatic arms of the female (A, bar= 250 μm) and male (B, bar= 250 μm) carpels at day 28 after bud burst. C. Ovules of the female flowers showing the differentiated tegument and micropyle (bar= 50 μm). D-E. Uninucleate (D, bar=20 μm) and binucleate (E, bar= 10 μm) embryo-sacs in the female flowers at day 41 after bud burst. F. Tanniferous inclusions (arrows) in the carpel of the male flowers (bar= 50 μm). G-H. Mature 8-nucleate embryo-sacs in the female flowers at day 48 after bud burst. G, complete embryo-sac surrounded by well-defined integumentary tapetum (bar=50 μm). H, Eight- nucleate embryo-sac in detail (bar= 10 μm)

4.2.2 Polyamine extraction and determination

Samples were homogenized in 5% (w/v) tricloroacetic acid (TCA), incubated 1 h on ice, centrifuged 10 min at 1,500 g and supernatants were separated and measured. Pellets were washed twice with 5% TCA and resuspended in TCA.

Three replicates of supernatant and pellet were hydrolysed with the same volume of 12 N HCl for 12 h at 110°C in flame sealed ampoules according to Slocum and Galston (1985). Following centrifugation at low speed to remove charred debris, the hydrolysates were taken to dryness under vacuum in a 60°C bath and resuspended in 200 μl of TCA. Replicates of both supernatant and hydrolysate were dansylated as described by Smith and Best (1977). Dansylated polyamines were separated and determined using TLC precoated plates of silicagel 60 with concentrating zone (Merck). Ethylacetate:cyclohexane 2:3 (v/v) was used as the solvent. Spots were scraped from the plates, resuspended in 2 ml of pure anhydrous acetone on a Vortex mixer and centrifuged. Fluorescence was measured using a Jasco FP-550 spectrofluorimeter (excitation 360, emission 505.5 nm) and the results compared with dansylated standards. Polyamine determinations through TLC separation were compared with HPLC separation according to Torrigiani et al. (1995) and values obtained were practically the same.

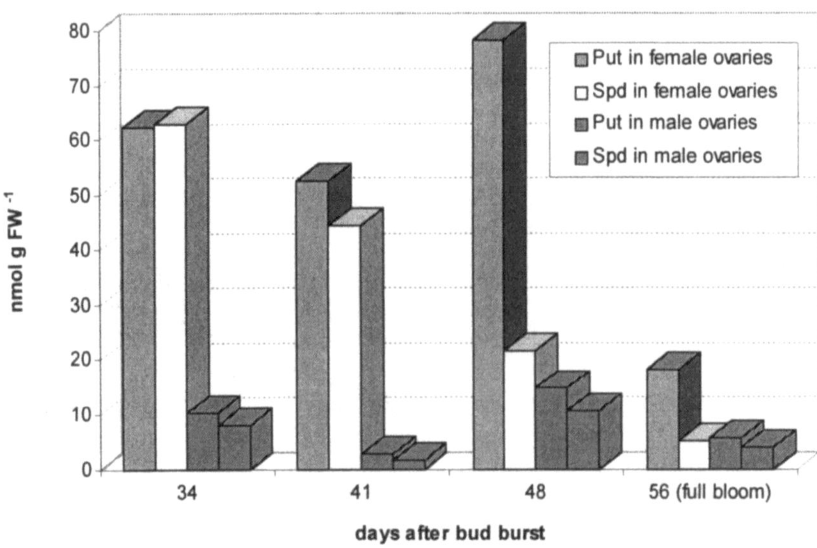

Fig. 4. Free polyamine content during the development of female and male kiwifruit ovaries

4.2.3
Enzyme extraction and assay

All enzyme procedures were carried out in an ice bath, unless otherwise indicated.

To determine arginine decarboxylase (ADC) and ornithine decarboxylase (ODC) activities isolated male and female ovaries were grounded in a mortar with liquid nitrogen and then homogenised with Tris-HCl buffer 100 mM (pH 8.5) containing pyridoxal phosphate 50 μM.

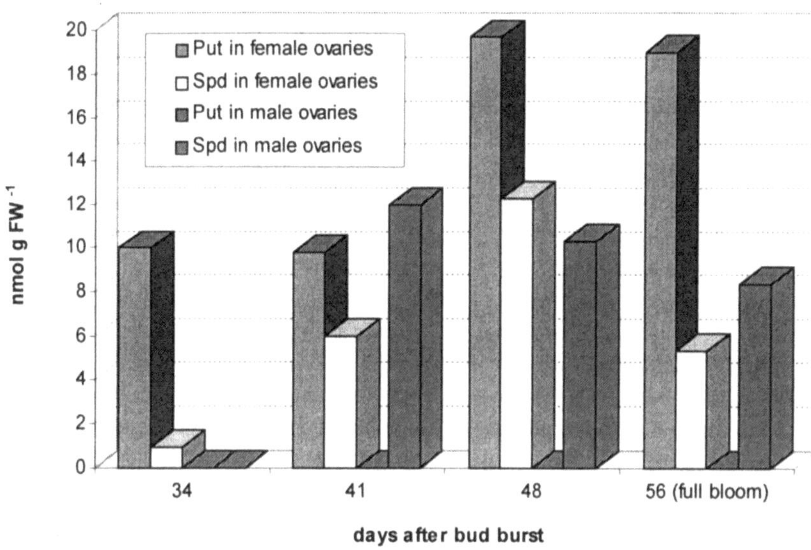

Fig. 5. TCA-insoluble polyamine content during. the development of female and male kiwifruit ovaries

The extracts were sonicated 3 times for 30 sec by means of an ultrasonic disintegrator (150W MSE, Crawley, UK, 20 kHz, amplitude 10 μm peak to peak) and then centrifugated at 20,000 g for 30 min at 0°C. The enzymatic activities were determined by incubating 0.5 ml of supernatant with 14.8 KBq in 20 μl of L-[U-^{14}C]arginine (12.43 GBq/mmol, Amersham, U.K.) or 14.8 KBq in 20 μl of L-[1-^{14}C]ornithine (1.93 GBq/mmol, Amersham, U.K.). The assay was carried out as previously described (Bagni et al. 1983), with minor modification (Beckman Tissue Solubilizer 450, Beckman, instead of Protosol, NEN).

Enzyme extraction for SAMDC assay was performed as described above with Tris-HCl buffer 100 mM (pH 7.6) containing pyridoxal phosphate 25 μM and EDTA 50 μM. No sonication was needed in this case and 7.4 KBq in 20 μl of S-adenosyl-L-[carboxyl-^{14}C]methionine (2.22 Gbq/mmol, Amersham, U.K) were used.

For ODC and ADC activities, tests performed with different concentrations of not labelled substrate (0 mM, 2 mM, 20 mM) showed that the trend was exactly the same, but the better results were obtained with 2mM cold substrate. For SAMDC activity the optimal concentration of cold substrate was 100 μM.

Protein content was measured by the protein-dye binding method according to Bradford (1976) using serum albumin as the standard.

4.3 Results

4.3.1 Carpel development

The histological investigation showed that ovary development occurred similarly both in the pistillate and staminate flowers up to 18 days after bud burst, when they showed comparable weight (7.11±0.72 mg and 5.15±0.59 mg for female and male ovary respectively) and size (2.90±0.22 mm of diameter and 2.25±0.16 mm for female and male ovary respectively). At the same date, the stigmatic surface was already present and showed a comparable developmental stage. Both pistillate and staminate ovaries exhibited regions of meristematic cells similarly located just below the stigmatic surface at the site of ovule formation (Figs. 1 A-C). Ovary development started to diverge between pistillate and staminate flowers almost one month before anthesis. The ovary of the pistillate flower was characterized by a rapid increase in fresh weight lasting up to anthesis, while the ovary of the male flower did not show any further significant increment in size (Fig. 2). One month before anthesis the pistillate ovaries showed well defined carpel loculi and still totally meristematic ovules (Fig. 1 D). The carpel loculi in the staminate ovaries were, instead empty, although occasionally in some loculi pseudo-parenchymatic ovule-like structures were visible (Figs. 1 E,F). At this time the stigmatic arms were still present in the male flower, even if less developed than in the pistillate one (Figs. 3 A,B). One week later (34 days after bud burst), in the pistillate ovaries the ovules showed tegument differentiation, micropyle end formation and onset of megasporogenesis (Fig. 3 C). Fourty-one days after bud burst, while the carpel loculi of staminate flowers continued to be empty, uninucleate and binucleate embryo-sacs were observed in the ovules of pistillate ovaries (Figs. 3 D,E). In both ovary types, high levels of tanniferous inclusions were observed (Fig. 3 F). One week before anthesis (48 days after bud burst), mature ovules, showing complete 8-nucleate embryo-sacs surrounded by a well-defined integumentary tapetum were observed in female flower ovaries (Figs. 3 G,H). The stigmatic arms were totally defined with well grown unicellular papillae. At this date, male ovaries were atrophic and the stigmatic surface was necrotic.

4.3.2 Polyamine content and biosynthesis

Polyamines (putrescine and spermidine) were always present in female and male ovaries, both in the free and TCA-insoluble fractions (Figs. 4-5), while spermine was only detectable in traces. No conjugated polyamines were found in the TCA-soluble fraction. The level of free polyamines in female ovaries was 3 to 22 folds higher than in male ovaries, with putrescine reaching a maximun one week before anthesis (48 days after bud burst). Both polyamines decreased at the anthesis in female and male ovaries (Fig. 4). Nonetheless, by expressing the data as nmol per ovary unit, free polyamines did not decrease.

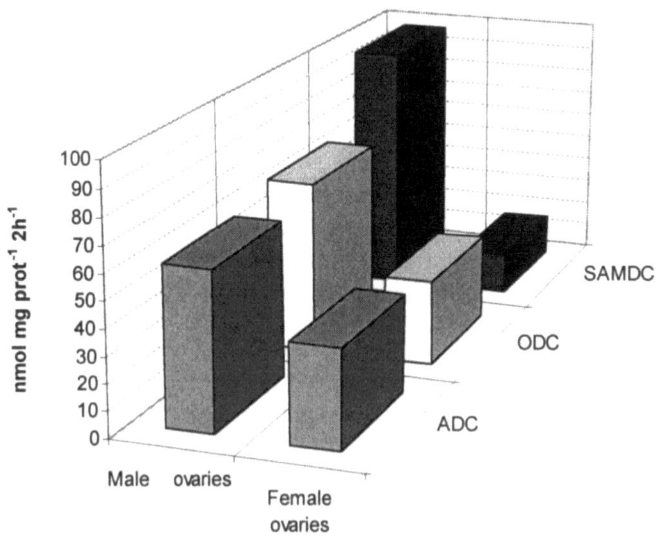

Fig. 6. Polyamine biosynthetic activities (arginine-, ornithine- and S-adenosylmethionine-decarboxylase) in female and male kiwifruit ovaries at 56 days after bud burst (full bloom)

Conjugated polyamines in TCA-insoluble fraction displayed a completely different pattern. In female ovaries they represented 1/10 of free polyamines at 34 days after bud burst and then increased until at anthesis free and conjugated polyamines reached similar values (Figs. 4-5). In the male ovaries spermidine was the only conjugated polyamine, present in approximately the same concentration as the free form. Putrescine was detectable only in trace amounts (Fig. 5).

In both female and male ovaries the trend of free and conjugated polyamines during development was different, the latter showing only a slight decrease at the anthesis (Fig. 5).

In all the developmental stages in female ovaries total putrescine/spermidine ratio was bigger than 1 and increased over 3 in the last two stages. In male ovaries on the other hand this ratio was always less than 1 due to the disappearance of putrescine in TCA-insoluble fraction (Figs. 4-5).

At anthesis higher polyamine biosynthetic activities (ODC, ADC and SAMDC) were detected in male ovaries in comparison with the female ones, especially with regard to SAMDC (Fig. 6).

4.4 Discussion

We have investigated the physiology of male and female flower development in kiwiplants as a first step to study sex determination in dioecious plants and the possibility to use polyamines as markers of sexual differentiation. Firstly, our results show that the level of free polyamines, the active form of these growth regulators, is correlated with cell proliferation as measured by increased fresh weight, ovary development, protein content and the occurence of histological events such as ovule differentiation and embryo sac formation in female in comparison to male ovaries. The decrease of free polyamine content in the last week before bloom could be due to cell enlargement as confirmed when data are expressed as nmol per ovary unit (data not shown). The free polyamine pattern in female ovaries is similar to that obtained in the entire flower buds (Biasi et al. 1997), while in male ovaries it is different because of the complete development of pollen. On the other hand no relationship with cell proliferation was observed for TCA-insoluble polyamines.

Secondly, a shift of free to TCA-insoluble conjugated polyamines was found just before full bloom in both female and male ovaries (in the latter one only for spermidine). The TCA-insoluble conjugated polyamines in anthers, pollen and entire flowers have been identified as bound to di-and tri-substituted hydroxycinnamic acids in several species (Bokern et al. 1995; Werner et al. 1995) and similar conjugated forms might also be present in ovaries of kiwi.

Diferulic acids and other dihydroxycinnamic acids cross-linked to hemicelluloses and/or lignins through ester-ether bonds have been identified in cell walls (Markwalder and Neukom 1976; Ishii 1997). Polyamines could therefore play a role in the cross linkage between hemicelluloses and lignin through their interaction with diferulic acids. This hypothesis is supported by the presence of conjugated polyamines in the TCA-insoluble fraction and not in the TCA-soluble one. In fact, with the used extraction procedure, the TCA-insoluble fraction contains mainly cell wall fragments. These interactions could be responsible for cell wall rigidity and thus hydroxycinnamic acids and polyamines might have a regulatory role in cell wall development.

The lack of conjugated polyamines in the TCA-soluble fraction in ovaries suggest that the conjugation of polyamines to dihydroxycinnamic acids occur after the conjugation of these compounds to hemicelluloses. We do not exclude however that a small fraction of polyamines present in the TCA-insoluble fraction can also be conjugated to cell wall proteins through transglutaminase activity as recently identified in the cytoplasm of ungerminated and germinated apple pollen (Del Duca et al. 1997).

The events occurring in the female ovaries during the last developmental stages, i.e. the thickening of ovule tegument walls and the formation of a conspicuous thickened endothelium (Figs. 3 C-E, G and Hopping 1976) correlate with the above-mentioned increase in conjugated polyamines. In the male ovaries, these histological events are not present, however the TCA-insoluble conjugated polyamines increase (Fig. 5). Since in the male ovaries tanniniferous cells are abundant (Fig. 3F and Schmid 1978) the increase in the TCA- insoluble conjugated polyamines in these ovaries might be correlated with tannin deposition. In fact, it is known that hydroxycinnamic acids have a central role as precursors in phenolic metabolism (Harborne 1980). However, it can not be ruled out that at least in the tanniniferous cells modification in the cell walls, undetectable at the light microscope, may also occur as reported for *Diospyros kaki L.* (Yonemori and Matsushima 1987).

Polyamine biosynthetic enzyme acitivity is very high in male ovaries compared with female ovaries although polyamine accumulation is lower. This suggest that the polyamine biosynthetic machinery is still functional even if male ovaries became atrophic. For an earlier developmental stage it was demonstrated that the percentage of necrotic male ovaries in *in vitro* culture can be reduced by supplying exogenous polyamines (Biasi et al. 1997), expecially spermidine which is able to stabilize membrane structure. This might show a sensitivity of male ovary to polyamines. The effect of senescence reduction and/or increase in viability due to spermidine was also shown in *Pinus parolini* pollen (Bagni and Bordoni 1980).

Contrary to the data obtained with tomato ovaries, in which ODC activity was the main polyamine biosynthetic activity (Heimer and Mizrahi 1983), in kiwi, female ovaries ADC and ODC activities are at approximately the same level, while SAMDC activity is lower. The relative contribution of ADC and ODC to polyamine synthesis may in fact vary upon various factors, such as tissue type and developmental stage (Slocum and Galston 1985).

In conclusion, the high level of free polyamines in female ovaries is responsible for cell division processes while it is still unclear whether the only role of insoluble conjugated polyamines is related to modifying cell wall rigidity/plasticity. The presence of TCA insoluble spermidine and the absence of putrescine in male ovaries, as well as high SAMDC activity compared to female ovaries, indicates that spermidine could be a good candidate as marker of sexual differentiation.

Acknowledgements

This work was supported by Consiglio Nazionale delle Ricerche (CNR), Italy to M.M. Altamura, N. Bagni and R. Biasi.

References

Altman A (1989) Polyamines and plant hormones. In: Bachrach U, Heimer YM (eds) The Physiology of Polyamines, vol 2. CRC Press, Boca Raton Florida, pp 121-159

Bagni N (1986) The role of polyamines in pollen germination. In: Cresti M, Dallai R (eds) Biology of Reproduction and Cell Motility in Plants and Animals, University of Siena Press, Siena, pp 113-118

Bagni N (1989) Polyamines in plant growth and development. In: Bachrach U, Heimer YM (eds) The Physiology of Polyamines, vol 2. CRC Press, Boca Raton Florida, pp 107-120

Bagni N, Bordoni A (1980) Aspetti biochimici nel processo di invecchiamento del polline. Inform Bot Ital 12: 393-398

Bagni N, Pistocchi R (1990) Binding, transport and subcellular compartmentation of polyamines in plants. In: Flores HE, Arteca RN, Shannon JC (eds) Polyamines and Ethylene: Biochemistry, Physiology and Interactions. Am Soc Plant Physiol, Rockville, Maryland, pp 62-72

Bagni N, Caldarera M, Moruzzi G (1967) Spermine and spermidine distribution during wheat growth. Experientia 23: 139-140

Bagni N, Serafini-Fracassini D (1979) Polyamines and plant tumors. It J Biochem 28: 392-394

Bagni N, Barbieri P, Torrigiani P (1983) Polyamine titer and biosynthetic enzymes during tuber formation of *Helianthus tuberosus*. J Plant Growth Regul 2: 177-184

Biasi R, Falasca G, Altamura MM, Fabbri S, Bagni N (1997) Reproductive morphogenesis in kiwifruit in relation to polyamines. Acta Hort 444: 479-484

Bokern M, Witte L, Wray V, Nimtz M and Meurer-Grimes B (1995) Trisubstituted hydroxycinnamic acid spermidine from *Quercus dentata* pollen. Phytochemistry 39: 1371-1375

Bradford MM (1976) A rapid and sensitive method for the quantification of microgram quantities of protein utilizing the principle of protein-dye binding. Ann Biochem 72: 248-254

Caporali E, Carboni A, Galli MG, Rossi G, Spada A, Marziani Longo GP (1994) Development of male and female flower in *Asparagus officinalis*. Search of point of transition from hermaphroditic to unisexual developmental pathway. Sex Plant Reprod 7: 239-249

Del Duca S, Bregoli AM, Bergamini C, Serafini-Fracassini D (1997) Transglutaminase-catalyzed modification of cytoskeletal proteins by polyamines during the germination of *Malus domestica* pollen. Sex Plant Reprod 10: 89-95

Egea-Cortines M, Mizrahi Y (1991) Polyamines in cell division, fruitset and development and seed germination. In: Slocum RD and Flores HE (eds) Biochemistry and Physiology of Polyamines in Plants. CRC Press, Boca Raton Florida, pp 143-158

Feuerstein BG, Marton LJ (1989) Specificity and binding in polyamine/nucleic acid interaction. In: Bachrach U, Heimer YM (eds) The Physiology of Polyamines, vol 1. CRC Press, Boca Raton, Florida, pp 109-124

Gerats AGM, Kaye C, Collins C, Malmberg RL (1988) Polyamine levels in Petunia genotypes with normal and abnormal floral morphologies. Plant Physiol 86: 390-393

Harborne JB (1980) Plant phenolics. In: Bell EA, Charlwood BV (eds) Encyclopedia of Plant Physiology, New Series,Vol. 8 Secondary Plant Products. Springer-Verlag, Berlin Heidelberg New York, pp 329-395

Heimer Y, Mizrahi Y (1982) Characterization of ornithine decarboxylase of tobacco cells and tomato ovaries. Biochem J 201: 373-376

Hopping ME (1990) Floral biology, pollination and fruit set. In: Warrington IJ, Weston GC for the NZSHS (eds) Kiwifruit: Science and Management. RAY Richards Publisher in association with the New Zealand Society for Horticultural Science, Auckland, pp 72-97

Hopping ME (1976) Structure and development of fruit and seeds in Chinese gooseberry (*Actinidia chinensis* Planch.). New Zeal J Bot 14: 63-68

Ishii T (1997) Structure and functions of feruloylated polysaccharides. Plant Sci 127: 111-127

Jensen WA (1962) Botanical histochemistry. Principles and Practice. Freeman and Co, San Francisco

Malmberg RL, McIndoo J (1983) Abnormal floral development of a tobacco mutant with elevated polyamine levels. Nature 305: 623-625

Markwalder HV, Neukom H (1976) Diferulic acid as a possible crosslink in hemicelluloses from wheat germ. Phytochem 15: 836-837

Marziani Longo GP, Rossi G, Scaglione G, Longo CP, Soave C (1990) Sex differentiation in *Asparagus officinalis* L. III. Hormonal content and peroxidase isoenzymes in female and male plants. Sex Plant Reprod 3: 236-243

Ombrello TM, Garrison SA (1987) Endogenous gibberellin and cytokinin in spear tips of *Asparagus officinalis* in relation to sex expression. J Amer Soc Hort Sci 112: 539-544

Quigley GJ, Teeter MM, Rich A (1978) Structural analysis of spermine and magnesium ion binding to yeast phenylalanine transfer RNA. Proc Natl Acad Sci USA 75: 64-68

Schmid R (1978) Reproductive anatomy of *Actinidia chinensis* (Actinidiaceae). Bot Jahrb Syst 100: 149-195

Slocum RD, Galston AW (1985) Changes in polyamine biosynthesis associated with postfertilization growth and development in tobacco ovary tissues. Plant Physiol 79: 336-343

Smith TA, Best GR (1977) Polyamines in barley seedlings. Phytochem 16: 841-843

Torrigiani P, Rabiti AL, Betti L, Marani F, Brizzi M, Bagni N, Canova A (1995) Improved method for polyamine determination in TMV, a rod-shaped virus. J Virol Meth 53: 157-163

Torrigiani P, Scaramagli S, Franceschetti M, Michael A, Bagni N (1998) S-adenosylmethionine decarboxylase (SAMDC) activity regulation in tobacco. In Bardocz S, White A, Tiburcio AF (eds) COST 917 "Biogenically active amines in food" Vol 1 "Biogenic Amines in Transgenic Plants". European Communities, pp 37-42

Werner C, Hu W, Lorenzi-Riatsch A, Hesse M (1995) Di-coumaroylspermidines and tri-coumaroylspermidines in anthers of different species of the genus *Aphelandra*. Phytochem 40: 461-465

Yonemori K, Matsushima J (1987) Changes in tannin cell morphology with growth and development of japanese persimmon fruit. J Amer Soc Hort Sci 112: 817-821

5 Protein targeting by ubiquitin during anther and pollen development in male and female flowers of kiwifruit (*Actinidia deliciosa*)

Scoccianti[1] V, Speranza*[2] A, Crinelli[3] R, Calzoni[2] GL, Teti[2] G and Bagni[2] N

5.1 Introduction

For the success of male reproductive processes, anther and pollen development are finely timed and choreographed (Koltunow et al. 1990; Scott et al. 1991). Gene expression is temporally and spatially regulated during anther development, producing highly differentiated cells and tissues responsible for non-reproductive and reproductive functions. Anther-specific mRNAs have been shown to encode a number of proteins (Goldberg et al. 1993; McCormick 1991). Multiple molecular events under tight gene control are also involved in the progression of microspores through the pollen developmental pathway (Bedinger and Edgerton 1990). Variations in the protein patterns of male gametophytes related to different stages of development have been widely demonstrated (Bedinger and Edgerton 1990; Delvallée and Dumas 1988; Hruba and Tupy 1998; Stinson et al. 1987; Tupy et al. 1983; Vergne and Dumas 1988). Therefore, the importance of protein degradation and turnover closely associated to the pathway of anther and pollen development can be hypothesized.

Ubiquitin is a small (8.6 kDa) highly conserved protein found in all eukaryotes (Vierstra 1993): a covalent ATP-dependent attachment of ubiquitin targets proteins for selective degradation. Ubiquitin-dependent proteolysis appears to act in the breakdown of many intracellular proteins in plants, since many ubiquitin-protein conjugates can be detected in crude protein extracts (Worrall and Twell 1994). A regulatory role of this degradative pathway in protein re-arrangement accompanying developmental processes has been suggested (Wunsch and Haas 1995). Ubiquitin-conjugated proteins have been detected by immunocytochemistry in developing anthers of *Nicotiana alata* and shown to be

[1] Institute of Botany; University of Urbino; Italy
*[2] for correspondence: Department of Biology; University of Bologna; Italy
[3] Institute of Biological Chemistry; University of Urbino; Italy

tissue- and time-specific (Li et al. 1995). Pollen maturation has been shown to be accompanied by a dramatic loss of free and conjugated ubiquitin in maize (Callis and Bedinger 1994); by contrast, high levels of ubiquitin and ubiquitinated proteins were detected in mature, mainly binucleate, pollen of several other species (Kulikauskas et al. 1995).

The aim of the present work was to investigate, using a biochemical approach, the presence and temporal changes of free ubiquitin and ubiquitin conjugates during anther and pollen development in a binucleate pollen species such as the dioecious vine, *Actinidia deliciosa* (kiwifruit). In this plant both staminate and pistillate flowers develop the reproductive structures, although they are not functional, of the opposite sex: empty loculi in the ovary of male flowers, unviable pollen in the anthers of the female ones.

5.2
Material and methods

Male (cv Tomuri) and female (cv Hayward) *Actinidia deliciosa* (A. Chev.) C. F. Liang and A. R. Ferguson growing in experimental plots near Faenza (Italy) were sampled during flower development starting from buds 1 cm in diameter (day 0). Either whole anthers or isolated microgametophytes of male and female flowers from the microspore tetrad stage (day 14) to mature pollen (day 35) were extracted in 50 mM Tris-HCl buffer pH 8 containing protease inhibitors. Protein content was determined according to Lowry et al. (1951) after TCA precipitation. SDS-PAGE was carried out according to Laemmli (1970). The gels were stained with Coomassie blue or electroblotted according to Towbin et al. (1979). Blots were probed with an affinity-purified rabbit polyclonal anti-ubiquitin antibody kindly provided by Prof. A. L. Haas, Medical College of Wisconsin, Milwaukee, USA. Horseradish peroxidase-conjugated goat anti-rabbit IgG was used as secondary antibody. The detection system used was that of enhanced chemiluminescence (ECL). The autoradiographic signal was analyzed densitometrically using the Phoretix 1D Plus program (Phoretix International, UK).

5.3
Results

The time-course of water content in anthers from male and female flowers of kiwifruit progressively decreased from day 14 until pollen maturation, showing a strong linear component (Fig. 1). Anthers of male and female flowers began to dehisce with very similar water contents (see intercepts of the corresponding equations of Fig. 1). However, the rate of dehydration was more than two-fold higher in anthers of male flowers (see angular coefficients).

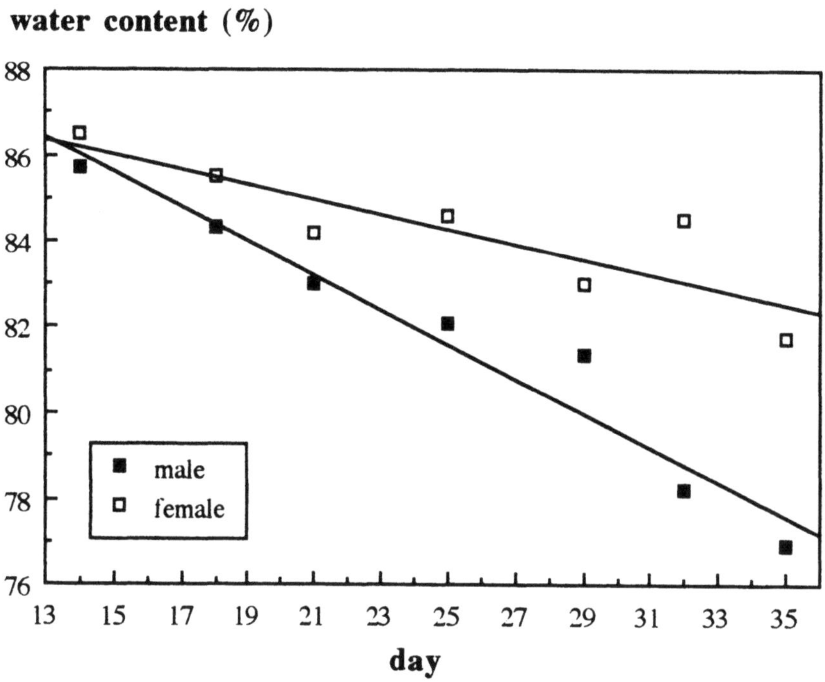

Fig. 1. Percent water content in anthers of male and female flowers of kiwifruit during development up to pollen maturity. Water content decreased according to a linear regression: y = 91.6 - 0.40 x (male flowers); y = 88.6 - 0.17 x (female flowers)

Crude protein extracts of anthers from male and female flowers were separated by SDS-PAGE and numerous polypeptide bands of different molecular mass were detected (Fig. 2A). In general, no particular changes in the protein pattern occurred during anther development in both male and female flowers. Some protein bands of low molecular mass were present at day 14, which later disappeared (Fig. 2A). By immunoblot experiments, several ubiquitin conjugates were detected in anther extracts (Fig. 2B). A large proportion of the ubiquitinate proteins were of high molecular mass (higher than 94 kDa), and they were always present in the three stages examined. At day 14, some ubiquitin immunoreacting bands of low molecular mass were detected in both male and female flowers. Their intensity decreased in the subsequent stages, and some of them disappeared. The presence of free ubiquitin was also demonstrated (Fig. 2C). In male flowers its content seemed to remain constant during anther development. By contrast, in the anthers from female flowers free ubiquitin content increased at day 21, then sharply declined dropping to 35 % of the initial value.

Changes of ubiquitin and ubiquitinated proteins were also studied during microgametophyte development (day 14, 21, 29) and in mature pollen (day 35) from male and female flowers. The Fig. 3 shows the densitometric profiles of ubiquitin-protein conjugates in the different stages. A large amount of high molecular mass (higher than 94 kDa) ubiquitin-conjugates was always present in the male from the tetrad stage until maturity; moreover, a peak corresponding to 67 kDa was observed at days 14 and 21. At 29 and 35 days a peak around 30 kDa was also detected. Finally, the content of high (higher than 67 kDa) and low molecular mass (lower than 21 kDa) ubiquitin conjugates increased during pollen maturation. Some of the latter proteins probably correspond to free ubiquitin. A different ubiquitination pattern characterized microgametophyte development in the female. The densitometric profile was similar to the male one only in the tetrad and early-uninucleate microspore stage. Later on, a dramatic loss of ubiquitination occurred at days 29 and 35, when only two small peaks were detectable. Free ubiquitin levels changed strongly during microgametophyte differentiation: two opposite trends marked the developmental path in male and female flowers (Fig. 4). While a continuous increase of free ubiquitin occurred in the male until maturity, in the microgametophyte of female flowers the largest amount was detected at the tetrad stage, then it rapidly decreased.

5.4 Discussion

Anther dehiscence is a complex process which starts when reproductive cells are at the tetrad stage (Goldberg et al. 1993). In tobacco, dehiscence has been shown to involve the formation of thickenings in the endothecial cell walls, degeneration of the circular cell cluster, breakdown of the tapetum and connective, rupture of the anther at the stomium and pollen release (Goldberg et al. 1993). Dehiscence is not a generalised desiccation of the anther tissues, but involves dehydration only of specific and localised areas of the anther, probably induced by programmed pumping of water from precise regions into other parts of the flower (Bonner and Dickinson 1989). Such a finely ordered series of events seems to have been somewhat modified in anthers of female flowers of kiwifruit. In fact, in female flowers the dehydration rate was markedly lower than that of male ones and this different behaviour could negatively affect the development of non-reproductive anther tissues.

Differences between the protein patterns of anthers at different stages of development have been reported (Abott et al. 1984; Delvallée and Dumas 1988; Wu and Murry 1985). It is known that many structural and enzymatic proteins are encoded by anther-specific mRNAs, which are in part exclusive of sporophytic tissues and in part in common with gametophytic cells (Golberg et al. 1993). It is reasonable to assume that a dynamic state of protein turnover could be related to the evolution of genomic expression and translation due to differentiation. In agreement with the immunocytochemical findings of Li et al. (1995) in *Nicotiana*, we have shown by a different experimental approach that the differentiation process in the anther of kiwifruit involves extensive protein ubiquitination. Moreover, a substantially different pattern of ubiquitin-protein conjugates and

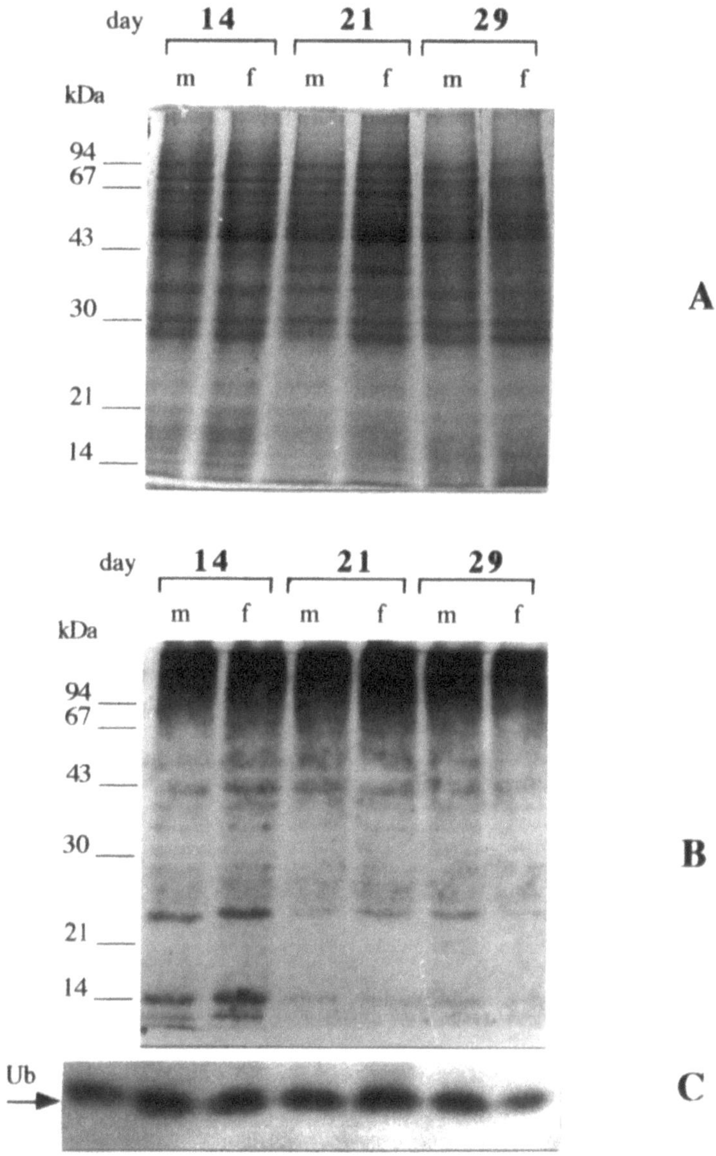

Fig. 2. Protein pattern, ubiquitin conjugates, and free ubiquitin in whole anthers of male (m) and female (f) flowers of kiwifruit during development. 20 µg of anther proteins were electrophoresed on 10% (w/v) polyacrylamide gel and stained with Coomassie blue (**A**) or electroblotted onto nitrocellulose membrane and probed with an affinity-purified polyclonal antibody against ubiquitin (**B**). For immunoblot analyses of free ubiquitin, 5 µg of the same protein extracts were fractionated on an SDS/15% polyacrylamide gel and immunoblotted as in B; in the first lane, purified bovine ubiquitin (Ub) was run as a standard (**C**). The position of molecular markers is shown on the left

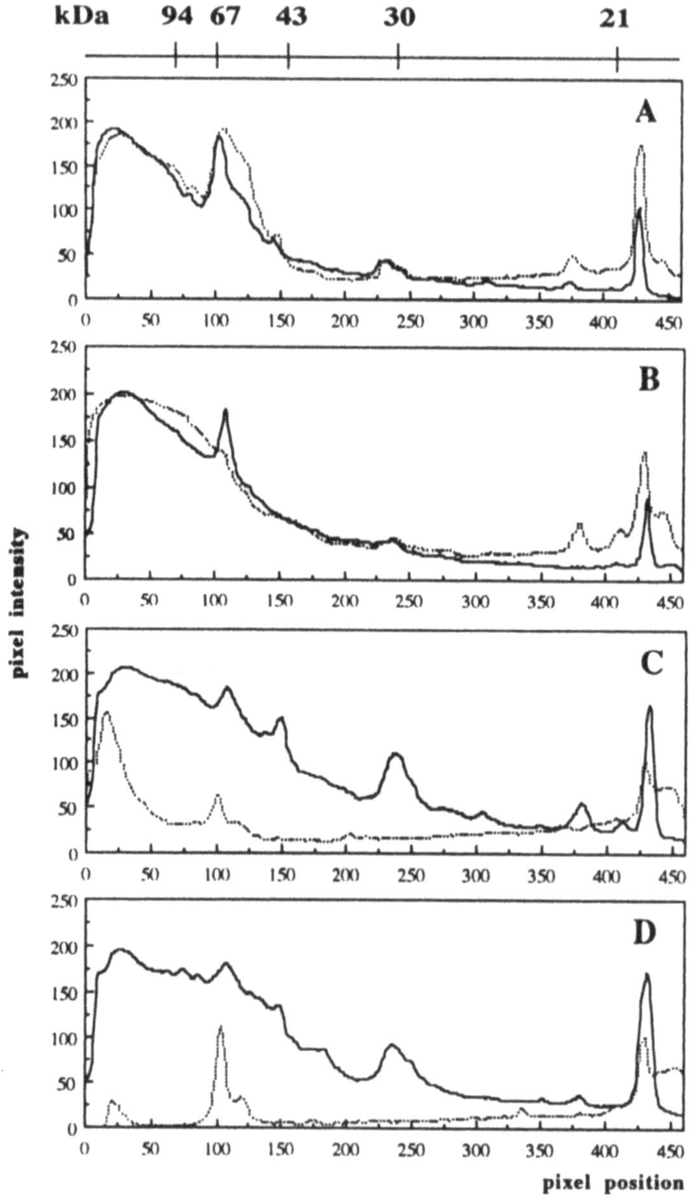

Fig. 3. Densitometric analysis of ubiquitin-protein conjugates in microgametophytes from male (unbroken line) and female (dotted line) flowers of kiwifruit during development. 15 μg of protein extracts were fractionated on an SDS/13% polyacrylamide gel and immunoblotted as described in fig. 2. **A**: day 14 (tetrads); **B**: day 21 (early-uninucleate microspore); **C**: day 29 (male: early-binucleate pollen; female: degenerating pollen at the uninucleate stage); **D**: day 35 (male: mature pollen; female: degenerated pollen). The position of molecular markers is shown at the top

changes in free ubiquitin were observed in developing anthers of male and female flowers.

We also studied ubiquitin involvement during the differentiation of microgametophytes in male and female kiwifruit flowers. Microgametophyte development in male flowers was characterized by a modulation of protein ubiquitination. By contrast, a progressive decrease of free and conjugated ubiquitin occurred in that of female ones due to the altered metabolic activity of this microgametophyte which is programmed to degenerate. A large amount of

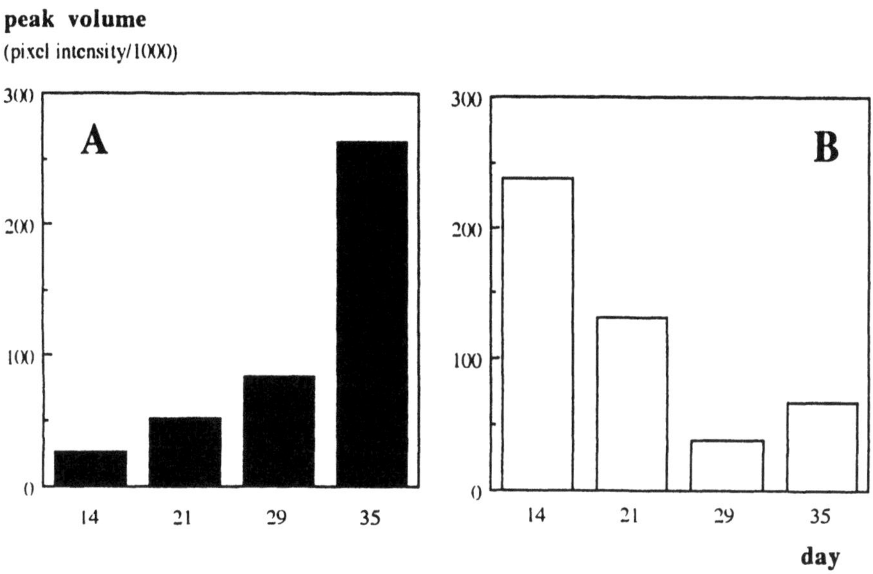

Fig. 4. Densitometric evaluation of free ubiquitin pattern during microgametophyte development in male (**A**) and female (**B**) kiwifruit flowers. 5 μg of protein extracts were fractionated on an SDS/15% polyacrylamide gel and immunoblotted as described in Fig. 2

high mass conjugates was present at all developmental stages of the microgametophyte of male flowers, until the time of pollen release. Free ubiquitin was shown to continuously increase during development, and an induction of free ubiquitin synthesis seemed to occur in mature kiwifruit pollen. Instead, a regulated loss of both free and conjugated ubiquitin was found in trinucleate maize pollen during maturation, so that 50-fold less conjugates than in young microspores were present (Callis and Bedinger 1994). It is known that trinucleate pollen has already performed the second haploid division prior to dehiscence and release, and is equipped with highly developed mitochondria and preformed proteins essential for tube emergence and growth (Hoekstra 1983). Due to its fully developed metabolic state at maturity, the grass pollen should not require a tight regulation of protein turnover after dehiscence, and could have therefore eliminated the ubiquitin system (Callis and Bedinger 1994). The presence and increasing levels of free ubiquitin were observed during spermiogenesis in chicken (Agell and Mezquita

1998). As in mature grass pollen, free ubiquitin was no longer detectable in mature chicken spermatozoa. Therefore, if a role of ubiquitin in differentiation processes can be hypothesized (Wunsch and Haas 1995), its loss seems to correlate with the completion of the processes. In the present paper we show that free and conjugated ubiquitin increases during kiwifruit microgametophyte development and is still high in the mature binucleate pollen. It is known that binucleate pollen is shedded in a more precocious developmental and slower metabolic state than the trinucleate one, and a number of activities (the second mitosis included) have yet to be performed after germination. Therefore, the different levels of both free ubiquitin and ubiquitinated proteins in binucleate and trinucleate microgametophytes could be related to their different metabolic state at maturity.

Acknowledgements

This work was supported by a CNR contribution to N.B. We thank Prof. M. M. Altamura (Università La Sapienza, Roma, Italy) for the microgametophyte developmental stage determination. We also thank the 'Istituto Professionale di Stato per l'Agricoltura' of Persolino (Faenza, Italy) for kindly allowing us to collect flowers from their kiwifruit plants.

References

Abott AG, Ainsworth CC, Flavell RB (1984) Characterization of anther differentiation in cytoplasmic male sterile maize using a specific isozyme system (esterase). Theor Appl Genet 67: 469-473

Agell N, Mezquita C (1998) Cellular content of ubiquitin and formation of ubiquitin conjugates during chicken spermatogenesis. Biochem J 250: 883-889

Bedinger P, Edgerton MD (1990) Developmental staging of maize microspores reveals a transition in developing microspore proteins. Plant Physiol 92: 474-479

Bonner LJ, Dickinson HG (1989) Anther dehiscence in *Lycopersicon esculentum* Mill. I. Structural aspects. New Phytol 113: 97-115

Callis J, Bedinger P (1994) Developmentally regulated loss of ubiquitin and ubiquitinated proteins during pollen maturation in maize. Proc Natl Acad Sci USA 91: 6074-6077

Delvallée I, Dumas C (1988) Anther development in *Zea mays*: changes in protein, peroxidase and esterase patterns. J Plant Physiol 132: 210-217

Goldberg RB, Beal TP, Sanders PM (1993) Anther development: basic principles and practical applications. Plant Cell 5: 1217-1229

Hoekstra FA (1983) Physiological evolution in angiosperm pollen: possible role of pollen vigour. In: Mulcahy DL, Ottaviano E (eds) Pollen: biology and implications for plant breeding. Elsevier Biomedical, New York, pp 35-41

Hruba P, Tupy J (1998) Developmental-related glycoproteins in tobacco pollen and microspores. In: University of Reims, Conference on 'Anther and pollen: from biology to biotechnology', Reims 9-10 April, p 40

Koltunow AM, Truettner J, Cox J, Wallroth KH, Goldberg RB (1990) Different temporal and spatial gene expression patterns occur during anther development. Plant Cell 2: 1201-1224

Kulikauskas R, Hou A, Muschietti J, McCormick S (1995) Comparisons of diverse plant species reveal that only grasses show drastically reduced levels of ubiquitin monomer in mature pollen. Sex Plant Reprod 8: 326-332

Laemmli UK (1970) Cleavage of structural proteins during the assembly of the head of bacteriophage T4. Nature 227: 680-685

Li YQ, Southworth D, Linskens HF, Mulcahy DL, Cresti M (1995) Localization of ubiquitin in anther and pistils of *Nicotiana*. Sex Plant Reprod 8: 123-128

Lowry OH, Rosebrough NJ, Farr AL, Randall RJ (1951) Protein measurement with the Folin phenol reagent. J Biol Chem 193: 265-275

McCormick S (1991) Molecular analysis of male gametogenesis in plants. Trends Genet 7: 298-303

Scott R, Hodge R, Paul W, Draper J (1991) The molecular biology of anther differentiation. Plant Sci 80: 167-191

Stinson JR, Eisenberg AJ, Willing RP, Pe ME, Hanson DD, Mascarenhas JP (1987) Genes expressed in the male gametophyte of flowering plants and their isolation. Plant Physiol 83: 442-447

Towbin H, Stahelin T, Gordon J (1979) Electrophoretic transfer of proteins from polyacrylamide gels to nitrocellulose sheets: procedure and some applications. Proc Natl Acad Sci USA 76: 4350-4354

Tupy J, Suss J, Hrabetova E, Rihova L (1983) Developmental changes in gene expression during pollen differentiation and maturation in *Nicotiana tabacum* L. Biol Plant 25: 231-237

Vergne P, Dumas D (1988) Isolation of viable wheat male gametophytes at different stages of development and variations in their protein patterns. Plant Physiol 88: 969-972

Vierstra RD (1993) Protein degradation in plants. Annu Rev Plant Physiol Plant Mol Biol 44: 385-410

Worrall D, Twell D (1995) Pollen maturation: where ubiquitin is not required ? BioEssay 16: 873-875

Wu FS, Murry LE (1985) Changes in protein and amino-acid content during anther development in fertile and cytoplasmic male sterile *Petunia*. Theor Appl Genet 71: 68-73

Wunsch AM, Haas AL (1995) Ubiquitin-protein conjugates selectively distributed during early chicken embryogenesis. Dev Dynam 204: 118-132

6 Mechanisms of Angiosperm anther dehiscence, a historical review

Keijzer* CJ

6.1 Morphological changes during anther dehiscence

Research on Angiosperm anthers dates back for over three centuries, when Grew (1682) published drawings of anthers and pollen in his "Anatomy of Plants" (Fig. 1). Special emphasis on the opening mechanism was not given in this first masterwork on plant anatomy. The first reports relating to the opening mechanism were produced more than one century later by Brisseau-Mirbel (1808), who pointed to the role of the connective tissue in anther dehiscence. Meyen (1828) and Von Mohl (1830) reported the presence of fibrillar structures in hypodermal cells of the anther. Purkinje (1830) introduced the names exothecium and endothecium for the anther epidermal and hypodermal cell layer, respectively, and related evaporation from the latter with anther dehiscence (Fig. 2). A main part of his book is dedicated to the species-dependent morphological variation of these cells (Fig. 2). Schleiden (1849) agreed with the name exothecium for the epidermis, but proposed mesothecium for the hypodermal layer (Purkinjes endothecium) and endothecium for the layer that nowadays is called the tapetum. Chatin (1870a;b) discussed this terminology, but finally followed Purkinjes terms. Goebel (1902) and Staedtler (1923) reserved the name exothecium for an epidermis containing (endothecium-like) wall thickenings, as could be found in gymnosperms and proposed the name epidermis for angiosperm anthers.

Chatin (1870a-d) gave his doubts on the uniform explication of Purkinje that the dehydrating endothecium caused the entire process of anther dehiscence and reported that anthers lacking wall thickenings in the endothecium dehisced as well. He was the first author to distinguish two different steps in the process: the disruption of the tissue between two adjacent locules (a narrow area where the endothecium lacks wall thickenings) on the one hand and the opening of the thus fused locules by one slit in

* Laboratory of Plant Cytology and Morphology; Department of Experimental Botany; Wageningen Agricultural University; Arboretumlaan 4; 6703 BD Wageningen; The Netherlands

the epidermis on the other (Fig. 3). Moreover he observed that these two steps took place before anthesis (i.e. flower opening). Years later these dissociating tissues were called the septum and the stomium, respectively (Coulter and Chamberlain 1903).

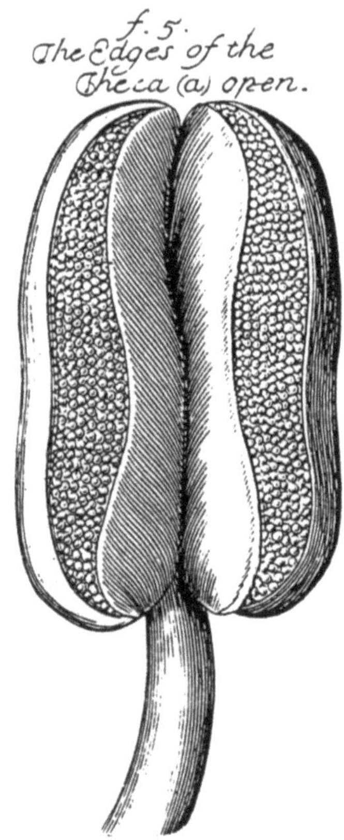

Fig. 1. The open anther of *Hyoscyamus* (Grew 1682)

Given this differentiation, much research around the beginning of our century focussed on the underlaying mechanism of stomium and septum opening. No doubt existed about the theory that the dehydrating endothecium cells caused the final recurving of the locule walls at anthesis. However, the idea of Schrodt (1885; 1901), Steinbrinck (1895) and Hannig (1910) that the dehydrating endothecium created the one and only force that disrupted the cells of both the septum and stomium was sharply attacted by Schneider (1908; 1909; 1911; Fig. 4), who demonstrated that dehydration of (premature) anthers containing wall thickenings in the endothecium could not open the septum and stomium. He claimed that the swelling pollen population produced the force for this disruption. Some years before, Strasburger (1902) already reported the idea that the growing pollen mass disrupted the septum and that wall degrading enzymes dissociated the cells of the stomium of water plants.

In contrast, Steinbrinck (1909) reported a normal opening process in anthers that were not completely filled with pollen. After this era of combatting german investigators, Namikawa (1919) and Bequerel (1932) reported the activity of cell wall degrading enzymes in both the septum and the stomium. The latter also reported the

Si loculum antherae *coniothecam* appellaveris membranae epidermidali externae *exothecii*, interno vero cellularum contextui nomen *endothecii* jure tribuetur.

Fibras organum principale endothecii esse duco, quibus functionem suam, dispersionem pollinis absolvit, cellulas vero earum quasi matrices in quarum cavo ex earum succis propriis efformantur. Vis, quae, dum antherarum loculi dehiscent, fibras contrahit, videtur esse proprietas quaedam communis physica fors hycroscopica, contractionem vero exsiccationis quodam processu atmosphaerae aëris ambientis influxu suboriri putaverim.

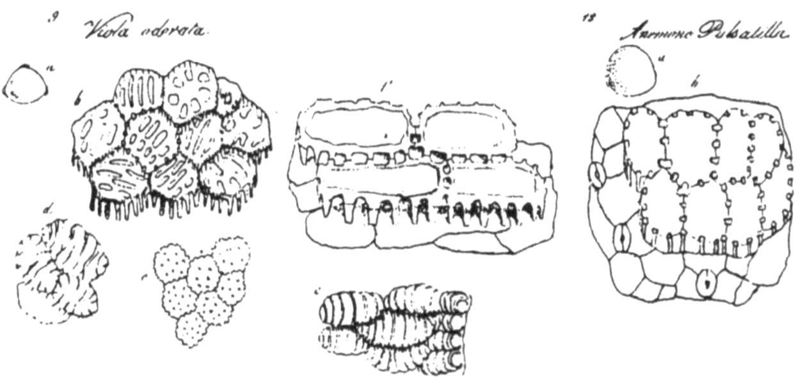

Fig. 2. The naming, functional explication and species-dependent morphological variation of exo- and endothecium (Purkinje 1830)

presence of a longitudinal vascular tissue in the septum of lily anthers, an observation that could never be confirmed in later years. Woycicki (1924a) found a normal opening process after preventing the pollen to produce pressure in the locules. Using lily anthers with their huge epidermis cells adjacent to the stomium, Woycicki (1924a) concluded that this latter site was mechanically opened by the locule walls when they were inward bending due to the swelling (i.e. hydration) of both these huge epidermis cells and the endothecium (Fig. 5). Given this mechanism, the epidermis bordering the stomium acted as a hinge, an observation that was supported by Staedtlers (1923) findings that anthers of many species lacked an epidermis with the exeption of some cells bordering the stomium (Fig. 6). Given Woycicki's (1924a, b) experiments, that could be succesfully repeated and supported by additional modern research (Keijzer 1987a), he was the first author to bring the vision that, apart from the outward bending action of the dehydrating endothecium, known for years, also the opposite action, i.e. inward bending due to water uptake, took place and even had a crucial function (stomium rupture) in the anther opening process. For the preceding opening of the septum he reported cell wall degrading enzymes to be responsible (Woycicki 1924a; Fig. 5). When testing the older theories, using different micromanipulation and modern microscopical techniques, Keijzer (1987a) found the

> membrane ou exothecium. Mais ce serait aller trop loin que de reporter à la membrane épidermique le rôle exclusif qu'attribuait Purkinje à son endothecium, et je ne vois en elle que l'un des agents de la déhiscence, phénomène complexe, le plus souvent subordonné à plusieurs causes, parmi lesquelles doivent être comptés, outre la membrane externe, la deuxième membrane ou endothecium de Purkinje, la destruction des cloisons des logettes, le décollement et la destruction des bords suturaux des valves, plus accessoirement la troisième membrane, peut-etre même le connectif.

Fig. 3. The first subdivision of the anther opening process (Chatin 1870b)

theories of Woycicki suitable for different plant species (Fig. 7) and came to the following sequence of lysigenic and schizogenic events:

1/ Deposition of (generally) U-shaped wall thickenings in the endodermis cells with the exception of a small area between each two bordering locules, the septum;

2/ Lysigenic opening of the septum due to local enzymatical cell wall degradation;

3/ Schizogenic rupture of the tapetal membranes lining the septum by the expanding pollen mass;

4/ Schizogenic rupture of the stomium by the inward bending locule walls due to water uptake by epi- and endothecium;

5/ Opening of the fused locules by the outward bending locule walls due to desiccation of epi- and endodermis at anthesis.

Recently Keijzer et al (1996) found enzymatic cell wall degradation in the stomium of maize in addition to the schizogenic process. Namikawa (1919), Oryol and Zhakova (1977), Horner and Wagner (1980), Wagner (1983) and Bonner and Dickinson (1989) found storage of crystals in the septum of Solanaceous anthers. The latter two authors identified these as calcium oxalate, relating this substance to the activation of cell wall degrading enzymes, necessary to open the septum. Oryol and Zhakova (1977) reported cellulases to be active in the septum. Sexton et al (1990) found that cellulases in the septum were similar to those found in abscission layers of flower buds, thus referring the septum to as an abscission layer.

Purkinje (1830), Kühn (1908) and Richter (1929) made extensive studies on distribution, shape and size of the endothecial wall thickenings in different plant species, revealing a wide variation among different species. On both the morphology of endothecial wall thickening and the localisation of the stomium in particular species, numerous reports are available (see Davis 1966). Depending upon the species, endothecial wall thickenings may contain pectins, cellulose and lignin (Oryol and Zhakova 1977; Keijzer 1987b).

durch Wind, Insekten usw. freigelegt wird." Ich kann STEINBRINCK nicht beipflichten. Die Erforschung der Antherenmechanik, an der die hervorragendsten Gelehrten gearbeitet haben, dauert nun schon hundert Jahre. Immer sprach man vorbehaltlos von der „Öffnungsmechanik", von „Öffnungsmechanismus", vom „Öffnungsproblem der Antheren"; die Pariser: „Causes de la déhiscence des anthères", Daß nun keiner der Autoren an die Hauptsache des Öffnungsmechanismus gedacht habe, ohne welche es überhaupt niemals eine Öffnung geben könnte, nämlich an die erste Entfernung, an die Lostrennung der Klappenränder vom Verbindungsgewebe (Connectiv), das wäre doch fast nicht denkbar. STEINBRINCKs Meinung wird auch dadurch als unrichtig dargetan, daß in den ersten vierzig Jahren der Antherenforschung von MIRBEL, LUDWIG, PURKINJE, TREVIRANUS, MEYEN, H. v. MOHL [1]), „der Druck der wachsenden Pollenmasse" als Öffnungsursache der Antheren in Betracht gezogen und allerdings meistens bekämpft worden war.

„zur Verteidigung der STEINBRINCKschen Erklärungsweise zu veröffentlichen" für gut fand, weil „in der letzten Zeit J. M. SCHNEIDER in einer ziemlich absprechenden Weise die zweifellos sehr sorgfältigen und verdienstvollen Arbeiten STEINBRINCKs bekämpfte und durch neue Behauptungen die Sachlage noch weiter zu verwirren drohte [1])". HANNIGs Arbeit enthält

Fig. 4. The public discussions among the German researchers (Schneider 1911)

6.2 The water dislocations

Apart from the studies mentioned above that focussed on the cell dissociations in septum and stomium, a number of scientists studied the underlaying mechanisms of water uptake and subsequent dehydration of the endothecium (and the epidermis).

Woycicki (1924b) and Keijzer and Willemse (1988) reported breakdown of starch preceding the swelling of the epidermis and endothecium cells. Keijzer (1987a) demonstrated an increasing osmotic pressure in these cells preceding stomium rupture.

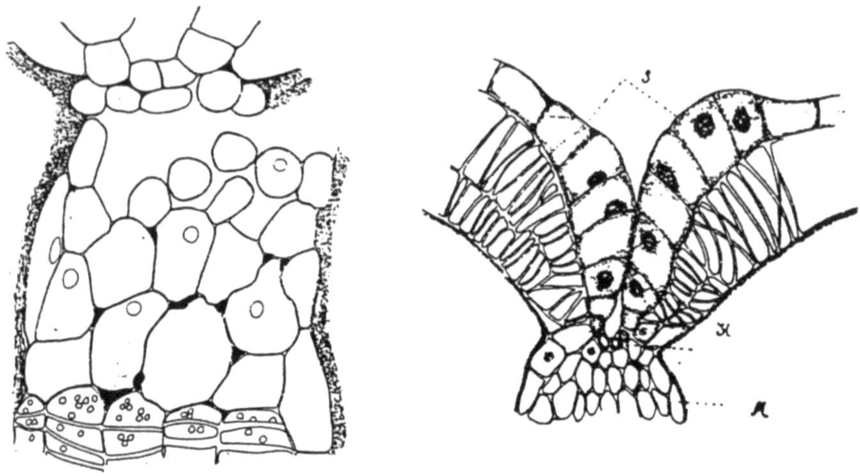

Fig. 5. The lysigenic opening of the septum (left) and the schizogenic opening of the stomium (right) (Woycicki 1924a)

Purkinjes (1830) initial view on the endothecium dehydration, i.e. evaporation, is still supported in our days, most recently by experiments in our lab (Keijzer 1987a). However, Burck (1906) denied evaporation and proposed retraction of water from the endothecium by the nectaries. However, Hannig (1910), Wolff (1924) and Schmid and Alpert (1977) could reject his thesis. Hannig (1910) demonstrated that pigments in the epidermis and endothecium transformed light into temperature increases, thus accelerating evaporation and accordingly anther dehiscence. More recently Keijzer (1983; 1987a) and Linskens and Cresti (1988) proved the importance of evaporation in endothecium desiccation. The role of the undulating cuticle, found on maturating anthers of many species is still unclear (Keijzer 1987a), although Cheng, Greyson and Walden (1986) assumed it to result from a tangential shrinkage of the epidermal cells, that in their turn would mechanically open the septum in this way. Bianchini and Pacini (1996) related variation in cuticle thickness to preferential evaporation from particular pre-determined epidermal cells, leading to a sudden, explosive anther dehiscence. Keijzer, Hoek and Willemse (1987) sealed stomates, found on the connective tissue of many species (Kenda 1950; 1952) and in this way demonstrated their role in controlling the evaporation rate and thus the time of anther dehiscence relative to the time of anthesis. Asynchronous opening of anthers and flowers, respectively, might result into cleistogamy, protandry or protogyny. In *Ipomoea* sp. Keijzer (unpublished) found the closed flower bud to grow enormously during the final 24 hours before anthesis. The accompanying decrease of the internal relative humidity made the anthers dehisce before flower opening, resulting into cleistogamy. In contrast, onions keep their anthers closed up to 24 hours after anthesis (Keijzer et al. 1987).

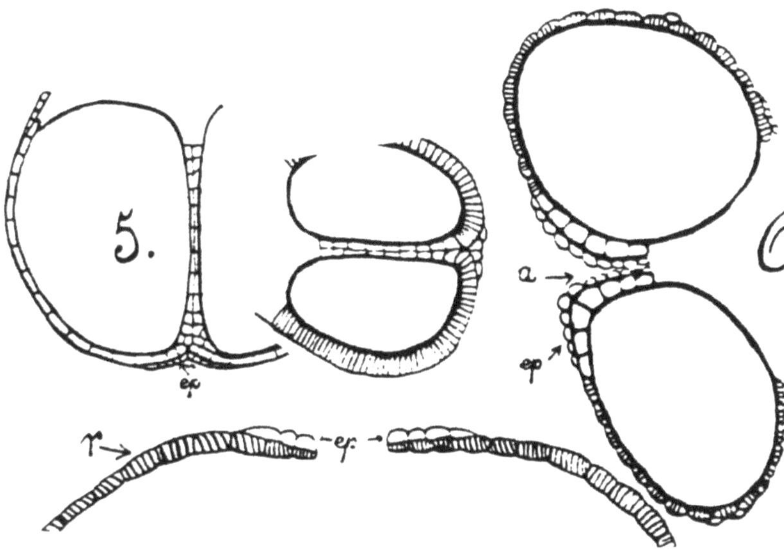

Fig. 6. Examples of epidermal reductions outside the stomium area (Staedtler 1923)

Recently Keijzer (in prep) demonstrated retraction of water from anthers to the (strongly) extending filaments of Poaceae, thus synchronously dehydrating the anthers on the one hand and filling the vacuoles of the filament cells on the other. Also Hannig (1910), using ingenious manipulations, and Schmid and Alpert (1977) presented interesting phenomena, showing that merely evaporation is too simple an explanation. The latter authors detached anthers from their filaments and found them to dehisce slower than those still attached to the flowers. Removal of the perianth appeared to delay anther dehiscence. Finally, contradicting to what should promote evaporation, they found that decreasing the relative humidity of the environment delayed anther dehiscence. In their thorough study on tomato anthers, also Bonner and Dickinson (1990) exclude evaporation as the mechanism behind the dehydration of the pre-determined anther tip and conclude environmentally-linked hygroscopic absorbtion.

Relating to the discussion whether the endothecium was the only tissue to cause the open bending force due to dehydration, Chatin (1870a;b) and Staedler (1923) (Fig. 5) found a normal opening process in anthers with a naturally reduced epidermis and also anthers from which the epidermis had been artificially removed (Gaston Bonnier and Leclerc du Sablon 1903) appeared to dehisce. Brodtmann (1898) and Schneider (1908) even isolated the wall thickenings from the endothecium cells and observed them closing their U-shape during desiccation. Apart from tracing the responsible cell layer, they also demonstrated a hygroscopic background for the underlaying process. A hygroscopic mechanism (located in the endothecial cell wall) was also proposed by Treviranus (1835), Von Mohl (1858), Schinz (1883), Leclerc du Sablon (1885), Schrodt (1885), Brodtmann (1898), Schwendener (1899), Colling

(1905) and Schips (1913). Following Kamerling (1898), others suggested the existence of cohesive forces (located in the cell lumen) to be responsible (Steinbrinck, 1898; 1899; 1901; 1903; Haberland 1924; Von Guttenberg 1926; Straka 1962). Pfeffer (1904), Hannig (1910), Canright (1952), Schmid (1976) and Keijzer (1987a) believed both forces to be operative, the main part being done by cohesion.

Fig. 7 to be reduced

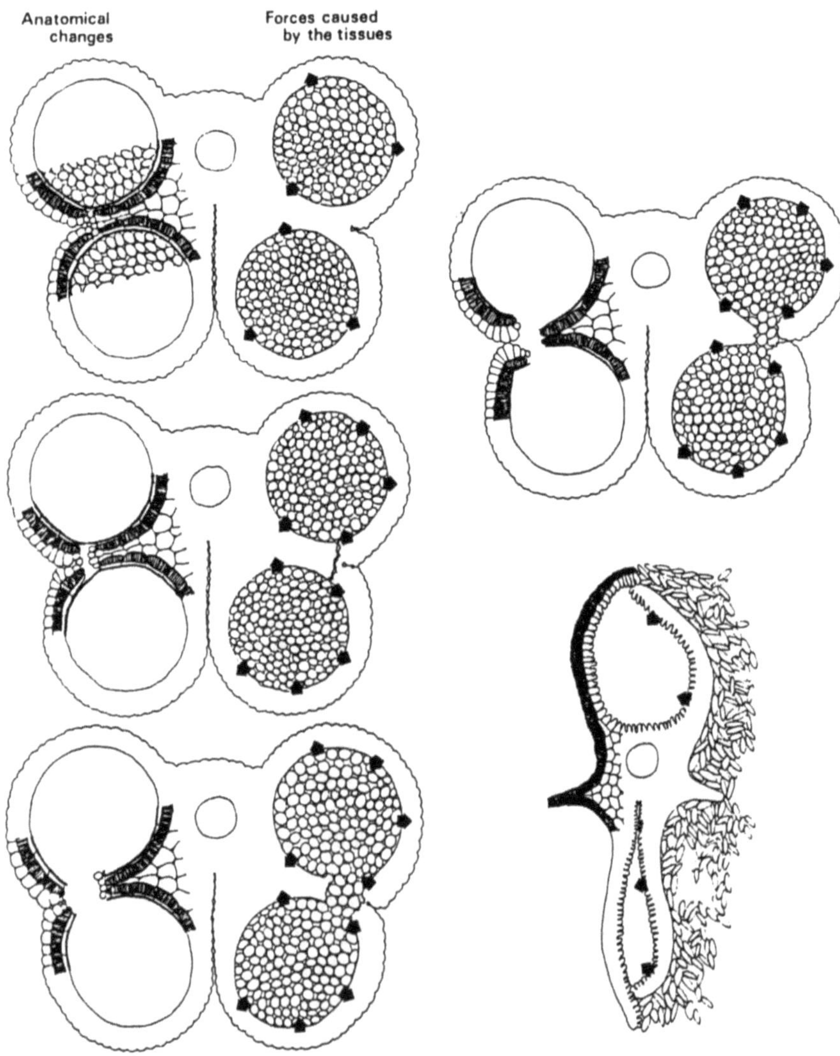

Fig.7. The main steps leading to anther dehiscence (Keijzer 1987a)

6.3 Triggering anther dehiscence and anthesis

In search for the triggering of anther dehiscence, Chaunan (1979) and Chaunan et al. (1982) observed that gametocides prevented both the development of the wall thickenings in the endothecium cells and the degeneration of the tapetum and hypothesized that any substance in the tapetum inhibited the wall thickenings to develop. Such an hypothesis was earlier ventilated by De Fossard (1969) and Gupta and Nanda (1973). In contrast, Keijzer and Cresti (1987) found that naturally male sterile anthers (that also missed locular fluid in parts of the anther) did develop endothecial wall thickenings. However, they did not pump their epidermis and endothecium (and accordingly kept their stomium closed) in the locular fluid-free areas, which might indicate the necessity of this fluid to open the stomium. Recently Keijzer (in preparation) demonstrated that the endogenous carbon dioxide concentration in the flower bud of maize determines the time of anthesis. Research is in course to prove that this gas originates from the respirating pollen, which would provide an additional explication why the septum and stomium should be open before anthesis. In favour of this, Tschermak (1890) had found that increasing temperatures brought rye ears into anthesis, describing himself messing around with a handful of flowers under his hat (Fig. 8).

beim Streichen durch den Mund, beim Tragen unter dem Hute,

Fig.8. The way Tschermak forced rye flowers into anthesis (Tschermak 1890)

6.4 Molecular biology

Recently also the molecular biologists made their first steps in the field of anther dehiscence.

Raghavan (1989) found the poly(A)RNA labelling in the endothecium of rice different from that in the epidermis and middle layers after in situ hybridizing anthers with 3H-poly(U). He associated this phenomenon with transcriptional activity required for the development of the wall thickenings.

Koltunow et al. (1990) found genes of a class that encodes enzymes for cell degeneration expressed preferentially in the stomium area (i.e. including septum) of tobacco anthers. Also stomium-(and septum-)specific antibodies have been raised, further supporting the view that these tissues are biochemically different from the rest of the anther wall (Trull et al. 1991; Cañas and Malmberg 1992).

6.5 Conclusions

The research on anther opening processes started in the very early days of experimental plant anatomy, presumably due to their clear visibility on the outside of the plant. Two centuries later, the main lines have become clear (Keijzer 1987a;b; Bonner and Dickinson 1989; 1990) but, as expected, a great species-dependent variation has been found, reflected in the extent (or abscence) of particular steps of the main line and in the pathways of water allocations. Using the words of Bonner and Dickinson (1990), the process is an "orchestrated programme of structural and physiological events". In the future, molecular biology might enable us to control this programme, thus opening ways to induce a handsome type of male sterility.

Acknowledgments

The author thanks Professor MTM Willemse for critically reading the manuscript.

References

Bequerel P (1932) La déhiscence de l'anthère du Lis blanc. C R Acad Sci Paris 195: 165-167

Bianchini M, Pacini E (1996) Explosive anther dehiscence in *Ricinus communis* L. involves cell wall modifications and relative humidity. Int J Plant Sci 157: 739-745

Bonner LJ, Dickinson HG (1989) Anther dehiscence in *Lycopersicon esculentum* Mill. 1. Structural aspects. New Phytol 113: 97-115

Bonner LJ, Dickinson HG (1990) Anther dehiscence in *Lycopersicon esculentum* Mill. 2. Water relations. New Phytol 115:367-375

Bonnier G, Leclerc du Sablon M (1903) Cours de Botanique. Libr Gen Enseign Paris

Brisseau-Mirbel CF (1808) Exposition et défense de ma théorie de l'organisation végétale publié par le docteur Bilderdijk. Van Cleef, The Hague

Brodtmann F (1898) Über die Funktion der mechanischen Elemente beim Farnsporangium und bei der Anthere. Diss Erlangen

Burck W (1906) On the influence of the nectaries and other sugar containing tissues in the flower on the opening of the anthers. Rec Trav Bot Neerl 3: 163-172

Cañas L, Malmberg R (1992) Isolation and characterization of three monoclonal antibodies specific for pistil and stamen tissues of tobacco (*Nicotiana tabacum* L.). Pl Sci 83: 195-203

Canright JE (1952) The comparative morphology and relationships of the Magnoliaceae. 1. Trends in specialization in the stamens. Am J Bot 39: 484-497
Chaunan SVS (1979) Development of endothecium in relation to tapetal behaviour in some male sterile plants. Phytomorph 29: 245-251
Chaunan SVS, Dingra RK, Kinoshita T (1982) Effect of phytogametocidal compounds on the development of endothecium in some plants. Jap J Breed 32: 139-145
Chatin A (1870a) De l'anthère: recherches sur le développement, la structure et les fonctions de ses tissus. Baillière et fils, Paris.
Chatin A (1870b) Causes de la déhiscence des anthères. CR Acad Sci Paris 70: 201-203
Chatin A (1870c). Rôle de la deuxième membrane. CR Acad Sci Paris 70: 410-413
Chatin A (1870d). Rôle de la troisième membrane, de la cloison et du connectif. Experiences sur l'action de l'eau. CR Acad Sci Paris 70: 644-648
Cheng PC, Greyson RI, Walden DB (1986). The anther cuticle of *Zea mays*. Can J Bot 64: 2088-2097
Coulter JM, Chamberlain CJ (1903) Morphology of Angiosperms, part 2. Appleton, New York, pp 40-45
Davis GL (1966) Systematic Embryology of the Angiosperms. John Wiley, New York
De Fossard RA (1969) Development and histochemistry of the endothecium in the anthers of in vitro grown *Chenopodium rubrum*. Bot Gaz 134: 125-129
Goebel K (1902) Über die Pollenentleerung bei einige Gymnospermen. Flora 91 suppl.
Grew N (1682) The Anatomy of Plants. London
Gupta SC, Nanda K (1973) Fibrous endothecium, tapetum and pollen development in *Belamcanda chinensis* Dc. Bot Gaz 134: 125-129
Haberland G (1924) Physiologische Pflanzenanatomie. Engelmann, Leipzig
Hannig E (1910) Über den Öffnungsmechanismus der Antheren. Jahrb Wiss Bot 47: 186-218
Horner HT, Wagner BL (1980) The association of druse crystals with the developing stomium of *Capsicum annuum* (Solanaceae) anthers. Am J Bot 67: 1347-1360
Kamerling Z (1898) Oberflächenspannung und Kohäsion. Bot Zentralbl 73
Keijzer CJ (1983) Hydration changes during anther development. In Mulcahy DL, Ottaviano E (eds) Pollen: Biology and Implications for Plant Breeding. Elsevier, New York, pp 197-201
Keijzer CJ (1987a) The processes of anther dehiscence and pollen dispersal. 1. The opening mechanism of longitudinally dehiscing anthers. New Phytol 105: 487-498
Keijzer CJ (1987b) The processes of anther dehiscence and pollen dispersal. 2. The formation and the transfer mechanism of pollenkitt, cell wall development in the locule tissues and a function of the orbicules in pollen dispersal. New Phytol 105: 499-507
Keijzer CJ, Cresti M (1987) A comparison of anther development in male sterile *Aloe vera* and male fertile *Aloe ciliaris*. Ann Bot 59: 533-542
Keijzer CJ, Hoek IHS, Willemse MTM (1987) The processes of anther dehiscence and pollen dispersal. 3. The dehydration of the filament tip and the anther in some monocotyledonous species. New Phytol 106:281-287
Keijzer CJ, Leferink-ten Klooster HB, Reinders MC (1996) The mechanics of the grass flower: anther dehiscence and pollen shedding in maize. Ann Bot 78: 15-21
Keijzer CJ, Willemse MTM (1988) Tissue interactions in the developing locule of *Gasteria verrucosa* during microgametogenesis. Acta Bot Neerl 37: 475-492
Kenda G (1950) Über Stomata an Antheren. Thesis, Graz
Kenda G (1952) Stomata an Antheren. Phyton 4: 83-96

Koltunow AM, Truettner J, Cox KH, Wallroth M, Goldberg RB (1990) Different temporal and spatial gene expression patterns occur during anther development. Plant Cell 2: 1201-1224
Kühn E (1908) Über den Wechsel der Zelltypen im Endothecium der Angiospermen. Diss Zurich
Leclerc du Sablon M (1885) Recherches sur la structure et la déhiscence des anthères. Ann Sci Nat Bot 7: 97-134
Linskens HF, Cresti M (1988) The effect of temperature, humidity and light on the dehiscence of tobacco anthers. Proc Kon Ned Akad Wet C 91: 369-375
Meyen FJF (1828) Anatomische und Physiologische Untersuchungen über den Inhalt der Pflanzenzellen. Joseephy, Berlin
Namikawa I (1919) Über das Öffnen der Antheren bei einige Solanaceen. Bot Mag (Tokyo) 33: 62-69
Oryol LJ, Zhakova MA (1977) The mechanism of anther dehiscence of tomato *Lycopersicon esculentum* Mill. (*Solanaceae*). Bot Zhur 62: 1720-1730
Pfeffer W (1904) Pflanzenphysiologie. Engelman, Leipzig
Purkinje JE (1830) De cellulis antherarum fibrosis nec non de granorum pollinarium formis. Bratislava.
Raghavan V (1989) mRNAs and a clone histone gene are differentially expressed during anther and pollen development in rice (*Oryza sativa* L.). J Cell Sci 92: 217-229
Richter S (1929) Über den Öffnungsmechanismus der Antheren bei einige Vertretern der Angiospermen. Planta 8: 154-184
Schinz H (1883) Untersuchungen über den Mechanismus des Aufspringens der Sporangien und Pollensäcke. Diss Zürich
Schleiden (1849) Grundzüge der Wissenschaftlichen Botanik. Leipzig
Schmid R (1976) Filament histology and anther dehiscence. Bot J Linn Soc 73: 303-315
Schmid R, Alpert PH (1977) A test of Burck's hypothesis relating anther dehiscence to nectar secretion. New Phytol 78: 487-498
Schips M (1913) Zur Öffnungsmechanik der Antheren. Beih Bot Zentralbl, Abt 1, 31: 119-208
Schneider JM (1908) Der Öffnungsmechanismus der Tulipa Anthere. Ber Deutsch Bot Ges 26: 394-398
Schneider JM (1909) Zur ersten und zweiten Hauptfrage der Antherenmechanik. Ber Deutsch Bot Ges 27: 196-201
Schneider JM (1911) Über das Öffnen des Nahtgewebes der Antheren. Ber Deutsch Bot Ges 29: 406-416
Schrodt J (1885) Das Farnsporangium und die Anthere. Regensburg
Schrodt J (1901) Zur Öffnungsmechanik der Staubbeutel. Ber Deutsch Bot Ges 19: 483-488
Schwendener S (1899) Über das Öffnungsmechanismus der Antheren. Sitzungsber Kön Preuss Akad Wiss, Berlin
Sexton R, Del Campillo E, Duncan D, Lewis LN (1990) The purification of an anther cellulase (ß(1:4) 4-glucan hydrolase) from *Lathyrus odorata* L. and its relationship to the similar enzyme found in abscission zones. Plant Sci 67: 169-176
Staedtler G (1923) Über Reduktionserscheinungen im Bau der Antherenwand von Angiospermen-Blüten. Flora 116: 85-108
Straka H (1962) Nicht durch Reize ausgelöste Bewegungen. In Ruhland W (ed) Handbuch der Pflanzenphysiologie. Springer, Berlin, pp 716-835

Strasburger E (1902) Ein Beitrag zur Kenntniss von Ceratophyllum submersum. Jahrb wiss Bot 37: 477-524

Steinbrinck C (1895) Grundzüge der Öffnungsmechanik von Blütenstaub- und einigen Sporenbehältern. Bot Jaarb Dod (Gent) 7: 26-27

Steinbrinck C (1898) Ist die Kohäsion des schwindenden Füllwassers der dynamischen Zellen die Ursache der Schrumpfungsbewegungen von Antherenklappen, Sporangien und Moosblättern? Ber Deutsch Bot Ges 16: 97-103

Steinbrinck C (1899) Zum Vorkommen und zur Physik der pflanzlichen Kohäsionsmechanismen. Ber Deutsch Bot Ges 17: 170-178

Steinbrinck C (1901) Zum Öffnungsproblem der Antheren. Ber Deutsch Bot Ges 19: 552-556

Steinbrinck C (1903) Kohäsions- oder hygroskopischer Mechanismus ? Ber Deutsch Bot Ges 21: 42-52

Steinbrinck C (1909) Über den ersten Öffnungsvorgang der Antheren. Ber Deutsch Bot Ges 27: 300-312

Steinbrinck C (1915) Über den Nachweis von Kohäsionsfalten in geschrumpfelten Antherengeweben. Ber Deutsch Bot Ges 33: 66-72

Treviranus LC (1835) Physiologie der Gewächse. Adolph Marcus, Bonn

Trull MC, Holaway BL, Friedman WE, Malmberg RL (1991) Developmentally regulated antigen associated with calcium crystals in tobacco anthers. Planta 186: 13-16

Tschermak E (1890) Über künstliche Auslösung des Blühens beim Roggen. Ber Deutsch Bot Ges 8: 445-449

Von Guttenberg H (1926) Bewegungsgewebe und Perzeptionsorgane. In Linsbauer K (ed) Handbuch der Pflanzenanatomie. Borntraeger, Berlin.

Von Mohl H (1830) Über die fibrösen Zellen der Antheren. Flora

Von Mohl H (1858). Die Untersuchung des Pflanzengewebes mit Hilfe des polarisierten Lichtes. Bot. Zeitung 16

Wagner BL (1983) Genesis of the vacuolar apparatus responsible for druse formation in *Capsicum annuum* L. anthers. Scan El Micr 2: 905-912

Wolff GP (1924) Zur vergleigende Entwicklungsgeschichte und biologischer Bedeutung der Blütennektarien. Bot Arch 8: 305-311

Woycicki Z (1924a) Recherches sur la déhiscence des anthères et le rôle du stomium. Rev Gen Bot 36: 196-212

Woycicki Z (1924b) La formation de la fente dans les anthères et les phénomènes qui l'accompagnent. Rev Gen Bot 36: 253-268

7 Anther carbohydrates during in vivo and in vitro pollen development

Clément* C and Audran JC

7.1 Introduction

The male gametophyte of flowering plants is formed within the anther, following a succession of complex cytological events (Bedinger 1992). The anther is composed of three compartments: 1/ the sporophytic compartment including the connective tissue and the anther wall, 2/ the gametophytic compartment represented by the microspores/pollen grains, and 3/ the locular fluid representing the chemical link between the sporophytic and the gametophytic compartments (Clément et al. 1998). It is admitted that the sporophytic compartment produces the locular fluid and is involved in the development of microspores (Scott 1993; Pacini 1994) and their protection against biochemical and environmental variations (Heberle-Bors 1989; Clément and Audran 1995). The locule in which the microspores and pollen grains grow is particularly rich in sugars (Kawaguchi et al. 1996), which composition varies during developement (Pacini 1996; Saini 1997). These sugars are crucial nutrients required for the achievement of functional pollen grains (Pacini 1994). Further, they influence the initiation of microspore embryogenesis (Touraev et al. 1997), as well as the success of anther and microspore culture (Jähne and Lörz 1995). Sugars present in the locule are formed according to specific pathways (Miller and Ranwala 1994) under the control of anther or pollen-grain specific genes (Davies et al. 1992; Scott 1993, Bucciaglia and Smith 1994; Mu et al. 1994; Schrauwen et al. 1996; Wu et al. 1996). Disturbance in sugar metabolism dramatically affects pollen formation, leading in some cases to pollen abortion and male sterility (Banga et al. 1984; Dorion et al. 1996). This phenomenon is most evident under water-stress conditions (Saini and Lalonde 1998).

The biochemistry and the biology related to the various compartments of the anther, especially the carbohydrates present in the intermediate locular fluid, are

* for correspondence: Université de Reims Champagne Ardenne; Biologie et Physiologie Végétales; BP 1039; 51687 Reims Cedex 2; France

still widely unknown. The knowledge on the metabolic activity of the anther deserves to be deepened, since this organ governs various processes involved in pollen development and, thus, in plant reproduction. Therefore, this paper aims to present the information available on the content and physiology of sugars detected in the anther of various species. It focuses on a number of important aspects related to these sugars: their origin, composition, variation and utilization during pollen development, as well as their influence on in vivo and in vitro pollen biology.

7.2 Origin of anther sugars

During flower growth, most of sugar needs in developing reproductive organs are covered by photosynthesis occurring in the vegetative leaves (Mc Conchie and Lang 1991). In *Lilium*, the flower bud has the strongest sink strength in the plant (Wang and Breen 1986) and attracts up to 85% of the photoassimilates produced in the L_5 leaf. However, it has been shown in a few species that a part of photoassimilates are produced in the flower bud itself (Bazzaz and Carlson 1979; Vu et al. 1985; Weiss et al. 1988; Vemmos and Goldwin 1993; 1994; Clément et al. 1997a), reaching up to 64.5% of the total carbon balance in *Acer platanoides* (Bazzaz et al. 1979). In *Lilium*, the most intense photosynthetic period in the flower bud corresponds to the anther growth phase which takes place during microsporogenesis (Clément et al. 1997a). It is therefore likely that a part of carbohydrates synthesized in the flower bud are destined to anthers, although it was not clearly demonstrated. Following leaf and/or flower bud photosynthesis, carbohydrates are transported towards the anther through the filament (Keijzer et al. 1987). Sucrose is the main form of sugar transport in the plant (Lawrence and Mayne 1991) and was found to be the major form of carbohydrate in young filaments (Schaeverbeke 1965). In *Lilium*, the period of high sucrose content in the filament corresponds to the anther growth phase, from the microspore mother cell stage to the vacuolated microspore stage. Later, from pollen mitosis, when the anther growth is completed, the amount of sucrose decreases in the filament which thus enters its own growth (Clément et al. 1996). Finally, the anther sink strength for sucrose is dramatically reduced during the anther maturation phase (Reznickova 1983). These data strongly suggest that, at least in *Lilium*, the anther is mainly supplied with exogenous sucrose during its growth phase and is quite autonomous during its maturation phase in terms of sugar supply.

An endogenous synthesis of sucrose occurs in the anther wall involving the photosynthetic process. Indeed, some photosynthetically active radiations penetrate the flower bud, reach the internal anthers and allow them to perform CO_2 fixation (Clément et al. 1997b). In numerous species, chloroplast-like plastids were found in the epidermis, the endothecium and the middle layers (Kirichenko et al. 1977; Mlodzianowski and Idzikowska 1978; Gori 1982; Keijzer and Willemse 1988a; Clément et al. 1997b). In *Triticum aestivum*, *Secale cereale*

(Kirichenko et al. 1992; 1993) and in *Lilium* (Clément et al. 1997b), these anther plastids contain photosynthesis compatible chlorophyll and perform CO_2 fixation with more efficiency than those in the leaf. CO_2 fixation reaches respectively 35 mg CO_2/mg FW/h in *Triticum aestivum*, 99 mg CO_2/mg FW/h in *Secale cereale* and 364.3 mg $CO_2/m^2/h$ in *Lilium*. Additionnally, it has been shown that stomatas are present in the anther epidermis of some species, with a double density when compared to the leaf, permitting gas exchanges with the internal atmosphere of the bud during pollen development (Keijzer et al. 1987; Clément et al. 1997b). It seems that CO_2 produced in the closed developing flower bud by internal organ respiration is used by chlorophyllous anther cells for photosynthesis (Kirichenko et al. 1992). In *Lilium*, the presence of chloroplast-like plastids in the anther wall and the occurrence of anther photosynthesis was mainly detected during microsporogenesis. After pollen mitosis and during pollen maturation, plastids develop into chromoplasts (Keijzer and Willemse 1988b; Clément et al. 1997b) and the photosynthetic activity disappeared as early as the young pollen grain stage. The proportion of anther autonomy in terms of sugar nutrition remains however to be determined.

7.3 Carbohydrate variations during pollen development

7.3.1 Sugars

The anther is the plant organ containing the highest concentration of soluble sugars. In *Oryza sativa*, it reaches 2.5% of dry weight against a maximum of 0.05% in the other organs (Kawaguchi et al. 1996). Similarly, in *Lilium* soluble sugar concentration is 45 mg/g FW at the young pollen grain stage, whereas a maximum of 15 mg/g FW was detected in the other floral organs (Clément et al. 1996).

In *Hordeum vulgare* (Roberts-Oehlschlager et al. 1990), *Oryza sativa* (Kawaguchi et al. 1996) and *Lilium* (Clément et al. 1996), the major forms of sugars detected in the anther are sucrose, maltose, glucose and fructose. In each species, the anther has a constant sucrose concentration during pollen development. In *Oryza sativa* and in *Lilium*, the anther contains similar amounts of glucose and fructose through pollen development, suggesting a permanent sucrose hydrolysis by invertases. In *Hordeum vulgare* anther, glucose concentration is higher than fructose, but maltose is the most abundant form of anther sugar in this species (Roberts-Oehlschlager et al. 1990), meaning that maltose hydrolysis leads to glucose release and enhanced the glucose/fructose ratio. Maltose was assayed at a low concentration during pollen maturation in *Lilium* and was not detected at all in *Oryza sativa* anthers. Galactose and raffinose were also found in the anther of *Lilium*, especially during mirospore vacuolation.

Their concentration however was low when compared to the major forms of sugars, reaching a maximum of 3 mg/g FW at the tetrad stage (Clément et al. 1996).

A novel tetrasaccharide was recently identified in the anther of *Oryza sativa* (Kawaguchi et al. 1996). It represents up to 1 % of the anther dry weight and its structure is β-L-Ara-(1-3)-α-L-Ara-(1-3)-β-D-Gal-(1-6)-D-Gal. It exhibits strong similarities with a glycan unit of the plant cell surface arabinogalactan protein (AGP). This sugar is specifically accumulated in the anther during microsporogenesis and seems to be involved in anther response to chilling.

The variation of sugar content during pollen development appears to be linked to anther growth. In *Lilium*, the completion of anther growth is accompanied by a peak of both glucose and fructose concentrations (Clément et al. 1996), meaning that these compounds are no longer consumed by the growth process and transiently accumulated in the anther. A similar peak occurs in the anther of *Hordeum vulgare* (Roberts-Oehlschlager et al. 1990), *Triticum aestivum* (Dorion et al. 1996) and *Lycopersicon esculentum* (Bhadula and Sawhney 1989), but the authors did not find a correlation with anther growth. In *Lycopersicon esculentum*, this peak was thought to provide carbohydrate monomers for pollen intine formation (Bhadula and Sawhney 1989).

In isolated mature pollen grains, sucrose is the major form of sugar identified in Angiosperm pollen (Hoekstra et al. 1992; Speranza et al. 1997). Sucrose concentration is considerable in some species, reaching 23 % of pollen dry weight in *Typha* (Hoekstra et al. 1992). It seems that the carbohydrate content of the mature pollen grain is correlated with the degree of pollen hydration: saccharose is mainly present in dehydrated pollen species and is associated with polysaccharides (Pacini 1996; Speranza et al. 1997); in the opposite, hydrated pollen grains do not contain saccharose i.e. in *Zea mays* and *Cucrubita pepo* (Pacini 1996). It has been demonstrated that the high concentration of sucrose in dehydrated pollen protects the integrity of cell membrane phospholipids against dessication (Hoekstra and van Roekel 1988; Hoekstra et al. 1992) and allow pollen to survive during a longer period (Hoekstra et al. 1989). Glucose and fructose were also detected in most species (Hoekstra and van Roekel 1988; Hoekstra et al. 1989; Hoekstra et al. 1991), but their presence was independent of pollen hydration (Speranza et al. 1997).

7.3.2 Starch

Starch is the universal form of carbohydrate reserve in the microsporangium. In most Angiosperms, the anther tapetum is of secretory type and starch can be detected in each anther cell type but is absent in the acellular locular fluid. The amount of anther starch strongly fluctuates during pollen development in both the sporophytic and the gametophytic compartment.

In most species, amylaceous reserves are maximal in the anther wall layers at the time of meiosis. Starch degradation thus occurs during microsporogenesis in

Pisum sativum (Biddle 1979), *Smilax aspera* (Pacini and Franchi 1983), *Lycopersicon esculentum* (Bhadula and Sawhney 1989) and *Lilium* (Reznickova 1978; Reznickova and Willemse 1980; Clément et al. 1994), from pollen mitosis in *Gasteria verrucosa* (Keijzer and Willemse 1988a; 1988b), or regularly throughout pollen development in *Zea mays* (Hourcade et al. 1986). In *Lycopersicon esculentum* (Bhadula and Sawhney 1989) and *Lilium* (Reznickova 1978; Reznickova and Willemse 1980; Clément et al. 1994), a second wave of amylogenesis/amylolysis occurs in the anther wall, including starch synthesis just before pollen mitosis and starch breakdown during pollen maturation. The endothecium is the most favourite place for starch accumulation in the anther (Reznickova 1978; Biddle 1979; Reznickova and Willemse 1980; Pacini and Franchi 1983). In the anther of *Lilium*, starch content is maximal at meiosis and endothecium amylaceous reserves then represent 45.8 % of total anther starch, against 36.3 % in the connective tissue, 15.4 % in the middle layers, 0.8 % in the epidermis and 1.7 % in the guard cells of stomata (Clément et al. 1994). It should be noticed that stomata starch is maintained throughout pollen development even when starch is entirely degraded in the neighbouring anther cell layers (Clément et al. 1994), suggesting that stomatas remain functional and are used either for photosynthesis or during anther dehydration just before dehiscence (Keijzer 1987, Keijzer 1999).

Usually, tapetal reserves are stored as polysaccharides in the vacuole or in the cytoplasm (see below). But a few species exhibit starch grains in tapetal plastids, namely *Pisum sativum* (Biddle 1979), *Austrobaileya maculata* (Zavada 1984), *Trillium kamtschaticum* (Takahashi 1987), *Gasteria verrucosa* (Keijzer and Willemse 1988a; 1988b), *Brassica oleracea* (Murgia et al. 1991), *Arachis hypogea* (Xi 1991), *Tilia platyphyllos* (Hesse 1993). In this case, starch reserves are present during the earliest stages of microsporogenesis and are degraded concurrently with the numerous syntheses occuring in the tapetum, providing carbon skeletons for polysaccharide and lipid synthesis, as well as substrate for energy.

Two types of pollen grains can be distinguished according to their behaviour during development in terms of starch synthesis/degradation (Pacini and Franchi 1988). In the first case, pollen grains accumulate starch only once during development. In *Arachis hypogea* (Xi 1991), *Gossypium hirsutum* (Wetzel and Jensen 1992), *Lycopersicon esculentum* (Polowick and Sawhney 1993), *Lilium* (Clément et al. 1994) and *Cucurbita pepo* (Nepi et al. 1996), amylogenesis occurs around pollen mitosis and is followed or not by starch degradation during pollen maturation. In the second case, pollen grains undergo two waves of amylogenesis/amylolysis, the first one occurring during meiosis or microspore vacuolation, and the second one during pollen maturation. Such behaviour has been described in *Smilax aspera* (Pacini and Franchi 1983), *Ophrys lutea* (Feijo and Pais 1988), *Lycopersicon peruvianum* (Pacini and Viegi 1995), *Nicotiana tabacum* (Olmedilla et al. 1991), *Oenothera* (Noher de Halac et al. 1990; Noher de Halac et al. 1992) and *Pterostylis plumosa* (Pandolfi et al. 1993).

In mature pollen grains, independently, the vegetative cell contains either starch or polysaccharides as carbohydrate reserves (Franchi and Pacini 1988), which may be linked to pollen hydration. Starchy mature pollen grains were found in *Hordeum vulgare* (Cass and Karas 1975), *Malus communis* (Cresti et al. 1983), *Smilax aspera* (Pacini and Franchi 1983), *Lagerstromia indica* (Pacini and Bellani 1986), *Zea mays* (Hourcade et al. 1986; Chang and Neuffer 1989; Mandaron et al. 1990), *Gasteria verrucosa* (Keijzer and Willemse 1988b), *Papaver dubium* (Hoekstra and van Roekel 1988), *Nicotiana tabacum* (Olmedilla et al. 1991), *Oenothera* (Noher de Halac et al. 1990; Noher de Halac et al. 1992), *Lolium perenne* (Pacini et al. 1992), *Felicia muricata* (Jordaan and Kruger 1993), *Triticum aestivum* (Dorion et al. 1996) and *Lavatera arborea* (Nepi and Pacini 1999). Starch is more economic for metabolism than lipids, but lipids are more attractive for pollinators. It is then amazing that starchy mature pollen grains are mainly found in anemophilous species, whereas lipid reserves containing pollen grains are more specific to entomophilous species (Baker and Baker 1979).

7.3.3 Other polysaccharides

Polysaccharides were mainly cytochemically detected in the anther, most often without further biochemical precision. They are present in the secretory tapetum, in the loculus and in the microspores/pollen grains.

In the tapetum, polysaccharides were described either in the tapetal vacuole as in *Smilax aspera* (Pacini and Franchi 1983), or in cytoplasmic vesicles of the cytoplasm as in *Olea europea* (Pacini and Juniper 1979), *Allium sativum* (Gori 1982), *Gasteria verrucosa* (Keijzer and Willemse 1988), *Calotropis procera* (Audran and Dan Dicko Zafimahova 1992), *Ledebouria socialis* (Hess and Hesse 1994), or in both places as in *Lilium* (Clément et al. 1994; Clément et al. 1998). These compounds are PAS or PATAg positive, meaning that they can be classified as neutral polysaccharides. In the cytoplasm, polysaccharide containing vesicles are transported towards the locular side of the tapetum, fuse with the undulated tapetal plasma membrane and release their content into the loculus (Hess and Hesse 1994; Clément et al. 1998). It is therefore admitted that a part of the locular fluid polysaccharides originate from tapetal metabolism and secretions. However, the production of polysaccharides by the tapetum is transient during pollen development, since it mainly occurs during the earliest steps of microspore vacuolation. After that period, the reduction of dictyosomes and the increase of smooth endoplasmic reticulum associated with plastid lipid synthesis reveal that the whole metabolism of tapetum is switched towards pollenkitt formation (Hess and Hesse 1994; Clément et al. 1998). In most species, tapetum metabolism can be decomposed into 2 phases: the first one extends from meiosis to the middle vacuolated microspore and is devoted to at least carbohydrate elaboration and secretion; the second one extends from the middle vacuolated microspore to the degenerescence of tapetal cells and leads to the formation and the release of pollenkitt into the loculus.

Many papers reported the presence of polysaccharides in the locular fluid (Table 1). Their reactivity to the PAS or to the PATAG test indicates that they are neutral polysaccharides, although acidic polysaccharides were also detected in *Saxifraga cymbalaria* (Abadie and Hideux 1979) and in *Lilium* (Clément et al. 1998). Esterified and unesterified pectins were even identified in the locular fluid of *Ledebouria socialis* (Hess and Frosch 1994). Polysaccharides of the locular fluid partly originated from the tapetum (see above), but the degradation of the radial walls in tapetal cells also provides the loculus with fibrillar polysaccharides in *Saxifraga cymbalaria* (Abadie and Hideux 1979), *Gasteria verrucosa* (Keijzer and Willemse 1988), *Brassica oleracea* (Murgia et al. 1991) and *Lilium* (Reznickova and Willemse 1980; Clément et al. 1998). Moreover, tetrad callose wall dissolution by ß 1,3 endoglucanase (Stieglitz 1977; Bucciaglia and Smith 1994) releases additional polysaccharides in the loculus when meiosis is completed (Rodriguez-Garcia and Garcia 1978; Pacini and Franchi 1983; Hess and Hesse 1994; Clément et al. 1998). Whatever their nature, the polysaccharides are present in the loculus from meiosis to the achievement of microsporogenesis in all species investigated and next disappear at the first pollen mitosis.

Table 1. Polysaccharides in the locular fluid during microsporogenesis

Species	Polysaccharides	References
Saxifraga cymbalaria	PATAg positive Acidic polysaccharides	Hideux and Abadie 1979
Lilium sp.	PAS and PATAg positive + Toluidine blue metachromasy	Reznickova and Willemse 1980, Clément et al. 1994, Clément et al. 1998
Allium sativum	PAS positive	Gori 1982
Smilax aspera	PAS positive	Pacini and Franchi 1983
Thillandsia pallidoflavens	PATAg positive	Hess 1991
Calotropis procera	PATAg positive	Audran and Dan Dicko Zafimahova 1992
Oryza sativa	PAS positive	Wada et al. 1992
Ledebouria socialis	Esterified and unesterified pectins	Hess and Frosch 1994
Ledebouria socialis	PATAg positive	Hess and Hesse 1994

In the pollen grain, polysaccharides are mainly localized in the pollen wall and in the cytoplasm as reserves. The intine of the pollen wall contains both cellulose and pectins (Cresti et al. 1983; Bedinger 1992; van Aelst and van Went 1992; Hess 1993; Li et al. 1995), but it is still not clear whether polysaccharides enter the architecture of the sporopollenic exine. Neutral polysaccharides (associated with sucrose) are present in the vegetative cell cytoplasm of mature pollen grains (Weber 1989), especially in dehydrated pollen (Pacini 1996; Speranza et al. 1997). Moreover, recently, the fructose polymer fructan was identified in dehydrated pollen grains (Pacini, personal communication) and may be involved in pollen resistance to dehydration during maturation. This should be further correlated with pollen grains containing fructose as the major form of sugar at maturity (Clark and Lintas 1992).

Most of carbohydrate supply, biosyntheses and movements in the anther occur during microsporogenesis. From pollen mitosis, the anther is no longer provided with sucrose from the filament, photosynthesis is arrested in the anther wall, starch is degraded in the sporophytic compartment and polysaccharides disappear in the tapetum as well as in the locular fluid. During pollen maturation, the anther seems to be autonomous from sugar nutrition and mostly devoted to carbohydrate transformation for further anther dehiscence or pollen tube germination.

7.4 Carbohydrate fate in the anther

Sucrose reaching the anther is cleaved into glucose and fructose by anther wall invertases (Miller and Ranwala 1994; Dorion et al. 1996), leading to identical amounts of both monosaccharides in the anther during all pollen development (Roberts-Oehlschlager et al. 1990; Clément et al. 1996; Dorion et al. 1996; Kawaguchi et al. 1996). These compounds are incorporated in various anther metabolic pathways. Actually, when the anther is supplied with marked sucrose, radioactivity is found in anther starch, lipids and hexoses (Reznickova 1983).

The most intense period of sucrose attraction in the anther and the degradation of starch in the anther wall are simultaneous and occur during microsporogenesis. Such events are correlated with 1/ cellulose wall thickenings in the endothecium; 2/ tapetal syntheses and secretions, 3/ the presence of polysaccharides in the locular fluid, 4/ callose formation around meiospores and 5/ microspore wall formation and vacuolation.

Callose synthesis/degradation represents a particular event in the anther. Callose degradation was identified and characterized at the tetrad stage in the anther of *Lilium* and in *Nicotiana tabacum* (respectively Stieglitz 1977; Bucciaglia and Smith 1994). The responsible enzyme is a ß 1-3 endoglucanase and the corresponding gene is transiently expressed in the anther tapetum of tobacco at the end of meiosis. This phenomenon is accurately regulated since tapetum also synthesizes intralocule callase inhibitors that regulate the period of callase action (Schrauwen et al. 1996). It is likely that callose breakdown products are disseminated in the loculus and supply the fluid with additional monomers of glucose.

Substantial information is available concerning pollen sugar physiology during development. The disappearance of the locular fluid, especially polysaccharides, in parallel with microspore vacuolation strongly suggests that the loculus compounds are metabolized by developing microspores. In particular, it is remarkable that pectins are detected in the fluid during microsporogenesis in *Ledebouria socialis* (Hess and Frosch 1994) and in *Lilium* (Clément et al. 1998) and that the genes coding for pectin esterase in *Brassica napus* (Albani et al. 1991) and pectate lyase in *Nicotiana tabacum* (Rogers et al. 1992) are intensely expressed during the same period. Regarding starch accumulation/mobilization, the expression of the gene coding for starch synthase occurs at the tetrad stage and from pollen mitosis in *Nicotiana tabacum* (Olmedilla et al. 1991), which coincides with starch accumulation in the microspores/pollen grains. Also, in *Triticum aestivum*, enzymes involved in starch synthesis are active along with starch accumulation in the pollen grains (Dorion et al. 1996). Starch can be stored in the vegetative cell up to anthesis for pollen germination or degraded during pollen maturation. For example, in the pollen grain of *Lilium*, starch is partially degraded during maturation and transformed into lipids (Miki-Hirosige and Nakamura 1983). This is coherent with the inhibition of the starch synthesizing enzyme ADP-G pyrophosphorylase in the mature pollen grain of this species (Dickinson and Davies 1971). At the end of development, various activities related to sugar physiology were detected in mature pollen grains, but they are most often linked to the germinating process. Enzymes involved include invertases (Singh and Knox 1984) and pectinases (Pressey and Reger 1989; Rogers et al.1992; Robert et al. 1993; Mu et al. 1994; Tebbutt et al. 1994; Wu et al. 1996), which have key roles in respectively the use of stigmatic exudate and the penetration of the pollen tube in the style tissues.

Pollen sugar nutrition is controlled by the anther wall which provides an appropriate environment for pollen ontogenesis (Maheshwari et al. 1982; Heberle-Bors 1989), in particular the anther wall regulates the amount of sugar reaching the pollen grain during development (Aruga and Nakajima 1985; Heberle-Bors 1989; Kao et al. 1991; Kao 1993). In *Lilium*, when the anther is provided in vitro with surplus of sucrose, starch accumulates in the anther wall whereas the maturing pollen grains is not affected. If the pollen is isolated and incubated in the same medium, it accumulates abnormal amounts of starch and becomes unable to germinate (Clément and Audran 1995). This shows that the anther wall has a buffer function for developing pollen grains, accurately regulating the concentration of sugars (at least) in the loculus.

7.5 Anther carbohydrates under water stress

Disturbances in anther sugar physiology leads to pollen abortion and subsequent sterility (Banga et al. 1984; Dorion et al. 1996). For instance, water stress affects carbohydrate metabolism in the whole plant by decreasing the rate of photosynthesis and the export of assimilates from the leaves (Hanson and Hitz 1982), thus depriving growing flowers of sucrose (Saini 1997; Saini and Lalonde 1998). Female organs are not affected by water deficit but anther and pollen are particularly sensitive to such modification of environment. Induced changes of sugar physiology in the anther represent the first signs of water stress injury to pollen development and can then be considered as a marker of sensitivity.

The most sensitive stage to drought is meiosis, although the anther and pollen remain sensitive during all pollen development (Dorion et al. 1996; Sheoran and Saini 1996; Lalonde et al. 1997a). Under water stress, the anthers of *Triticum aestivum* (Dorion et al. 1996) and *Oryza sativa* (Sheoran and Saini 1996) accumulate abnormal amounts of soluble sugars, especially sucrose, but starch decreases simultaneously except in the connective tissue (Lalonde et al. 1997a). The sucrose-cleaving enzyme, acid invertase is inhibited (Dorion et al. 1996; Sheoran and Saini 1996), so that metabolizable sugars are no longer available for developing microspores/pollen grains. In addition, the activity of sucrose synthase in *Triticum aestivum* and starch synthase in *Oryza sativa* are also affected. In the pollen of both species, the polysaccharidic intine is reduced and starch is lacking in the vegetative cytoplasm (Saini et al. 1984; Sheoran and Saini 1996; Lalonde et al. 1997a). Recently, it was further shown that anther gene expression is affected by drought. In *Triticum aestivum*, the ADP-glucose pyrophosphorylase gene involved in starch sytnhesis is not expressed under water stress, leading to a lack of starch accumulation during pollen maturation (Lalonde et al. 1997b). Such pollen grains have a low vigour and are unable to germinate.

7.6 Anther carbohydrates and pollen in vitro behaviour

The success of androgenesis can be correlated with sugar physiology and metabolism in the microspore/pollen grains. First, the presence of starch is an indicator of microspore competence to androgenesis. It has been shown that starch containing microspores are unable to enter the androgenetic program in most species (Sangwan and Sangwan-Norreel 1987; Kott et al. 1988; Zaki and Dickinson 1990). Moreover, in *Zea mays*, the appearance of starch in the microspore occurs at mitosis and is simultaneous with the synthesis of new proteins, which may be involved in the disappearance of the androgenetic potential in this species (Mandaron et al. 1990).

The second aspect is the microspore pretreatment. It is necessary to stress the microspore in order to derive the gametophytic program towards the alternative sporophytic program. Sugar starvation is one of the most effective way for the induction of the microspore embryogenic development in a few species including *Nicotiana tabacum* (Zarsky et al. 1992; Garrido et al. 1995; Rihova et al. 1996; Touraev et al. 1996a; Touraev et al. 1996b; Touraev et al. 1997), *Oryza sativa* (Ogawa et al. 1994) and *Triticum aestivum* (Touraev et al. 1996c). Under starvation, G1 phase blocked microspores enter the S and G2 phases, undergo the first symetrical division and then develop into haploid embryos (Zarsky et al. 1992; Rihova et al. 1996; Touraev et al. 1996a). The mechanism of carbohydrate influence on the microspore cell cycle remains to be determined but it has been shown recently that sugars act as signal molecules for gene regulation (Hattori et al. 1990; Hattori et al. 1991; Yu et al. 1996), indicating that they could be involved in microspore embryogenesis induction. However, sugars are not the sole compounds responsible for microspore embryogenesis induction. Acually, the association of both sugar starvation and heat shock treatment increases the efficiency of the pretreatment and has permitted to obtain androgenetic plantlets in recalcitrant cultivars of *Triticum aestivum* (Touraev et al. 1996c).

Table 2. Sugars used in the induction of microspore embryogenesis
(AC) anther culture, (MC) microspore culture

Species	Sugar used	References
Nicotiana tabacum	sucrose	Benito-Moreno et al. 1988, Touraev et al. 1996c
Hordeum vulgare	sucrose + cellobiose + melibiose	Finnie et al. 1989, Roberts-Oehlschlager et al. 1990, Pickering and Devaux 1992, Hoeckstra et al. 1993, Scott and Lyne 1994b
Medicago sativa	sucrose + melibiose	Tanner et al. 1990
Sinocalamus latiflora	sucrose	Tsay et al. 1990
Solanum carolinense	sucrose	Reynolds 1990
Asparagus officinalis	sucrose	Feng and Wolyn 1991, 1994, Zhang et al. 1994
Brassica oleracea	sucrose	Yang et al. 1992
Solanum tuberosum	sucrose	Tiainen 1992
	maltose	Tiainen 1992
Triticum aestivum	maltose	Luckett and Darvey 1992 Karsai et al. 1994
Oryza sativa	maltose	Deimling et al. 1992, Deimling and Fleminghaus-Roux 1997
Aesculus carnea	sucrose	Marinkovic and Radojevic 1992
Zea mays	sucrose	Pescitelli et al. 1994, Büter et al. 1997
Fagopyrum esculentum	maltose	Bohanec et al. 1993, Bohanec 1997
Cyclamen persicum	sucrose	Ishizaka and Uematsu 1993
Linum usitatissimum	sucrose	Nichterlein and Friedt 1993
Solanum sp.	sucrose	Calleberg and Johansson 1993, Rokka et al. 1995
Populus sp.	maltose	Baldursson et al. 1993
Triticale cereale	maltose	Karsai et al. 1994
Phleum pratense	maltose	Abdullah et al. 1994
Tulipa gesneria	sucrose	van den Bulk et al. 1994
Camellia japonica	sucrose or glucose	Pedroso and Pais 1994
Brassica rapa	sucrose	Ferric et al. 1995
Capsicum annuum	sucrose	Gonzalez-Melendi et al. 1995
Helianthus annuus (AC)	sucrose	Zhong et al. 1995, Friedt et al. 1997
Helianthus annuus (MC)	maltose	Coumans and Zhong 1995, Friedt et al. 1997
Raphanus sativus	sucrose	Takahata et al. 1996
Sorghum sp.	sucrose	Liang et al. 1997
Avena sativa	maltose	Rines et al. 1997
Lilium sp.	sucrose	Arzate-Fernandez et al. 1997, Han et al. 1997
Dactylis glomerata	maltose	Christensen et al. 1997
Avena sterilis	sucrose	Kiviharju et al. 1997
Quercus suber	sucrose	Bueno et al. 1997
Apium graveolens	sucrose	Dohya et al. 1997

The source of carbohydrates in the culture medium may be crucial for the completion of the androgenetic process in many species and scientists therefore paid special attention to this point from the beginning of the 90's. Maltose and sucrose are the most commonly sugars used in the induction medium for anther or microspore culture of various species (Table 2).

In most Poaceae sugar is particularly crucial: maltose is the most suitable sugar (Table 2), whereas saccharose, glucose and fructose are toxic for microspores (Kao 1993; Scott and Lyne 1994b; Jähne and Lörz 1995). In *Hordeum vulgare*, it was shown that sucrose is rapidly metabolized by microspores which accumulate high amounts of starch and do not develop embryos, whereas maltose is assimilated more slowly and allows pollen embryogenesis (Scott et al. 1995). Further, in this family, maltose is now recognized not only to stimulate the production of pollen embryos (Finnie et al. 1989; Chu et al. 1990; Deimling et al. 1992; Pickering and Devaux 1992) but also to enhance the green/albinos ratio of haploid plantlet regeneration (Sorvari and Schieder 1987; Scott and Lyne 1994a; Caredda and Clément 1999). Besides, *Sinocalamus latiflora* (Tsay et al. 1990), *Zea mays* (Pescitelli et al. 1994; Büter 1997), *Avena sterilis* (Kiviharju et al. 1997) and *Sorghum* sp. (Liang et al. 1997) belong to the Poaceae but sucrose is preferred to maltose in those species. Also, in *Solanum tuberosum*, microspores seem to be indifferent to the source of carbohydrates, giving rise to androgenetic embryos under both pure sucrose or pure maltose in the induction medium with almost the same efficiency (Tiainen 1992). In some species, the best sugar to be used in the induction medium depends on the explant. Indeed, in *Helianthus annuus*, sucrose is preferred in anther culture (Zhong et al. 1995; Friedt et al. 1997), whereas maltose is more suitable in isolated microspore culture (Coumans and Zhong 1995; Friedt et al. 1997). In this case, it is likely that the anther wall plays a key role in regulating the nature of sugar reaching the developing microspores as in *Nicotiana tabacum* (Aruga and Nakajima 1985).

Microspore/pollen behaviour regarding sugars depends upon the species investigated but it seems that it governs both in vivo and in vitro development. Therefore, fundamental studies on microspore/pollen sugar physiology are necessary in order to predict and control microspore development under artifical conditions. This may be of importance not only for androgenesis but also for in vitro pollen maturation (Benito-Moreno et al. 1988; Tanaka and Ito 1980; 1981; Pareddy and Petolino 1992; Dupuis and Pace 1993; Touraev et al. 1995).

7.7
Conclusions

Carbohydrate physiology is fundamental for the completion of pollen development in vivo as well as in vitro. The anther wall appears as a multifunctional compartment in this yield, acting as 1/ a photosynthetic tissue providing the anther with endogenous sugar, 2/ a sink tissue cleaving the filament derived sucrose in metabolizable glucose and fructose, 3/ a buffer for sugar variations in the whole anther and 4/ a place of sugar monomer transformation and secretion into the locular fluid.

In species with secretory tapetum, the acellular locular fluid is the microenvironment in which pollen develops. This anther compartment, in particular the sugars in the locular fluid, deserve to be better studied, because of their decisive role in controlling pollen development in vivo. Moreover, increasing the basic knowledge of the locular sap may be of help, especially using isolated microspore culture in which the culture medium represents a substitute of the locular fluid.

Despite the importance of sugar influence on in vitro microspore development and possible consequences on new cultivar obtention by androgenesis, only few works focus on the mechanism of sugar action on microspore physiology. According to the close links between the importance of sugar environment for microspores/pollen and to the recent evidences of sugar influence on the expression of genes involved in sugar metabolism (Hattori et al. 1990; Hattori et al. 1991; Yu et al. 1996), it should be fruitfull to investigate the impact of carbohydrates on microspore/pollen gene expression both in vivo and in vitro.

Acknowledgements

The authors would like to thank so much Prof Ettore Pacini and Dr Elisabeth Pierson for critically reading the manuscript.

References

Abadie M, Hideux M (1979) L'anthère de *Saxifraga cymbalaria* L. ssp huetiana (Boiss) Engl. et Irmsch. en microscopie électronique (M.E..B. et M.E.T.).1. Généralités. Ontogénèse des orbicules. Ann Sci Nat Bot, Paris 13: 199-203

Abdullah AA, Pedersen S, Andersen SB (1994)Triploid and hexaploid regenerants from hexaploid timothy (*Phleum pratense* L.) via anther culture. Plant Breed 112: 342-345

Albani D, Altosaar I, Arnison PG, Fabijanski SF (1991) A gene showing sequence similarity to pectin esterase is specifically expressed in developing pollen of *Brassica napus*. Sequences in its 5' flanking region are conserved in other pollen-specific promoters. Plant Mol Biol 16: 501-513

Aruga K, Nakajima T (1985) Role of anther on pollen embryogenesis in anther culture of *Nicotiana tabacum* L. Japan J Breed 35: 390-397

Arzate-Fernandez AM, Nakazaki T, Yamagata H, Tanisaka T (1997) Production of doubled-haploid plants from *Lilium longiflorum* Thunb. anther culture. Plant Sci 123: 179-187

Audran JC, Dan Dicko Zafimahova L (1992) Aspects ultrastructuraux et cytochimiques du tapis staminal chez *Calotropis procera* (Asclepiadaceae). Grana 31 : 253-272

Baker HG, Baker I (1979) Starch in Angiosperm pollen grains and its evolutionary significance. Am J Bot 66: 591-600

Baldursson S, Krogstrup P, Norgaard JV, Andersen SB (1993) Microspore embryogenesis in anther culture of three species of *Populus* and regeneration of dihaploid plants of *Populus trichocarpa*. Can J For Res 23: 1821-1825

Banga SS, Labana KS, Banga SK (1984) Male sterility in indian mustard (*Brassica juncea* (L.) Coss.). A biochemical characterization. Theor Appl Genet 67: 515-518

Bazzaz FA, Carlson RW (1979) Photosynthetic contribution of flowers and seeds to reproductive effort of an annual colonizer. New Phytol 82: 223-232

Bedinger P (1992) The remarkable biology of pollen. Plant Cell 4: 879-887

Benito-Moreno RM, Macke F, Alwen A, Heberle-Bors E (1988) In situ seed production after pollination with in vitro matured, isolated pollen. Planta 176 : 145-148

Bhadula SK, Sawhney VK (1989) Amylolytic activity and carbohydrate levels during the stamen ontogeny of a male fertile, and a "gibberellin-sensitive" male sterile mutant of tomato (*Lypercosicon esculentum*). J Exp Bot 40: 789-794

Biddle JA (1979) Anther and pollen development in garden pea and cultivated lentil. Can J Bot 57: 1883-1900

Bohanec B, Neskovic M, Vujicic R (1993) Anther culture and androgenetic plant regeneration in buckwheat (*Fagopyrum esculentum* Moench). Plant Cell Tiss Org Cult 35: 259-266

Bohanec B (1997) Haploid induction in buckwheat (*Fagopyrum esculentum* Moench). In Jain SM, Sopory SK, Veilleux RE (eds) In vitro haploid production in higher plants. 4. Kluwer Academic publishers, the Netherlands, pp 163-170

Bucciaglia PA, Smith AG (1994) Cloning and characterization of Tag 1, a tobacco anther ß-1,3 glucanase expressed during tetrad dissolution. Plant Mol Biol 24: 903-914

Bueno MA, Gomez A, Boscaiu M, Manzanera JA, Vicente O (1997) Stress-induced formation of haploid plants through anther culture in cork oak (*Quercus suber*). Physiol Plant 99: 335-341

Büter B (1997) In vitro haploid production in maize. In Jain SM, Sopory SK, Veilleux RE (eds) In vitro haploid production in higher plants. 4. Kluwer Academic Publishers, The Netherlands, pp 37-71

Calleberg EK, Johansson LB (1993) The effect of starch and incubation temperature in anther culture of potato. Plant Cell Tiss Org Cult 32: 27-34

Caredda S, Clément C (1999) Androgenesis and albinism in Poaceae: influence of genotype and carbohydrates. Ibid, pp 211-228

Cass DD, Karas I (1975) Development of sperm cells in barley. Can J Bot 53: 1051-1062

Chang MT, Neuffer G (1989). Maize microsporogenesis. Genome 32: 232-244

Christensen JR, Borrino E, Olesen A, Andersen SB (1997) Diploid, tetraploid and octoploid plants from anther culture of tetraploid orchard grass, *Dactylis glomerata* L. Plant Breed 116: 267-270

Chu CC, Hill RD, Brule-Babel A (1990) High frequency of pollen embryoid formation and plant regeneration in *Triticum aestivum* L. monosaccharide containing media. Plant Sci 66: 255-262

Clark CJ, Lintas C (1992) Chemical composition of pollen from kiwifruit vines. New Zeal J Crop Hort Sc 20: 337-344

Clément C, Audran JC (1995) Anther wall layers control pollen sugar nutrition in *Lilium*. Protoplasma 187: 172-181

Clément C, Burrus M, Audran JC (1996) Floral organ growth and carbohydrate content during pollen development in *Lilium*. Am J Bot 83: 459-469

Clément C, Chavant L, Burrus M, Audran JC (1994) Anther starch variations in *Lilium* during pollen development. Sex Plant Reprod 7: 347-356

Clément C, Laporte P, Audran JC (1998) The loculus content and tapetum during pollen development. Sex Plant Reprod 11: 94-106

Clément C, Mischler P, Burrus M, Audran JC (1997a) Characteristics of the photosynthetic apparatus and CO2-fixation in the flower bud of *Lilium*. 1. Corolla. Int J Plant Sci 158: 794-800

Clément C, Mischler P, Burrus M, Audran JC (1997b) Characteristics of the photosynthetic apparatus and CO2-fixation in the flower bud of *Lilium*. 2. Anther. Int J Plant Sci 158: 801-810

Coumans M, Zhong D (1995) Doubled haploid sunflower (*Helianthus annuus*) plant production by androgenesis: fact or artifact ? Part 2. In vitro isolated microspore culture. Plant Cell Tiss Org Cult 41: 203-209

Cresti M, Ciampolini F, Sarfatti G (1983). Ultrastructural features of *Malus communis* L. mature pollen. In Mulcahy DL, Ottaviano E (eds) Pollen : biology and implications for plant breeding. Elsevier Science Publisher, Biomedical, New York, pp 165-172

Davies SP, Singh MB, Knox RB (1992) Identification and in situ localization of pollen-specific genes. Int Rev Cytol 140: 19-34

Deimiling S, Flehinghaus T, Bohn M, Geiger HH (1992) Sugary aspects of androgenesis-alterations of carbohydrates during haploid induction in gramineous species. In Angers University (ed), Proceedings XIIIth Eucarpia Congress, pp 153-154

Deimling S, Flehinghaus-Roux T (1997) Haploidy in rye. In Jain SM, Sopory SK, Veilleux RE (eds) In vitro haploid production in higher plants. 4. Kluwer Academic Publishers, The Netherlands, pp 181-204

Dickinson DB, Davies MD (1971) Metabolism of germinating lily pollen: pollen enzymes. In Heslop-Harrison J (ed) Pollen: development and physiology. Butterworths, London, pp 190-193

Dohya N, Matsubara S, Murakami K (1997) Callus formation and regeneration of adventitious embryos from celery microspores by anther and isolated microspore cultures. J Japan Soc Hort Sci 65: 747-752

Dorion S, Lalonde S, Saini HS (1996) Induction of male sterility in wheat by meiotic stage water deficit is preceded by a decline in invertase activity and changes in carbohydrate metabolism in anthers. Plant Physiol 111: 137-145

Dupuis I, Pace GM (1993) Factors affecting in vitro maturation of isolated maize microspores. Plant Cell Rep 12: 564-568

Feijo JA, Pais MSS (1988) Ultrastructural modifications of plastids and starch metabolism during the microsporogenesis of *Ophrys lutea* (Orchidaceae). Ann Bot 61: 215-219

Feng XR, Wolyn DJ (1991) High frequency production of haploid embryos in *Asparagus* anther culture. Plant Cell Rep 10: 574-578

Feng XR, Wolyn DJ (1994) Recovery of haploid plants from *Asparagus* microspore culture. Can J Bot 72: 296-300

Ferrie AMR, Epp DJ, Keller WA (1995) Evaluation of *Brassica rapa* L. genotypes for microspore culture response and identification of a highly embryogenic line. Plant Cell Rep 14: 580-584

Finnie SJ, Powell W, Dyer AD (1989) The effect of carbohydrate composition and concentration on anther culture response in barley (*Hordeum vulgare* L.). Plant Breed 103: 110-118

Franchi GG, Pacini E (1988) Pollen polysaccharide reserves in some plants of economic interest. In Cresti M, Gori P, Pacini E (eds) Sexual Reproduction in Higher Plants. Springer-Verlag, Berlin, Heidelberg, New York, Tokyo, pp 47

Friedt W, Nurhidayah T, Rocher T, Kohler H, Bergmann R, Horn R (1997) Haploid production and application of molecular methods in sunflower (*Helianthus annuus* L.). In Jain SM, Sopory SK, Veilleux RE (eds) In vitro haploid production in higher plants. 5. Kluwer academic publishers, The Netherlands, pp 17-35

Garrido D, Vicente O, Hebverle-Bors E, Rodriguez-Garcia I (1995) Cellular changes during the acquisition of embryogenic potential in isolated pollen grains of *Nicotiana tabacum*. Protoplasma 186: 220-230

Gonzalez-Melendi P, Testillano PS, Ahmadian P, Fadon B, Vicente O, Risueno MC (1995) In situ characterization of the late vacuolate microspore as a convenient stage to induce embryogenesis in *Capsicum*. Protoplasma 187: 60-71

Gori P (1982) Accumulation of polysaccharides in the anther cavity of *Allium sativum*, clone Piemonte. J Ultrastr Res 81 : 158-162

Han DS, Niimi Y, Nakano M (1997) Regeneration of haploid plants from anther cultures of the Asiatic hybrid lily "Connecticut King". Plant Cell Tiss Org Cult 47: 153-158

Hanson AD, Hitz WD (1982) Metabolic responses of mesophytes to plant water deficits. Ann Rev Plant Physiol 33: 163-203

Heberle-Bors E (1989) Isolated pollen culture in tobacco: plant reproductive development in a nutshell. Sex Plant Reprod 2: 1-10

Hess MW (1993) Cell wall development in freeze-fixed pollen: Intine formation of *Ledebouria socialis* (Hyacinthaceae). Planta 189: 139-149

Hess MW, Frosch A (1994) Subunits of forming pollen exine and Ubisch bodies as seen in freeze substituted *Ledebouria socialis* Roth (Hyacinthaceae). Protoplasma 182: 10-14

Hess MW, Hesse M (1994) Ultrastructural observations on anther tapetum development of freeze-fixed *Ledebouria socialis* Roth (Hyacinthaceae). Planta 192: 421-430

Hesse M (1993) Pollenkitt development and composition in *Tilia platyphyllos* (Tiliaceae) analysed by conventional and energy filtering TEM. Pl Syst Evol 7: 39-52

Hideux M, Abadie M (1979) L'anthère de *Saxifraga cymbalaria* L. ssp huetiana (Boiss) Engl. et Irmsch. en microscopie électronique (M.E..B. et M.E.T.). 3. Dynamique des relations entre les cellules tapétales et les cellules sporales. Ann Sci Nat Bot, Paris 13: 247-263

Hoekstra FA, Crowe JH, Crowe LM (1989) Differential dessication sensitivity of corn and *Pennisetum* pollen linked to their sucrose contents. Plant Cell Environ 12: 83-91

Hoekstra FA, Crowe JH, Crowe LM (1991) Effect of sucrose on phase behaviour of membranes in intact pollen of *Typha latifolia* L., as measured with Fourier transform infrared spectroscopy. Plant Physiol 97: 1073-1079

Hoekstra FA, Crowe JH, van Roekel T, Vermeer E (1992) Do phospholipids and sucrose determine membrane phase transitions in dehydrating pollen species ? Plant Cell Env 15: 601-606

Hoekstra FA, van Roekel T (1988) Dessication tolerance of *Papaver dubium* L. pollen during its development in the anther. Plant Physiol 88: 626-632

Hoekstra S, van Zijderveld MH, Heidekamp F, van der Mark F (1993) Microspore culture of *Hordeum vulgare* L.: the influence of density and osmolality. Plant Cell Rep 12: 661-665

Hourcade DE, Bugg M, Loussaert DF (1986) The use of Gaspé variety for the study of pollen and anther development in maize. In Mulcahy DL, Ottaviano E (eds)

Biotechnology and ecology of pollen. Springer-Verlag, New York, Berlin, Heidelberg, Tokyo, pp 319-324

Ishizaka H, Uematsu J (1993) Production of plants from pollen in *Cyclamen persicum* Mill. through anther culture. Japan J Breed 43: 207-218

Jähne A, Lörz H (1995) Cereal microspore culture. Plant Sci 109: 1-12

Jordaan A, Kruger H (1993) Pollen wall ontogeny of *Felicia muricata* (Asteraceae: Astereae). Ann Bot 71: 97-105

Kao KN (1993) Viability, cell division and microcallus formation of barley microspores in culture. Plant Cell Rep 12: 366-369

Kao KN, Saleem M, Abrams S, Pedras M, Horn D, Mallard C (1991) Culture conditions for induction of green plants from barley microspores by anther culture methods. Plant Cell Rep 9: 595-601

Karsai I, Bedo Z, Hayes PM (1994) Effect of induction medium pH and maltose concentration on in vitro androgenesis of hexaploid winter triticale and wheat. Plant Cell Tiss Org Cult 39: 49-53

Kawaguchi K, Shibuya N, Ishii T (1996) A novel tetrasaccharide, with a structure similar to the terminal sequence of an arabinogalactan-protein, accumulates in rice anthers in a stage specific manner. Plant J 9: 777-785

Keijzer CJ (1987) The process of anther dehiscence and pollen dispersal. 1. The opening mechanism of longitudinally dehiscing anthers. New Phytol 105: 487-498

Keijzer CJ (1999) Mechanisms of Angiosperm anther dehiscence, a historical review. Ibid, pp 55-67

Keijzer CJ, Hoek IHS, Willemse MTM (1987) The process of anther dehiscence and pollen dispersal. III. The dehydration of the filament tip and the anther in three Monocotyledonous species. New Phytol 106: 281-287

Keijzer CJ, Willemse MTM (1988a) Tissue interactions in the developing locule of *Gasteria verrucosa* during microsporogenesis. Acta Bot Neerl 37: 475-492

Keijzer CJ, Willemse MTM (1988b) Tissue interactions in the developing locule of *Gasteria verrucosa* during microgametogenesis. Acta Bot Neerl 37: 493-508

Kirichenko AB, Kirichenko EB, Chebotar AA (1977) Ultrastructure of anther of *Hordeum vulgare* L. at the stage of bicellular pollen: characteristics of plastid differentiation. Physiol Rast 24: 751-755

Kirichenko E, Krendéléva T, Koukarskikh G, Nizovskaia N (1992) Structure et activité fonctionnelle des chloroplastes des anthères et du péricarpe des caryopses de blé et de seigle. C R Acad Sci Paris 314: 365-370

Kirichenko E, Krendéléva T, Koukarskikh G, Nizovskaia N (1993) Photochemical activity in chloroplasts of anthers and caryopsis in cereals. Russ Plant Physiol 40: 229-233

Kiviharju E, Puolimatka M, Pehu E (1997) Regeneration of anther-derived plants of *Avena sterilis*. Plant Cell Tiss Org Cult 48: 147-152

Kott LS, Polsoni L, Beversdorf WD (1988) Cytological aspects of isolated microspore culture of *Brassica napus*. Can J Bot 66: 1658-1664

Lalonde S, Beebe DU, Saini HS (1997a) Early signs of wheat anther development associated with the induction of male sterility by meiotic-stage water deficit. Sex Plant Reprod 10: 40-48

Lalonde S, Morse D, Saini HS (1997b) Expression of a wheat ADP-glucose pyrophosphorylase gene during development of normal and water-stress-affected anthers. Plant Mol Biol 34: 445-453

Lawrence DK, Mayne RG (1991) Plant growth regulators and photosynthesis. In Baker NR, Percival MP (eds), Herbicides. Elsevier Science Publishers, pp 299-335

Li YQ, Faleri C, Geitmann A, Zhang H-Q, Cresti M (1995) Immunogold localization of arabinogalactan proteins, unesterified and esterified pectins in pollen grains and pollen tubes of *Nicotiana tabacum* L. Protoplasma 189: 26-36

Liang GH, Gu X, Yue G, Shi ZS, Kofoid KD (1997) Haploidy in *Sorghum*. In Jain SM, Sopory SK, Veilleux RE (eds) In vitro haploid production in higher plants. 4. Kluwer Academic Publishers, The Netherlands, pp 149-161

Luckett DJ, Darvey NL (1992) Utilisation of microspore culture in wheat and barley improvement. Aust J Bot 40: 807-828

Maheshwari SC, Rashid A, Tyagi AK (1982) Haploids from pollen grains. Retrospect and prospect. Am J Bot 69: 865-879

Mandaron P, Niogret MF, Mache R, Monéger F (1990) In vitro protein synthesis in isolated microspores of *Zea mays* at several stages of development. Theor Appl Gen 80: 134-138

Marinkovic N, Radojevic L (1992) The influence of bud length, age of the tree and culture media on androgenesis induction in *Aesculus carnea* Hayne anther culture. Plant Cell Tiss Org Cult 31: 51-59

McConchie R, Lang S (1991) Carbohydrate depletion and leaf blackening in *Protea neriifolia*. J Am Soc Hort Sc 116: 1019-1024

Miki-Hirosige H, Nakamura S (1983) Growth and differentiation of amyloplasts during male gamete development in *Lilium longiflorum*. In Mulcahy DL, Ottaviano E (eds), Pollen: biology and implications for plant breeding. Elsevier Science Publisher, Biomedical, New York, pp 141-147

Miller WB, Ranwala AP (1994) Characterization of three soluble invertase forms from *Lilium longiflorum* flower buds. Physiol Plant 92: 247-253

Mlodzianowski F, Idzikowska K (1978) The ultrastructure of anther wall and pollen of *Hordeum vulgare* at the microspore stage. Acta Soc Bot Pol 47: 219-224

Mu JH, Stains JP, Kao TH (1994) Characterization of a pollen expressed gene encoding a putative pectin esterase of *Petunia inflata*. Plant Mol Biol 25: 539-544

Murgia M, Charzynska M, Rougier M, Cresti M (1991) Secretory tapetum of *Brassica oleracea* L.: polarity and ultrastructural features. Sex Plant Reprod 4: 28-35

Nepi M, Ciampolini F, Pacini E (1996) Plastid differentiation during *Cucurbita pepo* (Cucurbitaceae) pollen grain development. Sex Plant Reprod 9: 17-24

Nepi M, Pacini E (1999) What may be the significance of polysiphony in *Lavatera arborea*? Ibid, pp 13-20

Nichterlein K, Friedt W (1993) Plant regeneration from isolated microspores of linseed (*Linum usitatissimum* L.). Plant Cell Rep 12: 426-430

Noher de Halac I, Cismondi IA, Harte C (1990) Pollen ontogenesis in *Oenothera*: comparison of genotypically normal with the male-sterile mutant *sterilis*. Sex Plant Reprod 3: 41-53

Noher de Halac I, Fama G, Cismondi IA (1992) Changes in lipids and polysaccharides during pollen ontogeny in *Oenothera* anthers. Sex Plant Reprod 5: 110-116

Ogawa T, Fukuoka H, Ohkawa Y (1994) Induction of cell division of isolated pollen grains by sugar starvation in rice. Breed Sci 44: 75-77

Olmedilla A, Schrauwen JAM, Wullems GJ (1991) Visulaization of starch-sybthase expression by in situ hybridization during pollen development. Planta 184: 182-186

Pacini E (1994) Cell biology of anther development. In Williams EG, Clarke AE, Knox RB (eds), Genetic control of self incompatibility and reproductive development in flowering plants. Kluwer Academic publishers, Dordrecht, Boston, London, pp 289-308

Pacini E (1996) Types and meaning of pollen carbohydrate reserves. Sex Plant Reprod 9: 362-366

Pacini E, Bellani LM (1986) *Lagerstroemia indica* L. pollen: form and function. In Pollen and spores: form and function. Linnean Society, London, pp 347-357

Pacini E, Franchi GG (1983) Pollen grain development in *Smilax aspera* L. and possible functions of the loculus. In Mulcahy DL, Ottaviano E (eds), Pollen: biology and implications for plant breeding. Elsevier Science Publisher, Biomedical, New York, pp 183-190

Pacini E, Juniper BE (1979) The ultrastructure of pollen-grain development in the olive (*Olea europea*). 1. Proteins in the pore. New Phytol 83: 157-163

Pacini E, Taylor PE, Singh MB, Knox RB (1992) Development of plastids in pollen and tapetum of rye-grass, *Lolium perenne* L. Ann Bot 70: 179-188

Pacini E, Viegi L (1995) Total polysaccharide content of developing pollen in two angiosperm species. Grana 34: 237-241

Pandolfi T, Pacini E, Calder DM (1993) Ontogenesis of monad pollen in *Pterostylis plumosa* (Orchidaceae, Neottiaideae). Pl Syst Evol 186: 175-185

Pareddy DR, Petolino JF (1992) Maturation of maize pollen in vitro. Plant Cell Rep 11: 535-539

Pedroso MC, Pais MS (1994) Induction of microspore embryogenesis in *Camellia japonica* cv. Elegans. Plant Cell Tiss Org Cult 37: 129-136

Pescitelli SM, Johnson CD, Petolino JF (1994) Isolated microspore culture of maize. In Bajaj YPS (ed) Biotechnology in Agriculture and Forestry. 25. Maize. Springer-Verlag, Berlin, pp 186-200

Pickering RA, Devaux P (1992) Haploid production: approaches and use in plant breeding. In Shewry PR (ed) Barley: genetics, biochemistry, molecular biology and biotechnology. CAB International, Oxford, pp 519-547

Polowick PL, Sawhney VK (1993) An ultrastructural study of pollen development in tomato (*Lypercosicon esculentum*). II. Pollen maturation. Can J Bot 71: 1048-1055

Pressey R, Reger BJ (1989) Polygalacturonase in pollen from corn and other grasses. Plant Sci 59: 57-62

Reynolds TL (1990) Wild tomato (*Solanum carolinense* L.): anther culture and the induction of haploids. In Bajaj YPS (ed) Biotechnology in agriculture and forestry. 12 Haploids in crop improvement, Springer Verlag, Berlin, Heidelberg, pp 498-513

Reznickova SA (1978) Histochemical study of reseve nutrient substances in *Lilium candidum*. CR Acad Sc Bulg 31: 1067-1070

Reznickova SA (1983) Metabolism of reserve substances in the developing anther. In Erdelska O (ed), Fertilization and embryogenesis in ovulated plants.Veda, Bratislava, pp 57-62

Reznickova SA, Willemse MTM (1980) Formation of the pollen in the anther of *Lilium*. 2. The function of the surrounding tissues in the formation of pollen and pollen wall. Acta Bot Neerl 29: 141-156

Rihova L, Capkova V, Tupy J (1996) Changes in glycoprotein patterns associated with male gametophyte development and with induction of pollen embryogenesis in *Nicotiana tabacum* L. J Plant Physiol 147: 573-581

Rines HW, Riera-Lizarazu, Nunez VM, Davis DW, Phillips RL (1997) Oat haploids from anther culture and from wide hybridizations. In Jain SM, Sopory SK, Veilleux RE (eds) In vitro haploid production in higher plants. 4. Kluwer Academic Publishers, The Netherlands, pp 205-221
Robert LS, Allad S, Gerster JL, Cass L, Simmonds J (1993) Isolation and characterization of a polygalacturonase gene highly expressed in *Brassica napus* pollen. Plant Mol Biol 23: 1273-1278
Roberts-Oehlschlager SL, Dunwell JM, Faulks R (1990) Changes in the sugar content of barley anthers during culture on different carbohydrates. Plant Cell Tiss Org Cult 22: 77-85
Rodriguez-Garcia MI, Garcia A (1978) Differentiation of the plastid population during microsporogenesis and the development of the pollen grain in the Liliaceae. Biol Cell, 33: 63-70
Rogers HJ, Harvey A, Lonsdale DM (1992) Isolation and characterization of a tobacco gene with homology to pectate lyase which is specifically expressed during microsporogenesis. Plant Mol Biol 20: 493-502
Rokka VM, Valkonen JPT, Pehu E (1995) Production and characterization of haploids derived from somatic hybrids between *Solanum* brevidens and *S. tuberosum* through anther culture. Plant Sci 112: 85-95
Saini HS (1997) Effects of water stress on male gametophyte development in plants. Sex Plant Reprod 10: 67-73
Saini HS, Lalonde S (1998) Injuries to reproductive development under water stress, and their consequences for crop productivity. J Crop Prod 1: 223-248
Saini HS, Sedgley M, Aspinall D (1984) Developmental anatomy in wheat of male sterility indued by heat stress, water deficit or abscissic acid. Aust J Plant Physiol 11: 243-253
Sangwan RS, Sangwan-Norreel BS (1987) Ultrastructural cytology of plastids in pollen grains of certain androgenic and nonandrogenic plants. Protoplasma 138: 11-22
Schaeverbeke J (1965) Mise en évidence par marquage au 14C d'une pénétration sélective de saccharose dans les filets staminaux des Graminées, au cours de l'anthèse. CR Acad Sc Paris 261: 1081-1084
Schrauwen JAM, Mettenmeyer T, Croes AF, Wullems GJ (1996) Tapetum-specific genes: what role do they play in male gametophyte development ? Acta Bot Neerl 45: 1-15
Scott R (1993) Anther development: a molecular perspective. In Jordan BR (ed), The molecular biology of flowering. CAB International, Wallingford, Oxon, pp 141-184
Scott P, Lyne RL (1994a) The effect of different carbohydrate sources upon the initiation of embryogenesis from barley microspores. Plant Cell Tiss Org Cult 36: 129-133
Scott P, Lyne RL (1994b) Initiation of embryogenesis from cultured barley microspores: a further investigation into the toxic effects of sucrose and glucose. Plant Cell Tiss Org Cult 37: 61-65
Scott P, Lyne RL, Ap Rees T (1995) Metabolism of maltose and sucrose by microspores isolated from barley (*Hordeum vulgare* L.). Planta 197: 435-441
Sheoran IS, Saini HS (1996) Drought-induced male sterility in rice: changes in carbohydrate levels and enzyme activities associated with the inhibition of starch accumulation in pollen. Sex Plant Reprod 9: 161-169
Singh MB, Knox RB (1984) Invertases of *Lilium* pollen. Plant Physiol 74: 510-515
Sorvari S, Schieder O (1987) Influence of sucrsoe and melibiose on barley anther cultures in starch media. Plant Breed 99: 164-171

Speranza A, Calzoni GL, Pacini E (1997) Occurrence of mono- or disaccharides and polysaccharide reserves in mature pollen grains. Sex Plant Reprod 10: 110-115

Stieglitz HB (1977) Role of ß-1,3 glucanase in postmeiotic microspore release. Develop Biol 57: 87-97

Takahashi M (1987) Development of omniaperturate pollen in *Trillium kamtschaticum* (Liliaceae). Am J Bot 74: 1842-1852

Takahata Y, Komatsu H, Kaizuma N (1996) Microspore culture of radish (*Raphanus sativus* L.): influence of genotype and culture conditions on embryogenesis. Plant Cell Rep 16: 163-166

Tanaka I, Ito M (1980) Induction of typical cell division in isolated microspores of *Lilium longiflorum* and *Tulipa gesneriana*. Plant Sci Lett 17: 279-285

Tanaka I, Ito M (1981) Control of division patterns in explanted microspores of *Tulipa gesneria*. Protoplasma 108: 329-340

Tanner GJ, Piccirilli M, Moore AE, Larkin PJ, Arcioni S (1990) Initiation of non physiological division and manipulation of development pathway in cultured microspores of *Medicago* sp. Protoplasma 158: 165-175

Tiainen T (1992) The influence of culture conditions on anther culture response of commercial varieties of *Solanum tuberosum* L. Plant Cell Tiss Org Cult 30: 211-219

Tebbutt SJ, Rogers HJ, Lonsdale DM (1994) Characterization of a tobacco gene encoding a pollen-specific polygalacturonase. Plant Mol Biol 25: 283-297

Theiler-Hedtrich R, Hunter CS (1995) Regeneration of dihaploid chicory (*Cichorium intybus* L. var. *foliosum* Hegi) via microspore culture. Plant Breed 114: 18-23

Tsay HS, Yeh CC, Hsu JY (1990) Embryogenesis and plant regeneration from anther culture of bamboo (*Sinocalamus latiflora* (Munro) McClure). Plant Cell Rep 9: 349-351

Touraev A, Fink CS, Stöger E, Heberle-Bors E (1995) Pollen selection: a transgenic reconstruction approach. PNAS 92: 12165-12169

Touraev A, Ilham A, Vicente O, Heberle-Bors E (1996a) Stress-induced microspore embryogenesis in tobacco: an optimized system for molecular studies. Plant Cell Rep 15: 561-565

Touraev A, Indrianto A, Wratschko I, Vicente O, Heberle-Bors E (1996b) Efficient microspore embryogenesis in wheat (*Triticum aestivum* L.) induced by starvation at high temperature. Sex Plant Reprod 9: 209-215

Touraev A, Pfosser M, Vicente O, Heberle-Bors E (1996c) Stress as the major signal controlling the developmental fate of tobacco microspores: towards a unified model of induction of microspore/pollen embryogenesis. Planta 200: 144-152

Touraev A, Vicente O, Heberle-Bors E (1997) Initiation of microspore embryogenesis by stress. Trends Plant Sci 2: 297-302

van Aelst AC, van Went JL (1992) Ultrastructural immuno-localization of pectins and glycoproteins in *Arabidopsis thaliana* pollen grains. Protoplasma 168: 14-19

van den Bulk RW, de Vries-van Hulten HPJ, Custers JBM, Dons JJM (1994) Induction of embryogenesis in isolated microspores of tulip. Plant Sci 104: 101-111

Vemmos SN, Goldwin GK (1993) Stomatal and chlorophyll distribution of Cox's orange pippin apple flowers relative to other cluster parts. Ann Bot 71: 245-250

Vemmos SN, Goldwin GK (1994) The photosynthetic activity of cox‘s orange pippin apple flowers in relation to fruit setting. Ann Bot 71: 245-250

Vu JC, Yelenosky G, Bausher MG (1985) Photosynthetic activity in the flower buds of "Valencia" orange (*Citrus sinensis* (L.) Osbeck). Plant Physiol 78: 420-423

Wada T, Ito T, Tio M, Takeoka Y (1992) Light microscopic observations on pollen and anther in rice (*Oryza sativa* L.). Jap J Crop Sci 61: 136-144

Wang YT and Breen PJ (1986) Partitioning of 14C assimilate in Easter lily as affected by growth stage and flower removal. Sci Hortic 29: 273-281

Weber M (1989) Ultrastructural changes in maturing pollen grain of *Apium nodiflorum* L. (Apiaceae), with special reference to the endoplasmic reticulum. Protoplasma 152: 69-76

Weiss D, Schönfeld M, Halevy AH (1988) Photosynthetic activities in the *Petunia* corolla. Plant Physiol 87: 667-670

Wetzel CL, Jensen WA (1992) Studies of pollen maturation in cotton: the storage reserve accumulation phase. Sex Plant Reprod 5: 117-127

Wu Y, Qiu X, Du S, Erickson L (1996) PO149, a new member of pollen pectate lyase-like gene family from alfalfa. Plant Mol Biol 32:1037-1042

Xi XY (1991) Development and structure of pollen and embryo sac in Peanut (*Arachis hypogaea* L.). Bot Gaz 152: 164-172

Yu S-M, Lee Y-C, Fang S-C, Chan M-T, Hwa S-F, Liu L-F (1996) Sugars act as signal molecules and osmotica to regulate the expression of alpha-amylase genes and metabolic activities in germinating cereal grains. Plant Mol Biol 30: 1277-1289

Zaki MA, Dickinson HG (1990) Structural changes during the first divisions of embryos resulting from anther and free microspore culture in *Brassica napus*. Protoplasma 156: 149-162

Zarsky V, Garrido D, Rihova L, Tupy J, Vicente O, Heberle-Bors E (1992) Derepression of the cell cycle by starvation is involved in the induction of tobacco pollen embryogenesis. Sex Plant Reprod 5: 189-194

Zavada MS (1984) Pollen wall development in *Austrobaileya maculata*. Bot Gaz 145: 11-21

Zhang CJ, Wang HL, Ma Y, Kang YQ (1994) Regeneration of haploid plants from isolated microspores of asparagus (*Asparagus officinalis* L.). Plant Cell Rep 13: 637-640

Zhong D, Michaux-Ferrière N, Coumans M (1995) Assay for doubled haploid sunflower (*Helianthus annuus*) plant production by androgenesis: fact or artifact ? Part 1. In vitro anther culture. Plant Cell Tiss Org Cult 41: 91-97

8 Expression of oleosin genes in the olive (*Olea europaea* L.) anther

Alché* JD, Castro AJ and Rodríguez-García MI

8.1 Introduction

Flowering plants store lipids in organs such as seeds, pollen, flowers, roots and stems in the form of subcellular structures called oil bodies (Huang 1996). Oil bodies are formed by a matrix of triacylglycerol which is surrounded by a monolayer of phospholipids. Oleosins are one of the major proteins present on the surface of such bodies (Huang 1992). The most studied oleosins are those of seeds (Huang 1992; Murphy 1993; Herman 1995), although evidence of their presence in several tissues of the anther is emerging (Oliveira et al. 1993; Ruiter 1997; Wu et al. 1997).

8.1.1 Oleosins in seeds

Oleosins may represent up to 8-10% of the total weight in the seeds of some plant species. These proteins are 15-25 kDa and frequently appear as heteromers of two isoforms in a 1:1 molar ratio. Recent reports have indicated that oleosins consist of three distinct domains, of which the most significant is a conserved central hydrophobic region composed of approximately 70 amino acids, and believed to interact directly with the internal lipid domain of the oil body (Huang 1991; Murphy 1993). This domain is flanked by relatively hydrophylic N- and C-terminal domains which exhibit less conservation, except within a special plant species (Ross and Murphy 1996). An extension at the C- terminus is also present in some oleosins, and is responsible, along with the highly variable N- terminus, for the differences of size in oleosins (Huang 1992). Ultrastructural localization on

* for correspondence: Departamento de Bioquímica; Biología Celular y Molecular de Plantas; Estación Experimental del Zaidín CSIC; Profesor Albareda 1; 18008 Granada; Spain

oleosins in transgenic rapeseed shows that oleosins are equally distributed on the surface of each oil body and that ER is apparently involved in the targeting of oleosins to oil bodies (Sarmiento et al. 1997).

At present, the functions of oleosins are still not fully understood. Two main putative functions have been suggested for them: first to maintain the seed oil bodies as small entities in vivo and in vitro, preventing adjacent oil-bodies from coalescing, particularly under dehydrating conditions such as those of maturing seeds (Tzen et al. 1992). In addition to this structural role, oleosins may also serve as recognition signals on the surface of the oil-bodies for the binding of newly synthesized lipase during germination (Huang 1996).

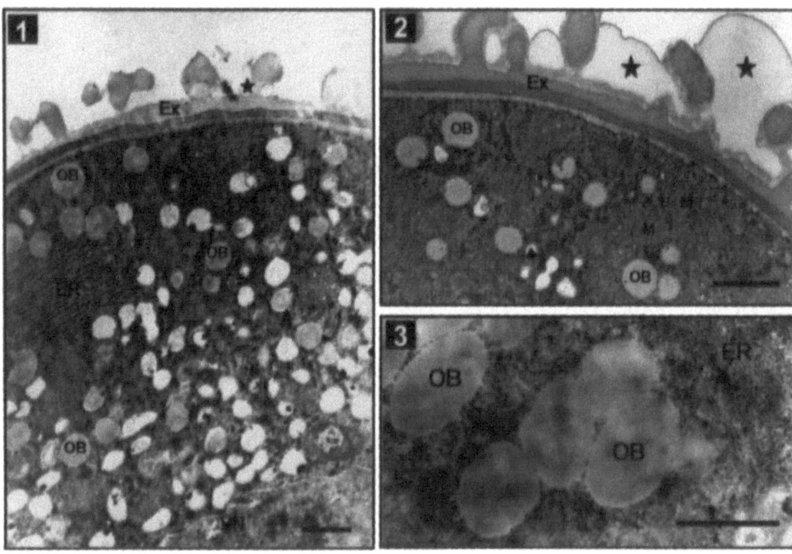

Fig. 1. Electron micrographs of sections from *Olea europaea* mature pollen grains fixed in 3% glutaraldehyde, postfixed in 2% OsO_4, embedded in Epon and contrasted with uranyl acetate and lead citrate. **(1.1)** General view of the vegetative cell cytoplasm showing oil bodies (OB), endoplasmic reticulum (ER), vegetative nucleus (VN), Exine (Ex) and pollenkit (star). **(1.2 and 1.3)** Enlarged views of lipid bodies inside the vegetative cytoplasm. Mitochondria (M). Bars represent 1 μm

8.1.2. Oleosins in anthers

Until now mRNA encoding putative oleosins have been detected in the tapetum of developing anthers of *Brassica* (Robert et al. 1994; Ross and Murphy 1996) and *Arabidopsis* (de Oliveira et al. 1993; Robert et al. 1993). The sizes of the predicted oleosins in the anther represent a broader range than those of seeds (from 14 to 60 kDa). However, the corresponding proteins had not yet been documented. Only very recently polypeptides encoded by cloned *Brassica* genes were identified as oleosins (Ruiter et al. 1997; Wu et al.1997). In addition, the subcellular localization of oleosins in the anther has been well documented by electron microscopy immunocytochemistry (Wang et al. 1997). The oleosins are located largely on the periphery of the plastoglobuli inside the tapetum plastids. They are also found on the surface of storage oil bodies inside the maturing microspores. According to these results, the gametophytic microspore oleosins share common epitopes with the sporophytic tapetum oleosins. These results also confirm the subcellular localization of oleosins on oil body surfaces reported for seed tissues (Sarmiento et al. 1997). In all cases, oleosins are situated so as to cover the oil bodies mantaining them as small non-coalescing entitities (Huang 1992). In the case of tapetum oleosins, they could also assist the globules in penetrating the lysed tapetum cell wall and /or in attaching the pollenkit to the pollen surface coat. In this way oleosins may interact with the stigma in order to guide water from the stigmatic papilae to the pollen grains facilitating the pollen hydration process before pollen tube germination (Ruiter et al. 1997). In mature pollen, oleosins may also be involved in the mobilization of oil bodies during pollen germination and pollen tube growth (Wang et al. 1997). The above-mentioned putative roles of oleosins in the tapetum/pollen do not have to be mutually exclusive. However, at this point no conclusive evidence yet exist. In order to find evidence supporting this hypothesis, and to elucidate other possible functions of oleosins in the anther, further studies in this direction are required.

8.2 Oleosins in the olive tree

The olive is one of the most important oil-storing plant species. It represents a key crop for the agroeconomy of Mediterranean countries, due to both the large production and the high quality of its oil. For these reasons, the study of oleosins in the olive plant and the physiological changes concomitant to pollen development, maturation of seeds and adaptation to dryness, can be considered topics of capital importance.

8.2.1. Oleosins are present in olive seeds

Oleosins have been studied in the olive seed (Ross et al. 1993). A 22 kDa polypeptide has been reported to be present in the long-term storage oil bodies from the embryo and endosperm tissues of the olive seed. Immunolocalization experiments exclusively localize the polypeptide on the surface of oil bodies, and therefore this polypeptide is likely to be an oleosin. The protein is apparently absent from oil bodies in the fruit of olive mesocarp.

8.2.2
The olive anthers contain oleosin transcripts in different tissues

On the basis of our studies during olive pollen development we can affirm that oil bodies of 0.2-1 µm diameter are present inside mature pollen grain (Fig. 1). They may be similar to seed oil bodies in structure and function. On the other hand, the tapetum of olive has been well studied by Pacini and Juniper (1979) and they also showed that there are lipid bodies and plastoglobuli within the plastids. Consequently, oleosins are likely to be found in the olive anther.

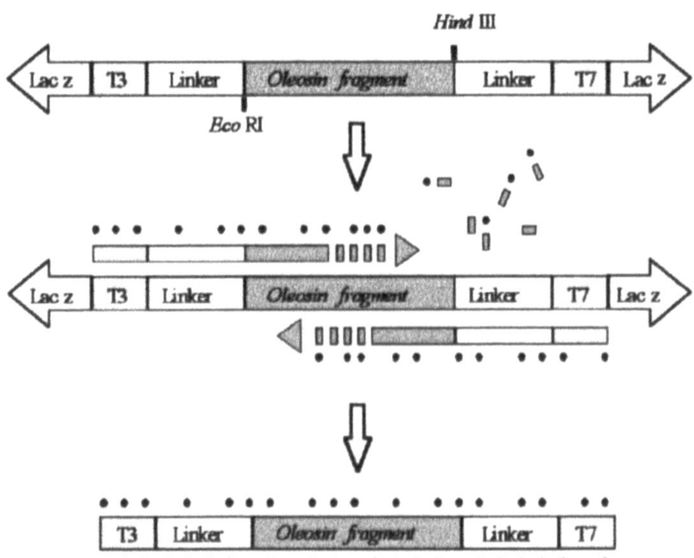

Fig. 2. Chart showing the strategy used for the preparation of dig-labelled probe. Ca. 200 bp corresponding to the central hydrophobic domain sequence from a *Brassica napus* oleosin anther cDNA subcloned into p-Bluescript were amplified by PCR using T3/T7 sequences as the oligos and digoxigenin (dots)-deoxi-nucleotides in the labelling mix. The amplified fragment was ethanol precipitated, checked by agarose gel electrophoresis and quantified

In order to obtain molecular evidence for the presence of oleosins in this organ, we have used a specific probe (about 200 bp long) to the central hydrophobic domain sequence from a *Brassica napus* oleosin anther cDNA kindly donated by Dr. Joanne Ross (John Innes Centre, Norwich, UK). The probe was synthesized by PCR amplification of the fragment subcloned into p-Bluescript plasmid using T3/T7 sequences as the oligos for amplification and digoxigenin-deoxinucleotides in the labelling mix (Fig. 2).

Genomic DNA for Southern blot analysis was obtained from both olive leaves and anthers by the CTAB standard protocol and cut with several restriction enzymes. The blot was probed and showed a number of bands when cut with *Eco* RI whose sizes ranged from between 3.5 to 7.0 Kpb. (3.47, 3.98, 4.47 and 6.99). The pattern coincided in both leaf and anther (Fig. 3). When double digestions were performed, bands shifted to lower sizes (Fig. 4).

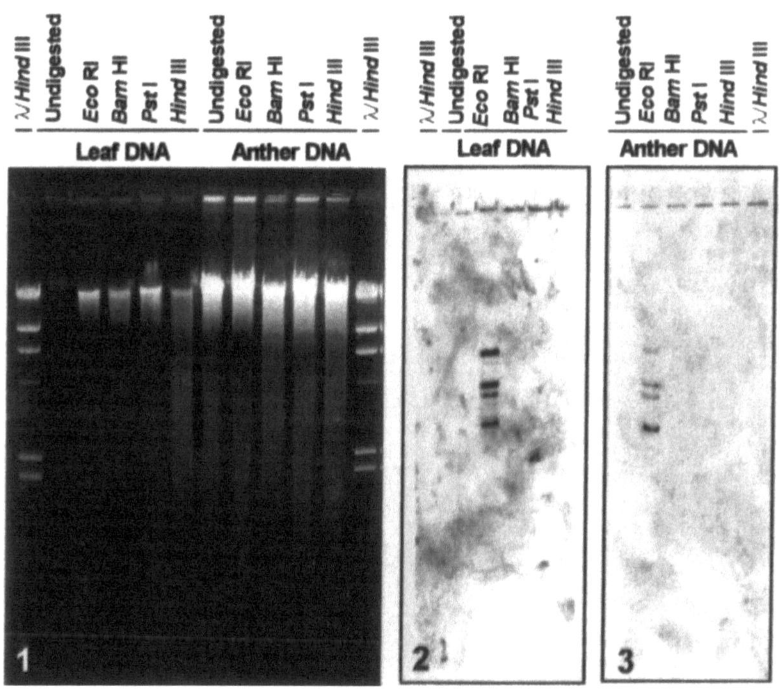

Fig. 3. Ethidium bromide stained agarose gel (1%) **(3.1)** and Southern blots of genomic DNA from leaves **(3.2)** and anthers **(3.3)**, single digested with different restriction enzymes. Filters were hybridized with the dig-labelled probe corresponding to the central hydrophobic domain of a *Brassica* oleosin. All lanes contain aproximately 20 μg (leafs) and 50 μg (anthers) of DNA. λ/*Hind* III was used as standard marker. The filters were hybridized for 12 h. at 40° C with 10 ng/ml of the probe in high SDS hybridization buffer (Boehringer Mannheim), washed with 2XSSC and 0.5XSSC, and detected using an anti-dig alkaline phosphatase conjugated antibody and chemiluminescent substrate (CSPD, Boehringer Mannheim) acording to the manufacturer's instructions

In order to detect the presence of oleosin transcripts, light microscopy (LM) in situ hibridization was performed using olive anthers at different developmental stages essentially as described by Coen et al. (1990) using the PCR-generated dig-labelled probe. Early floral buds (pre-meiotic stage) did not show labelling in any tisue (Fig. 5.1). Using this same probe, labelling was found to be present in both tapetum and microsporocytes of late-meiotic anthers (Fig. 5.2), tetrads (Figs. 5.3 - 5.4), early microspores (Fig. 5.5), late microspores (Fig. 5.6) and released pollen grains (Fig. 5.7).

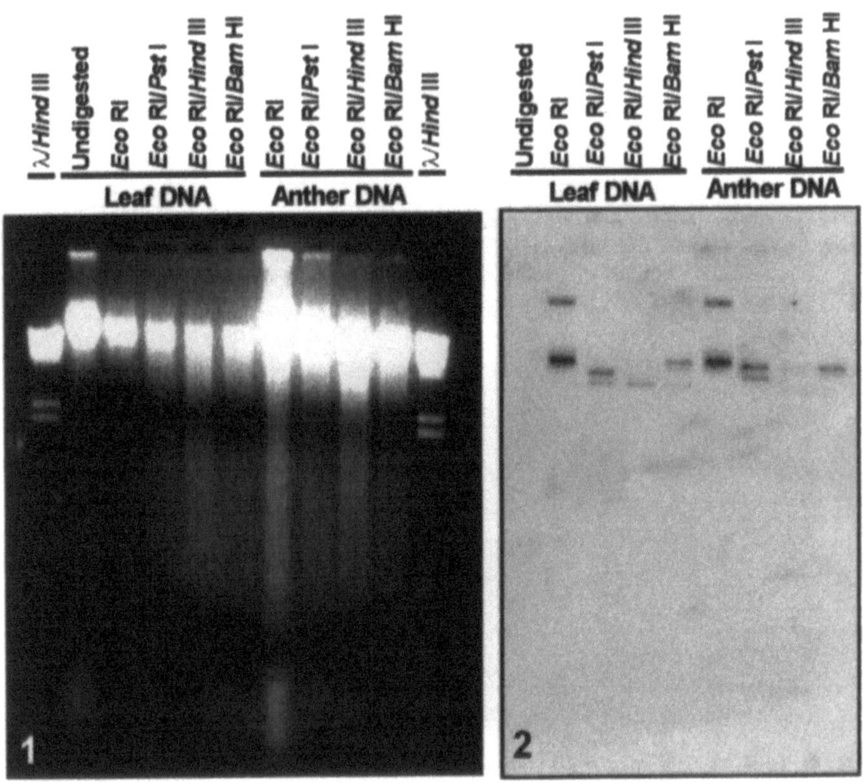

Fig. 4. Ethidium bromide stained agarose gel (1%) **(4.1)** and Southern blots of genomic DNA from leaves and anthers **(4.2)**, single/double digested with different restriction enzymes. Filters were hybridized and treated as described in Fig. 3

In conclusion, the results of Southern blot analysis of total genomic DNA suggest the presence of a multigene family encoding several oleosin isoforms in the olive, or at least multiple copies of oleosin genes. On the other hand, temporal and spatial analysis of oleosin gene expression show that oleosin mRNAs are

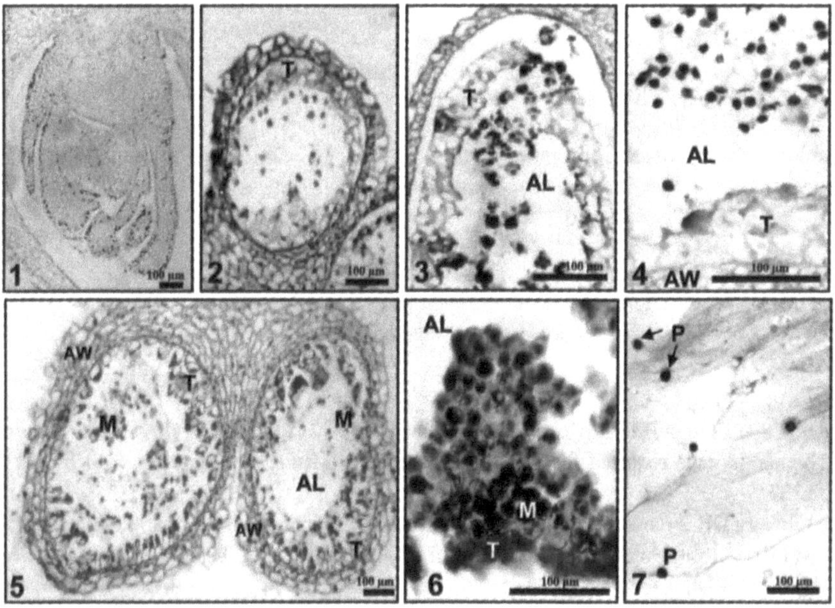

Fig. 5. LM *in situ* hybridization to oleosin mRNAs during olive anther development. **5.1**, whole young floral bud at the pre-meiotic stage. **5.2**, anthers at the late meiotic prophase stage. **5.3 and 5.4**, tetrads. **5.5**, early microspores. **5.6**, late microspores. **5.7**, mature (dehiscent) pollen. Samples were fixed overnight in 0.1 M cacodylate buffer, pH 7.2, containing 4% paraformaldehyde and 0.2 % glutaraldehyde. They were then washed, dehydrated in a graded ethanol series and xylene, and embedded in paraffin. Sections (7 µm thick) were attached to 3-aminopropyl triethoxy-silane-coated slides. After hybridization, slides were treated with 1:1000 anti-dig, alkaline phosphatase conjugated antibody (Boehringer Mannheim). Colour reaction (NBT/BCIP substrate) was left to develop for 16 h. Observations and photographs were made on permanent preparations. Negative control reactions (not shown) were set by omiting the probe from the hybridization mix and/or the anti-dig antibody. Anther locule (AL), anther wall (AW), microspores (M), pollen grains (P), tapetum (T). Bars represent 100 µm

present in both sporophytic (tapetum) and gametophytic (microsporocytes) tissues of anther during pollen development.

Northern analysis and RT-PCR are taking now underway in order to determine the expression of oleosin isoforms in the anther and whether these oleosin mRNAs are anther specific. Furthermore, oil bodies will be isolated after adapting the protocols designed for seeds, and antibodies will be raised to these proteins. The antibodies will be used in immunocytochemical studies to localize these proteins in the olive anther. Subcellular localization of oleosins in the anther is a prerequisite for understanding the physiological role of these proteins in the process of sexual reproduction.

Acknowledgements

This work was supported by the projects no. PB95-0080 (DGES, Spain) and OLI96-2160 (CICYT, Spain) and Junta de Andalucía (no. AGR160) with a grant to AC. We wish to thank Dr. Joanne Ross (John Innes Centre, Norwich, UK.) for kindly providing the cDNA probe.

References

Coen ES, Romero JM, Doyle S, Elliott R, Murphy G, Carpenter R (1990) *Floricaula*; a homeotic gene required for flower development in *Antirrhinum majus*. Cell 63: 1311-1322

De Oliveira DE, Franco LO, Simoens C, Seirinck J, Coppieters, Botterman J, Van Montagu M (1993) Inflorescence–specific genes from *Arabidopsis thaliana* encoding glycine-rich proteins. Plant Cell 2: 427-436

Herman EM (1995) Cell and molecular biology of seed oil body proteins. In: Kigel J, Galili G (eds). Seed development and germination. Marcel Dekker, New York, pp 195-214

Huang AHC (1992) Oil bodies and oleosins in seeds. Ann Rev Plant Physiol Mol Biol 43:177-200

Huang AHC (1996) Oleosins in seed and other organs. Plant Physiol 110: 1055-1061

Murphy DJ (1993) Structure, function and biogenesis of storage lipid bodies and oleosins in plants. Prog Lipid Res 31: 247-280

Pacini E, Juniper BE (1979) The ultrastructure of pollen-grain development in the olive (*Olea europaea*). 2. Secretion by the tapetal cells. New Phytol 83: 165-174

Robert LS, Allard S, Gerster JR, Cass L, Simmonds J (1993) Isolation and characterization of a polygalacturonase gene highly expressed in *Brassica napus* pollen. Plant Mol Biol 23: 1273-1278

Robert LS, Gerster JR, Allard S, Cass L, Simmonds J (1994) Molecular characterization of two *Brassica napus* genes related to oleosins which are highly expressed in the tapetum. Plant J 6: 927-933

Ross JHE, Sanchez J, Millan F, Murphy DJ (1993) Differential presence of oleosins in oleogenic seed and mesocarp tissues in olive (*Olea europaea*) and avocado (*Persa americana*). Plant Sci 93: 203-210

Ross JHE, Murphy DJ (1996) Characterization of anther-expressed genes encoding a major class of extracellular oleosin-like proteins in the pollen coat of *Brassicaceae*. Plant J 9: 625-637

Ruiter RK, Van Eldik GJ, Van Herpen RMA, Schrauwen JAM (1997) Characterization of oleosins in the pollen coat of *Brassica oleracea*. Plant Cell 9: 1621-1631

Sarmiento C, Ross JHE, Herman E, Murphy DJ (1997) Expression and subcellular targeting of a soybean oleosin in transgenic rapeseed. Implications for the mechanism of oil body formation in seeds. Plant J 11: 783-796

Tzen JTC, Lie GC, Huang AHC (1992) Characterization of the charged components and their topology on the surface of plant seed oil bodies. J Biol Chem 267:15626-15634

Wang TW, Balsamo RA, Ratnayake, Platt KA, Ting JTL, Huang AHC (1997) Identification, subcellular localization, and developmental studies of oleosins in the anther of *Brassica napus*. Plant J 11:475-487

Wu SSH, Platt KA, Ratnayake C, Wang TW, Ting JTL Huang AHC (1997) Isolation and characterization of neutral-lipid-containing organelles and globuli-filled plastids from *Brassica napus* tapetum. Proc Natl Acad Sci USA 94:12711-12716

9 Possible mechanisms of the "447" Cytoplasmic Male Sterility in *Vicia faba*, an unconventional CMS associated with a double-stranded RNA replicon

Pfeiffer* P

9.1 Introduction

Cytoplasmic male sterility (CMS) is a desirable trait for breeding purposes because it prevents self-pollination of hermaphroditic flowers in species where no self-incompatibility systems exist: it is therefore widely used for streamlining the production of hybrid seeds. Although a few interspecific crosses, *e.g. H. petiolaris* x *H.annuus* in sunflower and *T. aestivum x T. timopheevi* in wheat, give rise to a male-sterile progeny, there is to date no controlled and predictable way of obtaining CMS lines. Molecular studies in several plant species have revealed that CMS results from rearrangements of the mitochondrial DNA that lead to the production of aberrant, non-functional polypeptides that cause a collapse of mitochondrial biogenesis and energy production during pollen differentiation, at a time when energy demand peaks (Dewey et al. 1986; Young and Hanson 1987; Lewings 1993). The CMS trait of the "447" line of *Vicia faba* is however unusual in that it does not correlate with alterations of the mitochondrial DNA pattern, but rather with the presence of a linear double-stranded RNA of high molecular weight and of unknown origin (Grill and Garger 1981). This dsRNA is contained together with its specific replicase/transcriptase in membraneous vesicles, and is transmitted exclusively in a vertical mode. There is an absolute correlation of its presence with CMS: it is absent from maintainer lines, and is irreversibly lost after restoration of male fertility by crossing with a restorer line, or as a consequence of spontaneous reversion to fertility which occurs at a comparatively high rate (Scalla et al. 1981). Similarly, the presence or absence of the cytoplasmic vesicles correlates with that of the dsRNA, and we have previously established (Lefebvre

* Institut de Biologie Moléculaire des Plantes du CNRS; 12, rue du Général Zimmer; 67084 Strasbourg Cedex; France

et al. 1990) that these membranous vesicles do indeed contain the dsRNA described by Scalla et al. (1981) and Grill and Garger (1981). We also established that the structure of this dsRNA was unusual, with the (-) strand being continuous while the (+) strand was interrupted and could thus potentially define two subgenomic RNAs (Pfeiffer et al. 1993). Finally, "run-on" RNA synthesis performed with transcription/replication complexes contained in the cytoplasmic vesicles revealed that a small subgenomic RNA was preferentially synthetized.

We had however no information on the respective arrangement of the two subgenomic RNAs nor on the various proteins specified by the dsRNA in general. In addition, all attempts to convert fertile plants to male sterile by inoculation or by transfer of the dsRNA failed (Duc et al. 1984; Turpen et al. 1988), which precluded a direct demonstration of the involvement of the dsRNA in the male sterility together with potential extension of this CMS to other plants of agronomical interest.

The entire nucleotide sequence of the dsRNA was therefore determined (Pfeiffer 1998), and sequence analysis revealed that the dsRNA encodes a single long open reading frame featuring the typical RNA helicase and RNA-dependent RNA polymerase signatures that are common to single-stranded and double-stranded RNA viruses. This long ORF is interrupted by a nick located at a specific position on the coding strand, but it is not known whether this relates to the translational strategy used to express the genetic information encoded by the dsRNA. By analogy with similar dsRNA replicons and other positive-strand viruses, it is however very likely that the polypeptide(s) synthetized undergo(es) a series of proteolytic maturation steps that release the functional proteins required to complete the replication cycle of the dsRNA, and that one of these protein species is the determinant of the "447" CMS. The possible mechanisms of this non-conventional CMS will be discussed in the light of the specific features of the polyprotein encoded.

9.2 The "447" dsRNA replicon and its relatives

Many cultivars of plants contain high molecular weight, unencapsidated linear dsRNAs that have long remained undetected because their presence is symptomless (see e.g. Valverde et al. 1990). Their origin is unknown: they are not associated with any viral infection, are not derived from the genome of their host (contrary to a statement by Turpen et al. 1988) and were first supposed to be intramitochondrial; it was however subsequently recognized that they are probably contained within host-derived membranous vesicles, just like the replicative forms of positive-strand RNA viruses. They are generally strictly associated with a specific line or cultivar, cannot be transmitted horizontally by any of the classical inoculation procedures, but are transmitted efficiently in a vertical manner through both egg and pollen. For instance, a 13.2 kb dsRNA found in the "Barsoy" cultivar of barley could be traced to an ancestral progenitor imported from Japan at the

beginning of this century, and Zabalgogeazcoa et al. (1993) showed that this dsRNA had been maintained over 90 years during the whole selection procedure, indicating that it does not cause any adverse effects on the plant.

It had been assumed for some time that such dsRNA species present in the Black Turtle Soup bean (Wakarchuck and Hamilton 1985) and in the cultivated rice *Oriza sativa* ssp. *japonica* (Wang et al. 1990) was associated with cytoplasmic male sterility, but this hypothesis was later disproved (Mackenzie et al. 1988; Fukuhara et al. 1993). In addition, elucidation of the molecular mechanisms of cytoplasmic male sterility (for a review, see Lewings 1993) subsequently confirmed that this trait does not result from a viral infection, but rather is associated with mitochondrial DNA rearrangements that lead to the production of chimeric, non-functional polypeptides.

Previous studies had revealed that the "447" dsRNA itself has an unusual structure, with at least a nick in one of the strands. The specific association of the dsRNA with its cognate RNA-dependent RNA polymerase (RDRP) allowed to confirm by "run-on" experiments in which preinitiated complexes were allowed to pursue RNA synthesis in vitro, that the major product detected was a short transcript that remained associated with its template (Lefebvre et al. 1990). Subsequent studies, based on *in vitro* labelling of the 3' termini of the RNA strands using T4 RNA ligase-mediated addition of nucleotide derivatives or polyadenylation by poly(A) polymerase (Pfeiffer et al. 1993) confirmed the existence of a subgenomic messenger which was labelled preferentially by these enzymes that are able to recognize and modify a free 3'OH terminus. By analogy with the general mapping of subgenomic RNAs in viruses, it seemed therefore that the small subgenomic RNA mapped at the 3'OH end of the (+) strand. Complete sequencing of this subgenomic RNA revealed however that it was 2735 nts. long and mapped at the 5 'end of the positive strand in the dsRNA: why its 3'OH end, which is expected to be poorly accessible to modyfying enzymes is nevertheless highly reactive remains unexplained. Northern blot analysis performed with methylmercury hydroxide-denatured dsRNA revealed however that a minor population of dsRNA molecules (estimated at a fraction of a percent) had an uninterrupted (+) strand, a feature required for completion of the replication cycle of this dsRNA replicon (Pfeiffer et al. 1993). The role and coding capacity of this subgenomic RNA remained however unknown, and no sequence homologies of the "447" dsRNA with other described plant RNA viruses could be detected by hybridization studies (Turpen et al. 1988).

The closest relative to the "447" dsRNA is the dsRNA replicon found in cultivated rice *Oriza sativa* ssp. *japonica* (Fukuhara et al. 1993): this symptomless 14 kbp dsRNA is efficiently transmitted through both egg and pollen, and its copy number is tightly regulated but depends on the developmental stage, with a dramatic increase in developing pollen (though its viability is not affected), or in cell suspension cultures. Complete sequencing revealed that it encodes a single, long open reading frame of 4572 amino acid residues that contains the typical helicase and RDRP domains of RNA viruses (Moriyama et al. 1995). In addition, like in the "447" dsRNA, the coding strand features a nick at a specific position (Fukuhara et al. 1995) that interrupts the long open reading frame. An interruption

within the unique ORF was similarly identified at another location in the dsRNA resident in *O.rufipogon* lines which is related, but not identical to the *O. sativa* dsRNA: although the significance of such interruptions is not known, their unique character, together with the strong sequence similarities of the helicase and polymerase domains of the protein encoded by these two rice dsRNAs and the "447" dsRNA point to a common ancestry of these dsRNA replicons.

9.3
How a dsRNA replicon may affect its host: the example of the chestnut blight virus

The large size of the "447" dsRNA and its association with an RDRP within membranous vesicles (Lefebvre et al. 1990) were very reminescent of the virus CHV-1 found in the chestnut blight fungus *Cryphonectria (Endothia) parasitica.* Like the "447" dsRNA, this dsRNA replicon is not encapsidated in a coat protein but rather contained as a replicative complex in cytoplasmic membranous vesicles contributed by its fungal host (Hansen et al. 1985; Fahima et al. 1993). In the presence of CHV-1, this plant pathogenic fungus is converted into an hypovirulent form and, most interestingly, this virus and the associated phenotype can be further transmitted by hyphal anastomosis to virulent strains of the same vegetative compatibility group: this provided the basis for biological control of chestnut blight, a devastating disease that wiped out the American chestnut forests, and the availability of an infectious cDNA clone of this virus has now allowed extension of this therapy to other fungal strains (Choi and Nuss 1992).

It had been recognized for some time that the hypovirulent phenotype results from a down-regulation expression of a specific set of genes (Powell and Van Alfen 1987a,b) rather than from a general debilitation of the host. The complete sequence of the CHV-1 dsRNA was established by D.Nuss and his colleagues (Shapira et al. 1991): these authors have in particular analyzed the expression strategy and the sequence homologies of this dsRNA replicon, and concluded that it shares a common ancestry with potyviruses.

Potyviruses are a group of positive strand RNA plant viruses that express the complete genetic information encoded by their ca. 10 kb genomic RNA as a polyprotein that undergoes a series of self-maturation processing events: owing to the presence of two different papain-like cystein proteases, this unique precursor is converted into a series of discrete proteins endowed with specific functions, e.g. aphid transmission factor, genome-linked protein, helicase, polymerase and coat protein. Yet, in the case of CHV-1, the coding capacity is expressed not as a single, but rather as two ORFs arranged in a stop-restart mode and separated by the sequence UAAUG, where the underlined A is common to the stop codon terminating ORF A and to the start codon initiating ORF B; each ORF contains a papain-like protease activity which independently processes the two polyprotein precursors into specific products. Extensive mutational analysis and molecular biology studies allowed the correlation of various phenotypic effects exerted by this defective virus on its fungal host with specific regions or functions of the proteins expressed. Thus, Choi and Nuss (1992) demonstrated that expression of ORFA alone was sufficient to induce in the fungus, and in the absence of any replication of the dsRNA replicon, specific phenotypic effects that mimicked those observed in the virus-bearing strain, i.e. reduced pigmentation, suppressed fungal sporulation and reduced laccase accumulation. Fungal virulence itself was however not affected, demonstrating that specific viral domains, rather than replication of the virus itself, were responsible for corresponding phenotypic traits. This hypothesis of an uncoupling of hypovirulence and phenotypic effect could be further refined and tested by mutagenizing various domains of ORF A, and in particular the active site of the papain-like protease involved in the maturation of the 69 kD primary translation product into p29 and p40. The hypovirulence-associated traits could thus be assigned to the release of active p29 from its inactive polyprotein precursor, but their occurrence did not depend from an intrinsic protease activity (Craven et al. 1993). It is therefore likely that the effects exerted by CHV-1 on the phenotype and on the virulence of its fungal host do not result simply from the inactivation of critical cell factors by proteolytic cleavage, as happens for instance upon infection with poliovirus (Clark et al. 1991; Hellen et al. 1991), but that they are rather pleotropic in nature and must be mediated by signal transduction pathways. Indeed, the study of the fungal genes that are down-regulated following viral infection revealed that CPG-1, a G-protein a subunit, was repressed to nearly undetectable levels in the virus-infected fungus (Choi et al. 1995), which resulted in an elevation of intracellular cAMP levels: the resulting phenotypic effects could be mimicked by artificially elevating intracellular cAMP concentrations by phosphodiesterase inhibitors or by transgenic cosuppression of CPG-1, confirming the role of G-protein signalling in the interactions between the plant pathogen and its host and in virus-mediated hypovirulence (for a review on the perturbations of signal transduction processes in virus-infected chesnut blight, see Nuss 1996).

9.4
Coding capacity and gene expression strategy of the "447" dsRNA

Complete sequencing of the "447" dsRNA (EMBL accession number AJ000929) revealed that its organization is very similar to that of the cultivated rice dsRNA studied by T. Fukuhara and his colleagues. A single ORF was detected, spanning the entire length of the dsRNA with a short 5' untranslated sequence of 41 nt; the first AUG lies however in an unfavorable context, but it is not known whether translation initiates at this AUG or whether a mechanism of "leaky scanning" similar to that described in several viruses operates here, allowing initiation at a downstream AUG. Similarly, the nick in the (+) strand detected in over 99% of the dsRNA molecules is puzzling: it interrupts the single ORF after 898 amino acids without a termination codon, and only a small subpopulation of dsRNA molecules feature a continuous (+) strand from which the full-length 5,825 amino acid long polypeptide could be translated. It is unlikely that a "run-off" mechanism as occurs when in vitro translations are performed with truncated transcripts, operates in vivo as well on such interrupted mRNA. However, it must be kept in mind that for many viruses, self-processing mediated by a protease activity embedded in a precursor polypeptide is generally a co-translational process operating in cis before completion of translation and release of the full-length translation product.The main features of the "447" dsRNA are summarized in Figure 1.

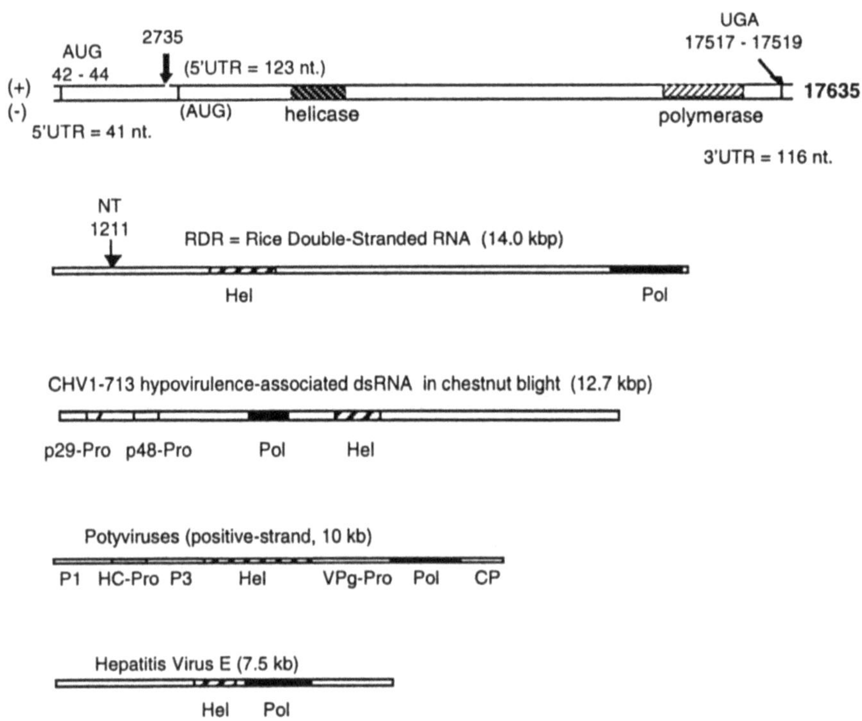

Fig. 1. Structure and coding capacity of the "447" dsRNA. The 17635 nucleotide long (+) strand is interrupted in more than 99% of the molecules, presumably by a nick that defines two potential subgenomic RNAs of respective length 2735 nts. and 14900 nts., with the smaller RNA being located on the 5' side. The (+) strand could in principle encode a single in-frame ORF starting at the first AUG at position 42-44 and ending with a TGA at 17517-17519. This potential polypeptide covering the entire length of the (+) strand would be 5,825 amino acids long and could be translated from the small subpopulation of dsRNA molecules in which the (+) strand is uninterrupted. It is likely that, in addition to the well-identified helicase and polymerase regions, the ORF(s) encode(s) protease functions that convert the precursor polyprotein(s) into functional polypeptides by a self-maturation process as occurs in many viruses.

9.5 Molecular basis of CMS

In all cases where a CMS-associated gene has been identified and shown to be actually responsible for CMS, the mitochondrial DNA had undergone rearrangements leading to chimeric open reading frames which frequently contained sequences of unknown origin. The corresponding polypeptides, which are often derivatives of subunits of the mitochondrial ATPase complex, may then compete with, or usurp the role of, their functional counterparts; this situation can be mimicked by transforming plants with nuclear copies of such aberrant ORFs equipped with adequate presequences that allow importation of the encoded polypeptide into these organelles. Similarly, incomplete or incorrect editing of mitochondrial transcripts can lead to deletion of start codons, wrong amino acid replacements and generation of premature stop codons, an hypothesis that has been verified by obtaining male sterility in plants transformed with nuclear-expressed constructs of unedited mitochondrial transcripts.

In several instances, a direct correlation could be established between the expression of a CMS-associated gene and alterations of mitochondrial functions using transgenic approaches. Thus, the well-studied sensitivity of maize with the T-sterile cytoplasm to the fungal T-toxin produced by the Texas race of southern corn leaf blight could be reproduced by expressing T-*urf13* in organisms as different as *E. coli* (Dewey et al. 1988), yeast (Glab et al. 1990; Huang et al. 1990) and tobacco (Chaumont et al. 1995). The recent isolation of several fertility restoration genes has allowed to directly confirm in specific cases what had so far been only correlative evidence and has helped to understand the various mechanisms that underly CMS (for a review, see Schnable and Wise 1998). Restoration of fertility in *Phaseolus vulgaris* is for instance straightforward since it results from the physical loss of a mitonchondrial DNA fragment that eliminates the CMS-associated sequence (Mackenzie and Chase 1990), but more subtle mechanisms operate in many other instances. In the case of maize with the T-sterile cytoplasm, fertility restoration mediated by the *Rf1* gene correlates with a reduced accumulation of the T-*urf13*-specific transcripts responsible for CMS, while the *Rf2* restorer gene was found to encode an aldehyde dehydrogenase that probably helps detoxify acetaldehyde: indeed, this compound is produced in large amounts due to aerobic fermentation in developping pollen and may cause degeneration of mitochondria weakened by the presence of the URF13 protein at a time where these organelles have to meet a peak in energy production demand (Tadege and Kuhlmeyer 1997). Other restorer genes have been found to reduce the level of transcription or to alter qualitatively the transcription pattern of the CMS-associated gene, to affect post-transcritional events such as RNA processing, editing, stability and translation of the transcripts derived from the CMS-associated gene, all of which generally lead to a reduced accumulation of the CMS-associated protein in mitochondria. It should be reminded, however, that mitochondria are not necessarily the sole target of CMS-associated proteins in developing pollen: in *Phaseolus* for instance, the ORF239 protein was found associated with the cell walls of developing microspores (Abad et al. 1995), and nuclear copies of the CMS-associated gene induced male sterility even in the absence of a mitonchondrial targeting sequence (He et al. 1996). Another important point is the species specificity of CMS-associated genes: although the T-toxin sensitivity associated with T-*urf13* expression could be reproduced in several systems, cytoplasmic male sterility was not obtained e.g. in transgenic tobacco (Chaumont et al. 1995), indicating that a gene product or a compound specific to maize anther tissue must be present for full expression of the phenotypic effects of this CMS-associated gene, possibly by allowing a conformational transition or triggering an allosteric change of the polypeptide chains inserted in the membrane.

9.6 Possible molecular mechanisms of the "447" CMS

Computer analysis of the single 5,825 amino acid long protein encoded by the "447"dsRNA revealed, as expected for an RNA replicon, the typical helicase and RDRP domains in the "ORF B" region with the same respective arrangement as in the rice dsRNA. No clear cut protease signature was identified so far, but viral papain-like proteases show only very modest degree of conservation: the search for specific protease function(s) inferred by the sheer size of the protein encoded will be best addressed by following an eventual self-maturation of in vitro translation products and by identifying the corresponding cleavage site(s). As in other positive-strand RNA viruses, the "house-keeping" genes required for completion of the replication cycle seem to be concentrated in the ORF B region which shows the highest sequence similarities with the rice dsRNA. In addition, a potential ß 1-3 glucanase signature (VLMGLEN<u>E</u>VL) was identified, in which the underlined conserved glutamic acid is an active site residue in glycosyl hydrolases (Py et al. 1991). This is reminiscent of the male sterility in petunia where mistimed microsporocyte callose wall degradation by callase, a ß 1-3 glucanase normally expressed at a specific stage in the anther locule, results in male sterility (Izhar and Frankel 1971). This could be reproduced experimentally by expressing a modified ß 1-3 glucanase in transgenic tobacco where premature dissolution of the callose deposits induced male sterility (Worall et al. 1992).

The "ORF A" region does not share any sequence homologies with other viral RNAs, and can be considered as the specific hallmark of the "447" dsRNA. Indeed, its 5' location and apparently preferential transcription suggests that it may contain the CMS determinant: a fumarate-lyase signature spanning residues 492 to 501 (GSTNMFSKNN) was identified that matched the short conserved sequence (GSx2Mx2KxN) (Bairoch and Bucher 1994) centered around a methionine residue probably involved in catalytic activity; further, a thioredoxin signature was also identified in this region. Expression studies will however be required to find out whether any enzymatic functions correspond to these signatures or not.

In addition to the variety of mechanisms that underly CMS, the unconventional nature of the "447" CMS in *Vicia faba*, associated with the presence of the dsRNA rather than with mitochondrial DNA alterations, makes it difficult to follow specific molecular signals associated with sterility and restoration of fertility such as mitochondrial DNA RFLPs or differential transcript accumulation or processing patterns. The "447" dsRNA could for instance encode a protein sequence that, once released from the polyprotein precursor, would be targeted to mitochondria and weaken them. Such a situation would be reminescent of that in maize with the T cytoplasm where the URF13 protein exerts a deleterious effect exclusively in the anther tissues, and in particular in the tapetum cells: this is due to the extreme solicitation to which these weakened mitochondria are submitted, both in terms of biogenesis and energy production, in these tissues.

The viral origin of the dsRNA suggests additional possible mechanisms for this CMS: indeed, several dsRNA viruses infecting fungi, e.g. the killer virus of yeast and of *Ustilago maydis*, encode protein toxins that are able to assemble in the plasma membrane of susceptible cells to form pores that kill them by causing efflux of ions and small electrolytes (virus-harboring cells are immune to the toxin). It is tempting to speculate that the "447" dsRNA may encode such a protein toxin that would either be expressed selectively in anther tissues, or activated by anther-specific compounds such as flavonoids.

9.7 Perspectives

One of the obvious organisms in which to test the molecular mechanisms of the "447" CMS is yeast, in which expression of a functional CMS determinant should result in a "petite" phenotype corresponding to impaired respiration. As mentioned earlier, even though the cleavage pattern of the polyprotein precursor remains unknown, self-processing by an embedded protease occurs also in heterologous systems and should allow release of functional polypeptides. Expression of the cDNA copies of the dsRNA in model systems such as tobacco may however not necessarily result in male sterility if species-specific compounds are required for full activity of the CMS determinant, as has been observed in transgenic tobacco expressing T-*urf13*.

References

Abad AR, Mehrtens BJ, Mackenzie S (1995) Specific expression in reproductive tissue and fate of a mitochondrial sterility-associated protein in cytoplasmic male-sterile bean. Plant Cell 7: 271-285

Chaumont F, Bernier B, Buxant R, Williams ME Levings CSIII, Boutry M (1995) Targeting the maize T-urf13 product into tobacco mitochondria confers methomyl sensitivity to mitochondrial respiration. Proc Natl Acad Sci USA 92: 1167-1171

Choi GH, Nuss DL (1992) Hypovirulence of chestnut blight fungus conferred by an infectious viral cDNA. Science 257: 800-803

Choi GH, Pawlyk DM, Nuss DL (1991a) The autocatalytic protease p29 encoded by a hypovirulence-associated virus of the chestnut blight fungus resembles the potyvirus-encoded HC-Pro. Virology 183: 747-752

Choi GH, Shapira R, Nuss DL (1991b) Cotranslational autoproteolysis involved in gene expression from a double-stranded RNA genetic element associated with hypovirulence of the chestnut blight fungus. Proc Natl Acad Sci USA 88: 1167-1171

Choi GH, Chen B, Nuss DL (1995) Virus-mediated or transgenic suppression of a G-protein a subunit and attenuation of fungal virulence. Proc Natl Acad Sci USA 92: 305-309

Clark ME, Hammerle J, Wimmer E, Dasgupta A (1991) Poliovirus proteinase 3C converts an active form of transcription factor IIIC to an inactive form: a mechanism for inhibition of host cell polymerase III transcription by poliovirus. EMBO J 10: 2941-2947

Craven MG, Pawlyk DM, Choi H, Nuss DL (1993) Papain-like protease p29 as a symptom determinant encoded by a hypovirulence-associated virus of the chestnut blight fungus. J Virol 67: 6513-6521

Dewey RE, Levings CSIII, Timothy DH (1986) Novel recombinations in the maize mitochondrial genome produce a unique transcriptional unit in the Texas male-sterile cytoplasm. Cell 44: 439-449

Dewey RE, Timothy DH, Lewings CSIII (1988) A 13-kilodalton maize mitochondrial protein in *E. coli* confers methomyl sensitivity to *Bipolaris maydis* toxin. Science 239: 292-295

Duc G, Scalla R, Lefebvre A (1984) New developments in cytoplasmic male sterility in *Vicia faba* L. In: Hebblethwaite PD, Dawkins TCK, Heath MC, Lockwood G (Eds). *Vicia faba:* Agronomy, Physiology and Breeding, pp. 254-260, Martinus Nijhoff/Dr. W. Junk Publishers, The Hague/Boston/Lancaster

Fahima T, Kazmierczak P, Hansen DR, Pfeiffer P, Van Alfen NK (1993) Membrane-associated replication of an unencapsidated double-strand RNA of the fungus, *Cryphonectria parasitica.* Virology 195: 81-89

Fukuhara T, Moriyama H, Pak JY, Hyakutake H, Nitta T (1993) Enigmatic double-stranded RNA in Japonica rice. Plant Mol Biol 21: 1121-1130

Fukuhara T, Moriyama H, Nitta T (1995) The unusual structure of a novel RNA replicon in rice. J Biol Chem 270: 18147-18149

Glab N, Wise RP, Pring DR, Jacq C, Slonimski P (1990) Expression in *S. cerevisiae* of a gene associated with cytoplasmic male sterility from maize: respiratory dysfunction and uncoupling of yeast mitochondria. Mol Gen Genet 223: 24-32

Grill LK, Garger SJ (1981) Identification and characterization of double stranded RNA associated with cytoplasmic male sterility in *Vicia faba.* Proc Natl Acad Sci USA 78: 7043-7046

Hansen DR, Van Alfen NK, Gillies K, Powell WA (1985) Naked dsRNA associated with hypovirulence of *Endothia parasitica* is packaged in fungal vesicles. J Gen Virol 66: 2605-2614

He S, Abad AR, Gelvin SB, Mackenzie SA (1996) A cytoplasmic male sterility-associated mitochondrial protein causes pollen disruption in transgenic tobacco. Proc Natl Acad Sci USA 93: 11763-11768

Hellen CUT, Fäcke M, Kräusslich HG, Lee CK, Wimmer E (1991) Characterization of poliovirus 2A proteinase by mutational analysis: residues required for autocatalytic activity are essential for induction of cleavage of eukaryotic initiation factor 4F polypeptide p220. J Virol 65: 4226-4231

Huang J, Lee S.H., Lin C, Medici R, Hack E, Myers AM (1990) Expression in the yeast of the T-URF13 protein from Texas male-sterile mitochondria confers sensitivity to methomyl and to Texas cytoplasm-specific fungal toxins. EMBO J 9: 339-347

Izhar S, Frankel R (1971) Mechanism of male sterility in *Petunia*: the relationship between pH, callase activity in the anthers, and the breakdown of microsporogenesis. Theor Appl Gen 41: 104-108

Koonin EV, Choi GH, Nuss DL, Shapira R, Carrington JC (1991) Evidence for common ancestry of a chestnut blight hypovirulence-associated double-stranded RNA and a group of positive-strand RNA plant viruses. Proc Natl Acad Sci USA 88: 10647-10651

Lefebvre A, Scalla R, Pfeiffer P (1990) The double-stranded RNA associated with the '447' cytoplasmic male sterility in *Vicia faba* is packaged together with its replicase in cytoplasmic membranous vesicles. Plant Mol Biol 14: 477-490

Lewings CS III (1993) Thoughts on cytoplasmic male sterility. Plant Cell 5: 1285-1290

Mackenzie SA, Pring DR, Bassett MJ (1988) Large double-stranded RNA molecules in *Phaseolus vulgaris* L. are not associated with cytoplasmic male sterility. Theor Appl Genet 76: 59-63

Mackenzie SA, Chase C (1990) Fertility restoration is associated with loss of a portion of the mitochondrial genome in cytoplasmic male-sterile common bean. Plant Cell 2: 905-912

Moriyama H, Fukuhara T, Nitta T (1995) Double-stranded RNA in rice: a novel RNA replicon in plants. Mol Gen Genet 248: 364-369

Pfeiffer P (1998) Nucleotide sequence, genetic organization and expression strategy of the double stranded RNA associated with the „447" cytoplasmic male sterility trait in *Vicia faba*. J Gen Virol 79: 2349-2359

Pfeiffer P, Jung JL, Heitzler J, Keith G (1993) Unusual structure of the double-stranded RNA associated with the '447' cytoplasmic male sterility in *Vicia faba*. J Gen Virol 74: 1167-1173

Powell WA, Van Alfen NK (1987a) Differential accumulation of Poly(A)$^+$ RNA between virulent and double-stranded RNA-induced hypovirulent strains of *Cryphonectria (Endothia) parasitica*. Mol Cell Biol 7: 3688-3693

Powell WA, Van Alfen NK (1987b) Two non-homologous viruses of *Cryphonectria (Endothia) parasitica* reduce accumulation of specific virulence-associated polypeptides. J Bact 169: 5324-5326

Py B, Bortoli-German I, Haiech J, Chippaux M, Barras F (1991) Cellulase EGZ of *Erwinia chrysanthemi*: structural organization and importance of His98 and Glu133 residues for catalysis. Protein Eng 27: 325-333

Scalla R, Duc G, Rigaud J, Lefebvre A, Meignoz R (1981) RNA-containing intracellular particles in cytoplasmic male sterile faba bean (*Vicia faba* L.). Plant Sci Lett 22: 269-277

Schnable P and Wise RP (1998) The molecular basis of cytoplasmic male sterility and fertility restoration. TIPS 3: 175-180

Shapira R, Nuss DL (1991) Gene expression by a hypovirulence-associated virus of the chestnut blight fungus involves two papain-like protease activities. J Biol Chem 266: 19419-19425

Shapira R, Choi GH, Nuss DL (1991) Virus-like genetic organization and expression strategy for a double-stranded RNA genetic element associated with biological control of chestnut blight. EMBO J 10: 731-739

Tadege M, Kuhlemeier C (1997) Aerobic fermentation during tobacco pollen development. Plant Mol Biol 35: 343-354

Turpen T, Garger SJ, Grill LK (1988) On the mechanism of cytoplasmic male sterility in the 447 line of *Vicia faba*. Plant Mol Biol 10: 489-497

Valverde R, Nameth S, Abdallha O, Al-Musa O, Desjardins P, Dodds JA (1990) Indigenous double-stranded RNA from pepper (*Capsicum annuum*). Plant Sci 67: 195-201

Wakarchuk DA, Hamilton RI (1985) Cellular double-stranded RNA in *Phaseolus vulgaris*. Plant Mol Biol 5: 55-63

Wang, B, Li YN, Zhang XW, Hu L, Wang JZ (1990) Double-stranded RNA and male sterility in rice. Theor Appl Genet 79: 556-560

Worall D, Hird DL, Hodge R, Paul W, Draper J, Scott R (1992) Premature dissolution of the microsporocyte callose wall causes male sterility in transgenic tobacco. Plant Cell 4: 759-771

Young EG, Hanson MR (1987) A fused mitochondrial gene associated with cytoplasmic male sterility is developmentally regulated. Cell 50: 41-47

Zabalgogeazcoa I, Cox-Foster DC, Gildow FE (1993) Pedigree analysis ot the transmission of a double-stranded RNA in barley cultivars. Plant Sci 91: 45-53

10 The location of the pollen-specific NTP303 protein

Schrauwen* JAM, Derksen J, Knuiman B, Wittink F and Wullems GJ

10.1 Pollen development

In higher plants fertilization starts with the fusion of a haploid male sperm cell nucleus and the haploid nucleus of a female egg cell. The male gametophyte, or pollen, is the mobile component in this process and its function is to deliver two male gametes into the embryo sac (Heslop-Harrison 1987; McCormick 1991; 1993).

The development of the male gametophyte takes place in the anther and starts with the formation of archesporial cells. Differentiation of the archesporial cells results in a primary parietal layer and an inner primary sporogenous layer. The cells of the primary sporogenous layer undergo mitotic divisions and diploid meiocytes are formed. These meiocytes or pollen mother cells enter meiosis. After post-meiotic development free microspores are formed. In these micropspores a generative cell is formed at pollen mitosis. Pollen maturation is accompanied with an increase of volume and weight and by an accumulation of a large number of new transcripts (Schrauwen et al. 1990). After dehiscence of the anther pollen grains are released. An arrest in the development of one of the anther tissues, like stomium cells (Beals and Goldberg 1997) effects pollen development and results in the production of sterile pollen. This shows that for the development of fertile pollen, occurring in the sporophytic anther, a coordinated gene expression is required in the gametophytic cells and the surrounding sporophytic tissues (Schrauwen et al. 1996).

* for correspondence: Department of Experimental Botany; University of Nijmegen; Toernooiveld 1, 6525 ED Nijmegen; The Netherlands

10.2 Gene regulation during pollen development

Although the morphological and chemical aspects of pollen development have been studied in great detail, knowledge of the underlying molecular processes is relatively limited (Muschietti et al 1994; Touraev et al 1995; Xu et al 1995). The developmental events leading to anther and pollen formation, and pollen release are precisely timed and regulated (Koltunow et al. 1990; Scott et al. 1991). These processes occur in a strict, chronological order that correlates with floral bud size. This opens an easy way to collect homogenous populations of pollen at discrete stages of development.

The high transcriptional and translational activity during pollen development leads to a considerable increase in the total amount of RNA, mRNA and dry weight from the late unicellular microspore stage up to mature pollen. A detailed study of the stage-related expression of mRNAs in the gametophyte during four developmental stages showed that some mRNAs are only transiently present. Others are present during more than one stage but their concentrations vary. A great number of new transcripts arise and accumulate during the late phase of pollen development (Bedinger and Edgerton 1990, Schrauwen et al. 1990). In addition, there are also RNAs present at a constant level during all stages investigated.

Table 1. The accumulation of NTP303 RNA and protein during pollen development and pollen tube growth

	Unicellular	Bi-cellular				
		early	mid	mature	germ 2h	germ 6h
RNA	-	(+)	+	++	++++	++++
Protein	-	-	+	+	++	++++

-, not detectable; + , relative amounts

Pollen formation can be divided in two developmental periods, with different sets of genes expressed (Mascarenhas 1990; 1993). Transcripts of the 'early' genes appear soon after meiosis and are reduced or undetectable in mature pollen. Transcripts of 'late' genes are first detected after microspore mitosis and continue to accumulate as pollen matures. Progress in the isolation of pollen expressed genes has been made by differential screening of cDNA libraries from both anthers and pollen. In this way genes have also been characteriszed with a less distinct temporal expression pattern than the 'early' and 'late' genes (Twell et al. 1993).

The knowledge of genes involved in pollen development has been focused on 'early' genes which are almost all expressed in the tapetum, and 'late' genes which are maximally expressed in mature pollen.

10.3 Expression of the late pollen specific gene ntp303 during development and pollen tube growth

The late pollen-specific gene *ntp303* (Weterings et al. 1992) was isolated by differential screening from a cDNA library prepared with RNA from mature pollen. The gene is fully characterized concerning its sequence, its promoter composition and its spatial and temporal expression kinetics (Weterings et al. 1992; 1995). Northern blot analysis confirmed that the *ntp303* RNA is only detectable in bicellular and germinating pollen (Table 1).

In tobacco pollen, transcription of *ntp303* starts after pollen mitosis I and continues during pollen tube growth. The ratio level of the accumulated transcript varies between 0.6:1:3 for mid-bicellular, mature and 2h germinated pollen respectively. *Ntp303* is regulated concerted with a group of other genes in pollen of tobacco. Their transcripts appear after the first haploid mitosis and accumulate during the maturation of the microgametophyte (Schrauwen et al. 1990). Northern blot analysis in our lab have proven that the pattern of *ntp303* gene expression during in vitro ripening of pollen is similar to the expression pattern of *ntp303* during in vivo maturation (VanHerpen and de Groot, personal communication). This shows that the expression of the gene is regulated by the haploid gametophyte itself, without any interaction with sporophytic factors. In addition, pulse labelling experiments *in vitro* RNA labelling during germination have shown that the gene is also actively expressed in the growing pollen tube (Weterings et al. 1992). This active transcription of a pollen specific gene during germination is in contrast to previous reports on an anther specific clone from *Lycopersicon esculentum* (*lat52*), which suggests that no transcription occurs during the first hours of pollen tube growth (Twell et al. 1989; Ursin et al. 1989).

In situ hybridization experiments showed that transcription starts at mid-bicellular pollen stage and was only detectable in the vegetative cell. *Ntp303* transcripts are still present near the tip of pollen tubes 72 hours after pollination of the pistil (Reijnen et al. 1991; Weterings et al. 1995). These data indicate a function of the gene during pollen tube growth or during the process of fertilization.

Dissection of the *ntp303* promoter for the *cis*-regulatory elements was assayed by 5'deletion analysis and the micro-projectile facilitated transient expression technique. These experiments showed that the minimal pollen-specific promoter is located between -103 and -51 nucleotides upstream of the transcriptional start site. In this region two hexameric sequences AAATGA, of which one was inverted, proved to have transcription activating properties. Using gel mobility shift assays, we demonstrated that this minimal functional promoter interacts with a leaf nuclear GT-1 binding activity. These data suggest that the ubiquitous transcription factors are involved in the pollen-specific expression of the gene (Hochstenbach et al. 1996).

10.4 The protein encoded by ntp303 and its location

A major goal in pollen biology is to determine the function of the proteins encoded by gametophytic mRNAs exclusively present in pollen. Protein gel blot and immunolocalization analysis will distinguish between cytoplasmic and membranous bound proteins. Such an approach will show the moment of translation which might be an indication for the involvement of the protein in a specific process. More information on the function of a protein will be obtained by a loss-of-function analysis.

Here we present results obtained with the antibody approach. Two coding regions of the *ntp303* were overexpressed in *E.coli*. Both recombinant proteins were isolated, purified and used to raise antisera in separate immunization procedures (Wittink et al. 1998). Both antisera reacted only with a 69kD protein from pollen and not with proteins from other organs.

Protein gel blot analysis with the NTP303 protein showed the appearance of the protein in germinated pollen and the accumulation thereafter in accordance to the size of the pollen tube. These results revealed that the protein is mainly associated with the pollen tube wall and is not detectable in the cytoplasm. Electron microscopic immuno-localization analysis showed the arise of the protein in mid-bicellular pollen (Table 1). Its location is in the vegetative membrane surrounding the generative cell and sperm cells, and not in their generative membranes of developing respectively germinating pollen. Besides the location of NTP303 at the vegetative plasma membranes, the protein was also present in callose plugs of pollen tubes, but not in the tip of the tube.

The location of NTP303 in all vegetative membranes except the pollen tube tip and its strong accumulation during development and pollen tube growth raised the question of its particular function. Since NTP303 is strictly expressed in pollen and homologous transcripts are found in many species, its function is expected to be related to a pollen-specific event, like the interaction process between female and male cells.

In conclusion, we have shown that the late pollen gene ntp303 is translated during pollen development, germination and pollen tube growth in a very specific manner. The protein is specifically present in the plasma membrane of the vegetative cell, the vegetative membrane that envelopes the gametes and in callose plugs. The protein is present in the microgametophyte from bicellular pollen to germinated pollen tubes, but not in the pollen tube-tip. The involvement of NTP303 with callose biosynthesis cannot sustained by the results that showed a different temporal appearance of callose and NTP303 expression.

The loss of function analysis of the pollen specific gene ntp303, both by an anti sense and co-suppression approach, to examine the effect of abolishing the production of NTP303 on pollen development and pollen tube growth, may unravel the function of the gene ntp303. Such analysis are in progress.

References

Beals TP, Goldberg RB (1997) A novel cell ablation strategy blocks Tobacco anther dehiscence. The Plant Cell 9: 1527-1545

Bedinger PA, Edgerton MD (1990) Developmental staging of maize microspores reveals a transition in developing microspore proteins. Plant Physiol 92: 474-479

Heslop-Harrison J (1987) Pollen germination and pollen-tube growth. Int Rev Cytol 107: 1-78

Hochstenbach R, De Groot P, Jacobs J, Schrauwen J, Wullems G (1996) The promoter of a gene that is expressed only in pollen interacts with ubiquitous transcription factors. Sex Plant Reprod 9: 197-202

Koltunow AM, Truettner J, Cox KH, Wallroth M, Goldberg RB (1990) Different temporal and spatial gene expression patterns occur during anther development. Plant Cell 2: 1201-1224

Mascarenhas JP (1990) Gene activity during pollen development. Ann Rev Plant Phys Mol Biol 41: 317-338

Mascarenhas JP (1993) Molecular mchanisms of pollen tube growth and differentiation. Plant Cell 5: 1303-1314

McCormick S (1991) Molecular analysis of male gametogenesis in plants. Trends Genet 7: 298-303

McCormick S (1993) Male gametophyte development. Plant Cell 5:1265-1275

Muschietti J, Dircks L, Vancanneyt G, McCormick S (1994) LAT 52 protein is essential for tomato pollen development: pollen expression antisense LAT52RNA hydrates and germinates abnormally and cannot achieve fertilization. Plant J 6: 321-338

Reijnen WH, Vanherpen MMA, De Groot PFM, Olmedilla A, Schrauwen JAM, Weterings KAP, Wullems GJ (1991) Cellular localization of a pollen-specific mRNA by in situ hybridization and confocal laser scanning microscopy. Sex Plant Reprod 4: 254-257

Schrauwen JAM, De Groot PFM, Vanherpen MMA, Vanderlee T, Reynen WH, Weterings KAP, Wullems GJ (1990) Stage-related expression of mRNAs during pollen development in lily and tobacco. Planta 182: 298-304

Schrauwen JAM, Mettenmeyer T, Croes AF, Wullems GJ (1996) Tapetum-specific genes: what role do they play in male gametophyte development ? Acta Bot Neerl 45:1-15

Scott R, Hodge R, Paul W, Draper J (1991) The molecular biology of anther differentiation. Plant Sci 80: 167-191

Touraev A, Fink CS, Stöger E, Heberle-Bors E (1995). Pollen selection: A transgenic reconstruction approach. PNAS 92: 12165-12169

Twell D, Wing R, Yamaguchi J, McCormick S (1989). Isolation and expression of an anther-specific gene from tomato. Mol Gen Genet 217: 240-245

Twell D, Patel S, Sorensen A, Roberts M, Scott R, Draper J, Foster G (1993) Acitvation and developmental regulation of an Arabidopsis anther-specific promoter in microspores and pollen of *Nicotiana tabacum*. Sex Plant Reprod 6: 217-224

Ursin VM, Yamaguchi J, McCormick S (1989) Gametophytic and sporophytic expression of anther-specific genes in developing tomato anthers. Plant Cell 1: 727-736

Weterings K, Reijnen W, Vanaarssen R, Kortstee A, Spijkers J, vanHerpen M, Schrauwen J, Wullems G (1992) Characterization of a pollen-specific DNA clone from *Nicotiana tabacum* expressed during microgametogenesis and germination. Plant Mol Biol 18: 1101-1111

Weterings K, Schrauwen J, Wullems G , Twell D (1995) Functional analysis of the pollen-specific gene NTP303 reveals a novel pollen-specific, and conserved cis-regulatory element. Plant J 8: 55-63

Wittink FRA (1998) The pollen-specific NTP303protein. Characterization, localization and its involvement in fertilization. Thesis University of Nijmegen

Xu H, Knox RB, Taylor PE, Singh MB (1995). *Bcp1*, a gene required for male fertility in *Arabidopsis*. PNAS 92: 2106-2110

11 The special callose wall: a new insight

Majewska-Sawka*[1] A and Rodriguez-Garcia[2] MI

11.1 Introduction

Callose appears as a transient material in the cell wall during various steps of plant growth and development. In the course of microsporogenesis, callose begins to be synthesized during the first meiotic prophase, with initial deposits being visible at the corner of the cells, and then gradually spreading over the whole microsporocyte surface. During cytokinesis, which completes meiotic division, callose is also secreted between the four adjacent cells of the newly formed tetrad, contributing to the formation of the so-called „special callose wall" (Bhandari 1984).

Because the synthesis of this special wall in early phases of meiosis is temporally related to the transition from the diploid to the haploid phase of development, is thought to play several special functions. The structure and possible physiological significance of the callose wall have been widely studied since the early 1960s, and insights into its role in microsporogenesis have been accumulated since that time.

Callose is a highly homogenous layer composed of linear polymers of β-1,3-glucose (Rowley 1959), but thc pathway of its synthesis and secretion have remained a matter of debate for some time (Iiyama et al. 1993). After β-1,3-glucan synthase was identified, the plasma membrane was shown to be the site of the subcellular location and activity of this enzyme (Fink et al. 1987; Schlüpmann et al. 1993).

The special callose wall was initially assumed to be totally impermeable (Heslop-Harrison and McKenzie 1967), and was thus considered to have important functions in the isolation of microsporocytes from each other and from surrounding tapetal cells. It was therefore thought to act as barrier that prevented

---

[1]* for correspondence: Institute for Plant Breeding and Acclimatization; Powstanców Wielkopolskich 10; 85-950 Bydgoszcz; Poland

[2] Experimental Station of Zaidin (CSIC); 18008 Granada; Spain

the passage of molecules and protected the microsporocytes against drought or mechanical damage (Bhandari, 1984). Several lines of evidence have demonstrated, however, that callose is permeable to low- and high-molecular weight substances including colloidal iron (Rowley and Dunbar 1970), glucose, sodium acetate (Southworth 1971) and cerium ions/cerium perhydroxide (Rodriguez-Garcia and Majewska-Sawka 1992).

Previous light and electron microscopic observations indicated that under certain conditions the callose surrounding microsporocytes does not form a homogenous wall, but displays a layered structure (Waterkeyn 1962). In particular, the outermost part of the callose wall was noted to be less electron-dense than its internal part (Polowick and Shawney 1993), so the possible presence of other polysaccharides at the outer surface of callose has been postulated. The latter view was confirmed by several cytochemical methods that demonstrated the presence of pectocellulosic compounds in this area (Bindle 1979; Bhandari et al. 1981; Pacini 1994). It remains to be determined whether these components represent the remnants of the microcporocyte primary wall, and exact composition of the external callose layer have yet to be identified.

We studied the possible presence of pectins on the surface of the callose wall during subsequent stages of microsporogenesis in sugar beet anthers (*Beta vulgaris* L.). Immunocytochemical methods with monoclonal antibodies to esterified (JIM7) and deesterified (JIM5) pectins (Knox et al. 1990; Casero and Knox 1995) were used in combination with the anti-β-1,3-glucan antibody.

11.2 Material and methods

Anthers of sugar beet (*Beta vulgaris* L.) containing microsporocytes and tetrads of microspores enclosed by the special callose wall were fixed with a mixture of 4% paraformaldehyde and 0.25% glutaraldehyde in 0.05 M Pipes buffer, pH 7.3, for 2 h at room temperature. Then the samples were washed several times in buffer, dehydrated in a series of ethanols and embedded in LR Gold resin (London Resin, UK).

The anthers were sectioned on a Reichert-Jung Ultracut E microtome and ultrathin sections were collected on Formvar-coated nickel grids. The immunocytochemical procedure was started by incubating the sections on drops of 0.1 M PBS, pH 7.2, containing 5% BSA for 45 min. Then the grids were placed directly on the drops of JIM5 (JIM7) antibodies diluted 1:1 in PBS supplemented with 0.3% BSA, and incubated at room temperature for 2 h. Both antibodies were a gift from Dr. Paul Knox, University of Leeds, United Kingdom.

After several washings in PBS, the sections were placed on drops of goat anti-rat monoclonal antibody (1:50 in PBS) coupled to 15-nm colloidal gold particles (Sigma Chemical Co., St. Louis, USA) for 2 h. Controls were run by omitting the primary antibody. After the immunocytochemical procedure all grids were rinsed several times in buffer and double-distilled water, and air dried.

In double labeling procedures the grids were initially incubated in JIM5 antibody (1:2 in PBS supplemented with 0.5% BSA) for 2 h, then thoroughly washed in several changes of buffer and placed on drops of secondary antibody (1:50 in PBS supplemented with 0.5% BSA) for 1 h. After washing, the grids were

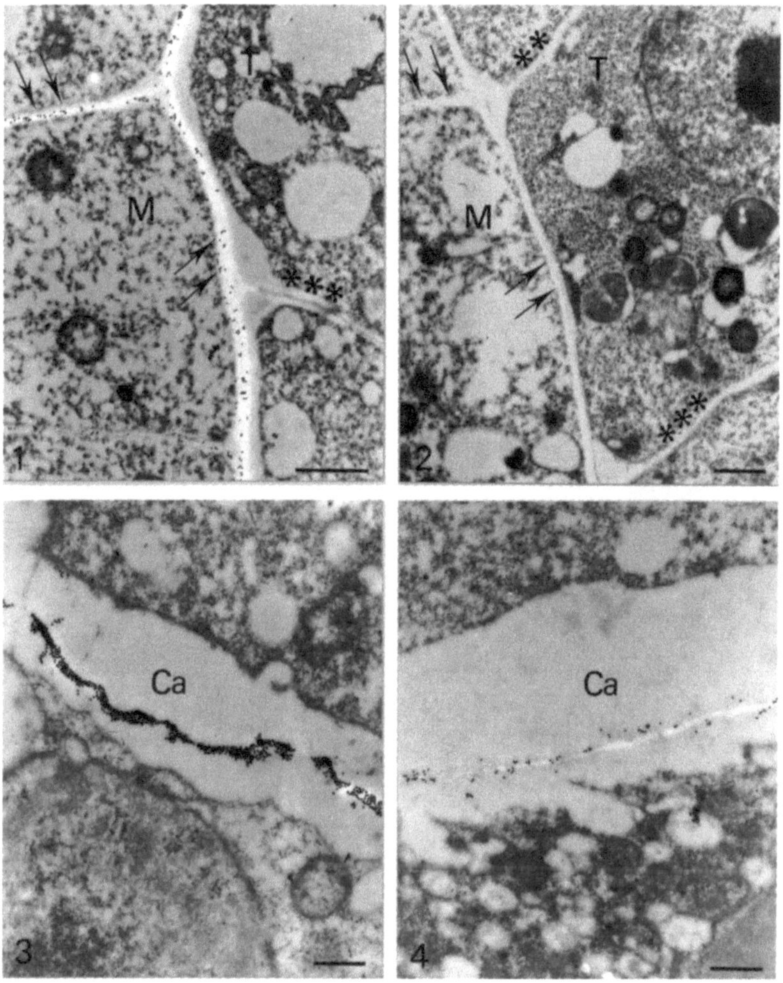

Figs. 1-2. Sections of sugar beet anther in early meiotic prophase I. Immunogold labeling with JIM5 antibody was seen in cell walls surrounding meiocytes (M, arrows), whereas the tapetal walls (T, asterisks) were free of gold particles (1). JIM7 antibody produced no labeling in the meiocyte or in tapetal walls (2). Bars: 1 μm. **Figs. 3-4.** Meiocytes in pachytene with visible incipient deposits of special callose wall (Ca). Primary walls were strongly labeled with both JIM5 (3) and JIM7 (4). Bars: 0.5 μm

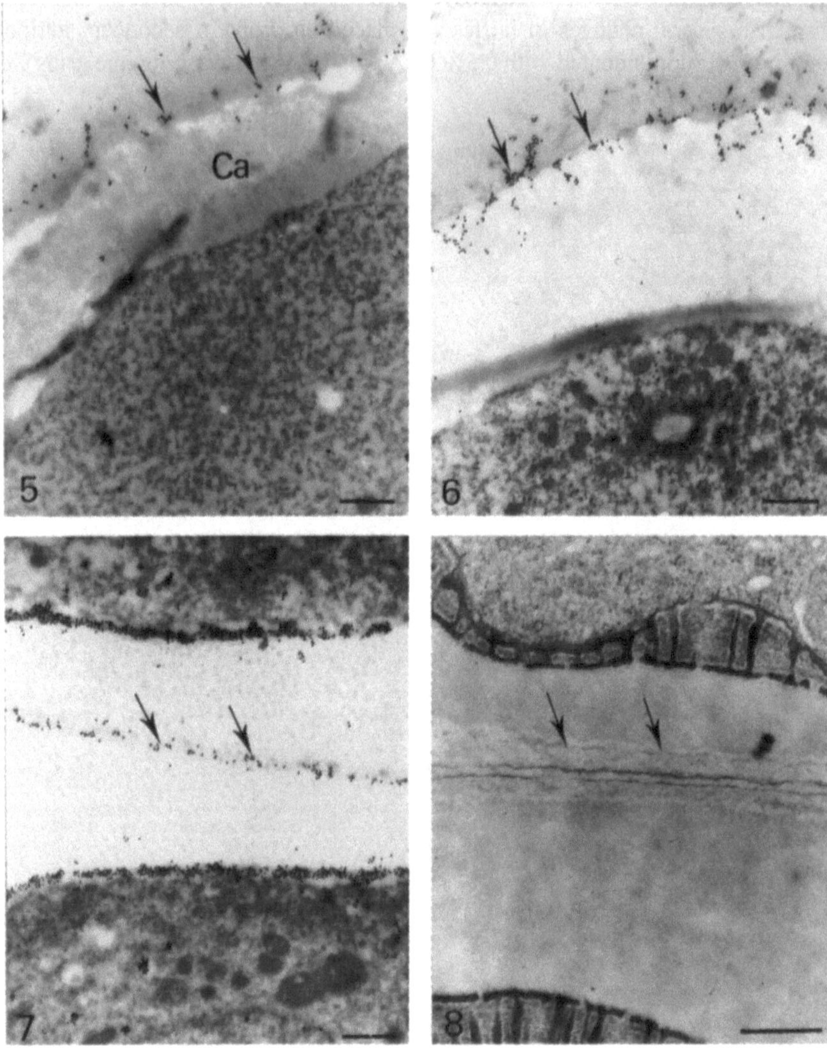

Figs. 5-8. External layer of the special callose wall (Ca, arrows) labeled with JIM5 at metaphase I (5), telophase II (6), and in the tetrad stage (7). This layer (arrows) was also distinguishable in conventional electron micrographs (8). Bars: 0.5 μm

incubated with anti-β-1,3-glucan antibody (Australia Biosupplies, Parkville Victoria) diluted 1:60 in PBS for 2 h, and then in rabbit-anti-mouse antibody (Sigma Chemical Co., St. Louis, USA) coupled to 20-nm gold particles. All further steps in the procedure were done as described above for experiments with the single antibody. After reactions, the sections were lightly stained in 5% uranyl acetate for 10 min and observed under Zeiss 10C electron microscope at 60 kV.

Fixation of anthers for conventinal electron microscopic observations was carried out as described previously (Majewska-Sawka et al. 1993).

11.3 Results

In very early stages of meiotic prophase, the JIM5-responsive pectin epitope was distributed inside the walls surrounding the meiocytes and in the walls between meiocytes and the tapetum, whereas the tapetal walls were completely devoid of the signal (Fig. 1). At this stage there was no visible labeling after immunoreaction with JIM7 antibody in meiocyte walls or in tapetal cell walls (Fig. 2).

After the callose wall had started to be deposited outside the plasmalemma, both esterified and unesterified pectins were detected in the primary cell walls (Figs. 3, 4). As meiosis proceeded the microsporocytes became less angular and started to separate from each other as a result of their spatial displacement in the enlarging anther cavity and of the progressive thickening of the special callose wall. During all stages of meiosis, as well as after completion of telophase II, the labeling from JIM5 and JIM7 binding continued to be observed at the external surface of the callose wall (Figs. 5, 6), with the deesterified epitope clearly being more abundant. A similar pattern was observed in subsequent stages of tetrad growth and maturation (Figs. 7-12).

Labeling with anti-β-1,3-glucan antibody revealed a homegenous distribution of gold particles across the special callose wall, whereas JIM5-responding pectin formed a thin layer at the outermost zone of this wall, as well as in the incipients of microspore exine (Fig. 11).

11.4 Discussion

The fate of the components that form the primary cell wall of microsporocytes has been followed in anthers of only few plant species by light microscopic observations of cytochemically stained sections. The techniques used previously involved staining with PAS or PTA to detect polysaccharides (Genevés and Rutin 1975; Bhandari et al. 1981), with zinc chloride or JKJ-H_2SO_4 to demonstrate cellulose (Biddle 1979; Bhandari et al. 1981), and also with ruthenium red to locate pectins (Biddle 1979). The results obtained to date have been contradictory, demonstrating that this primary wall was either fully degraded concurrently with the deposition of the special callose wall, or persisted as the outermost layer of callose surrounding the meiocytes and microspore tetrads (Bhandari 1984).

Among studies supporting the latter view, discrepancies as to the composition of this outer layer have been reported. Whereas in some species only cellulose could

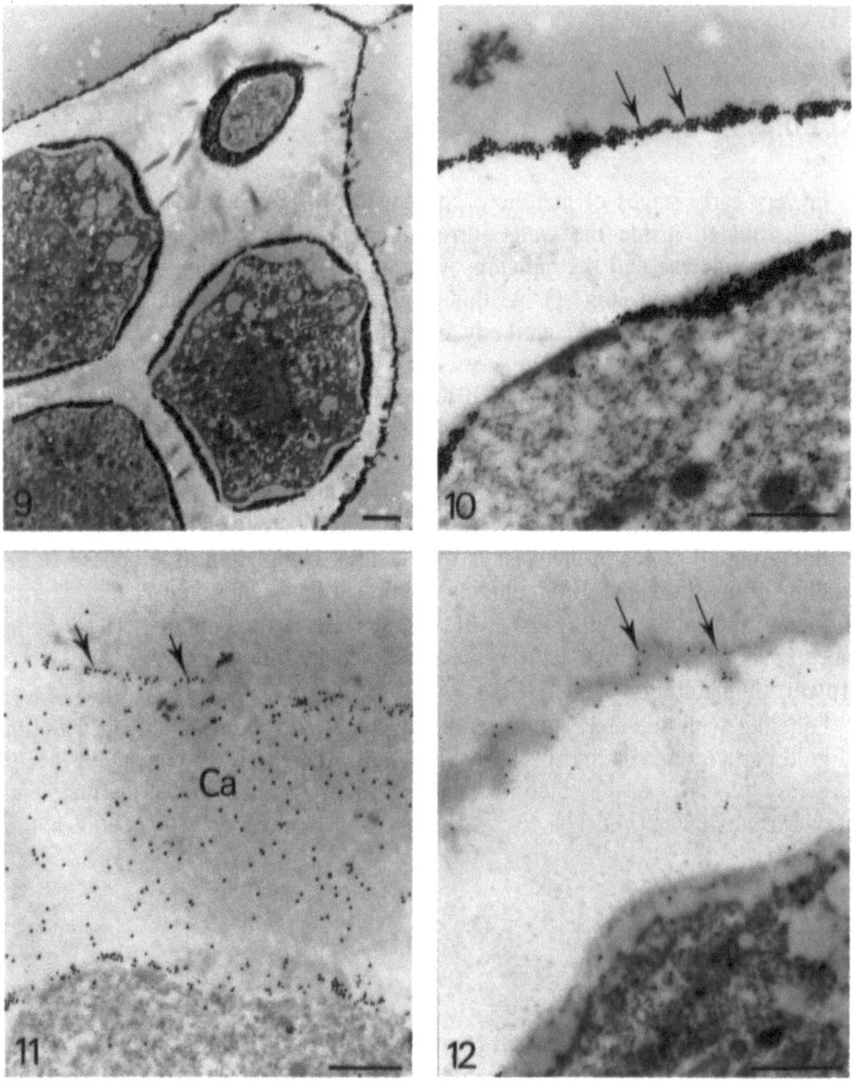

Figs. 9-12. The outermost part of the callose wall (Ca, arrows) in mature tetrads was strongly labeled with JIM5 (9-11), and more weakly labeled with JIM7 (12). Double labeling with JIM5 and anti-β-1,3-glucan antibody revealed a distinct pattern of distribution of small (arrows) and large gold particles (12). Bars: 1 μm (Fig. 9) and 0.5 μm, (Figs. 10-12)

be located (Bhandari et al. 1981), others displayed the presence of both cellulose and pectins (Biddle 1979).

To our knowledge, the exact composition of this pectin fraction has not been examined. We therefore undertook immunocytochemical studies designed to localize pectin epitopes in the walls of developing sugar beet microsporocytes. The presence of pectins was traced in successive developmental stages from the early prophase I to the formation of mature tetrads. We detected unesterified and methyl-esterified epitopes of pectin by immunogold electron microscopy with JIM5 and JIM7 monoclonal antibodies. JIM5 is reactive with pectin containing up to 50% randomly esterified groups, and reaction diminishes conspicuously with increasing esterification, whereas JIM7 is reported to produce equal reactions with degrees of esterification ranging from 35 to 90% (Knox et al. 1990).

The results of our immunocytochemical studies lead us to conclude that very young meiocytes which have just entered meiotic prophase are enclosed by a wall containing mainly deesterified pectin, which reacts with the JIM5 antibody. As meiosis proceeded and the meiocytes started to grow (Majewska-Sawka and Rodriguez-Garcia 1996), both pectin epitopes were detected in the primary walls and were also present in the area between two adjacent callose sheaths. This observation confirms conventional EM studies that showed that the rapid growth of meiocytes is accompanied by secretion of new wall materials which help to thicken the middle lamella (Polowick and Shawney 1992). Moreover, the appearance of JIM7-reactive epitope in the walls of rapidly expanding meiocytes is in agreement with the known pattern of pectin synthesis and modification in different plant tissues. Newly synthesized pectins are largely methyl-esterified in reactions that occur in the trans-Golgi membranes before secretion to the cell surface, but the degree of esterification may be modified by wall-bound pectin methyl esterases. In rapidly growing cells, the proportion of methyl-esterified pectins increases markedly (Carpita and Gibeaut 1993).

During further development of sugar beet microsporocytes, both pectin epitopes remained present on the external surface of the callose wall. The persistence of both pectin epitopes in successive stages of microsporocyte development thus suggests that these compounds may originate from the primary wall. The much stronger labeling with JIM5 than with JIM7 in all developmental stages indicates that the pectins inside the walls surrounding the microsporocytes of sugar beet are present mainly in deesterified form.

Acknowledgements

The study was supported by the Polish Committee for Scientific Research and the Spanish Ministry of Education and Science through the Polish-Spanish Bilateral Action Program, and by project PB95-0080 (DGES, Spain). The authors thank Dr. P. Knox for the supply of antibodies JIM5 and JIM 7, and K. Shashok for correcting the English version of the manuscript.

References

Bhandari NN (1984) The microsporangium. In Johri BM (ed) Embryology of angiosperms, Springer-Verlag, Berlin Heidelberg New York Tokyo, pp 53-121

Bhandari NN, Bhargava M, Geier T (1981) A persisting cellulosic wall of microspore mother cells during microsporogenesis in *Allium tuberosum* Rottl. and *Cyclamen persicum* Mill. Ann Bot 48: 425-431

Biddle JA (1979) Anther and pollen development in garden pea and cultivated lentil. Can J Bot 57: 1883-1900

Carpita NC, Gibeaut DM (1993) Structural models of primary cell walls in flowering plants: consistency of molecular structure with the physical properties of the walls during growth. Plant J 3: 1-30

Casero PJ, Knox JP (1995) The monoclonal antibody JIM5 indicates patterns of pectin deposition in relation to pit fields at the plasma- membrane-face of tomato pericarp cell walls. Protoplasma 188: 133-137

Fink J, Jeblick W, Blaschek W, Kauss H (1987) Calcium ions and polyamines activate the plasma membrane-located 1,3-β-glucan synthase. Planta 171: 130-135

Genevés L, Rutin J (1975) Réactivité a l'acide phosphotungstique (APT) des parois et du plasmalemme, dans les cellules staminales en méiose d'une angiosperme (*Ribes rubrum*, Grossulariacées). J Microsc Biol Cell 24: 391-396

Heslop-Harrison J, MacKenzie A (1967) Autoradiography of 2-[^{14}C]thymidine derivative during meiosis and microsporogenesis in *Lilium* anthers. J Cell Sci 2: 387-400

Iiyama K, Lam TBT, Meikle PJ, Ng K, Rhodes DI, Stone BA (1993) Cell wall biosynthesis and its regulation. In Jung H, Buxton D, Hatfield R, Ralph J (eds) Forage cell wall structure and digestibility, American Society of Agronomy, Madison, pp 621-683

Knox JP, Linstead PJ, King J, Cooper Ch, Roberts K (1990) Pectin esterification is spatially regulated both within cell walls and between developing tisses of root apices. Planta 181: 512-521

Majewska-Sawka A, Rodriguez-Garcia MI (1996) rRNA distribution during microspore development in anthers of *Beta vulgaris* L.: quantitative in situ hybridization analysis. J Cell Sci 109: 859-866

Majewska-Sawka A, Rodriguez-Garcia MI, Nakashima H, Jassem B (1993) Ultrastructural expression of cytoplasmic male sterility in sugar beet (*Beta vulgaris* L.). Sex Plant Reprod 6: 22-32

Pacini E (1994) Cell biology of anther and pollen development. In Williams EG (ed.) Genetic control of self-incompatibility and reproductive development in flowering plants. Kluwer Academic Publishers, pp 289-308

Polowick PL, Sawhney VK (1992) Ultrastructural changes in the cell wall, nucleus and cytoplasm of pollen mother cells during meiotic prophase I in *Lycopersicon esculentum* (Mill.). Protoplasma 169: 139-147

Polowick PL, Sawhney VK (1993) An ultrastructural study of pollen development in tomato (*Lycopersicon esculentum*). Tetrad to early binucleate microspore stage. Can J Bot 71: 1039-1047

Rodriguez-Garcia MI, Majewska-Sawka A (1992) Is the special callose wall of microsporocytes an impermeable barrier ? J Exp Bot 257: 1659-1663

Rowley JR (1959) The fine structure of the pollen wall in the *Commelinaceae*. Grana Palynol 2: 3-31

Rowley JR, Dunbar A (1970) Transfer of colloidal iron from sporophyte to gametophyte. Pollen et Spores 12: 305-328

Schlüpmann H, Bacic A, Read SM (1993) A novel callose synthase from pollen tubes of *Nicotiana*. Planta 191: 470-481

Southworth D (1971) Incorporation of radioactive precursors into developing pollen walls. In Heslp-Harrison J (ed.) Pollen: development and physiology, Butterworths, London, pp 115-120

Waterkeyn L (1962) Les parois microcytaires de nature callosique chez *Helleborus* et *Tradescantia*. Cellule 62: 225-255

12 Characterization of P26, a putative PLIM-1 partner in the pollen grain of *Helianthus annuus*

Evrard* JL, Vantard M, Baltz R and Steinmetz A

12.1 Introduction

Our group is investigating the role of different proteins in sunflower reproduction including pollen development and germination. This last process is highly complex, involving cell-cell interactions, signal transduction pathways as well as cytoskeletal reorganizations and cytoplasmic movements. P26, a protein of an apparent molecular weight of 26 kDa, is one of the major proteins of the mature sunflower pollen. Its function in pollen development is not known. A phosphorylatable protein of the same molecular weight which appears to be involved in the self-incompatibility process has been described in *Papaver rhoeas* (Rudd et al. 1997). In preliminary protein interaction studies the sunflower protein P26 was found to bind PLIM-1, a pollen-specific LIM protein (Baltz et al. 1992). P26 is also present in protein complexes immunoselected using anti-PLIM-1 antibodies. Since PLIM-1 is structurally related to the animal protein CRP shown to interact with the actin cytoskeleton (Beckerle 1997), we hypothesize that P26 functions in connection with cytoskeletal structures in the pollen grain. The present work describes the isolation of partial p26 cDNAs as well as a preliminary biochemical characterization of the protein.

12.2 Isolation of p26 cDNA clones

In order to isolate the P26 coding sequence, two complementary approaches were used: 1/ the use of antibodies against P26 to screen an expression pollen cDNA library and 2/ microsequencing of P26 to design degenerated oligonucleotides

---

* for correspondence: Institut de Biologie Moléculaire des Plantes du CNRS; 12, rue du Général Zimmer; 67084 Strasbourg Cedex; France

used in subsequent screening of a cDNA library or in amplification of a p26 cDNA using the RT-PCR technique. Due to its abundance in sunflower pollen the protein could be purified in a single step by preparative SDS-PAGE of protein extracts obtained after lysis of the pollen in a French Press. The protein was revealed by precipitation in a 0.1M KCl solution, electroeluted in order to be injected into rabbits for antibody production, or extracted from gel by diffusion for chymotrypsin digestion.

Fifty residues from the N-terminus could be unambiguously determined by the standard Edman degradation. To obtain internal information, a partial chymotrypsin digestion of P26 was performed and four fragments were isolated after SDS-PAGE (P26.1 to p26.4). Only P26.4 showed a new internal sequence (Fig.1), the three other fragments corresponding to the P26 N-terminus. Comparison with sequence databases did not reveal significant homologies with known proteins at this stage. From the peptide sequence data we designed three oligonucleotides: one using inosine for degeneration and two that were mixes from all possible codon combinations.

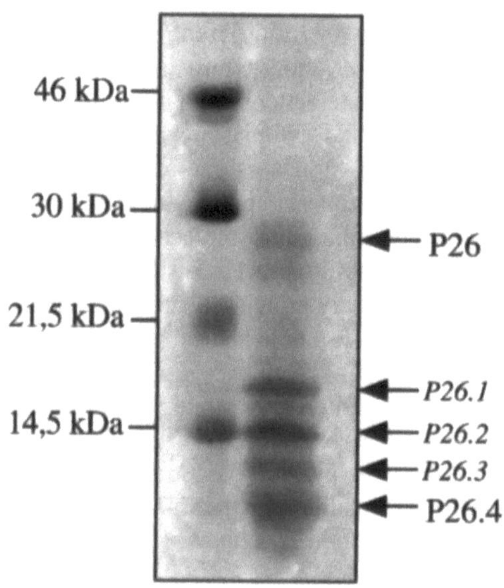

Fig.1. Chymotrypsin digestion of P26 leads to the production of four major fragments namely P26.1 to P26.4 here separated on SDS-PAGE. The microsequencing of these four fragments reveals only an internal sequence for P26.4. P26.1 to P26.3 correspond to the P26 N-terminus

To isolate a cDNA corresponding to p26 we have tried immunoscreening of an expression cDNA library from mature sunflower pollen using unpurified as well as purified antibodies. While several clones reacted positively with the two antibodies, the sequences of their inserts were found to be unrelated to that of P26 (Fig. 2).

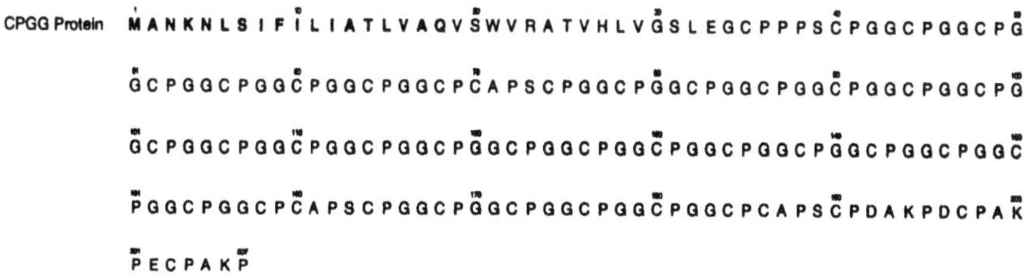

Fig. 2. Sequence of a putative cell wall protein from sunflower pollen. This same sequence was found in 10 identical cDNAs selected with anti-P26 antibodies. The bold sequence is a probable signal peptide. The CPGG motif is repeated thirty three times

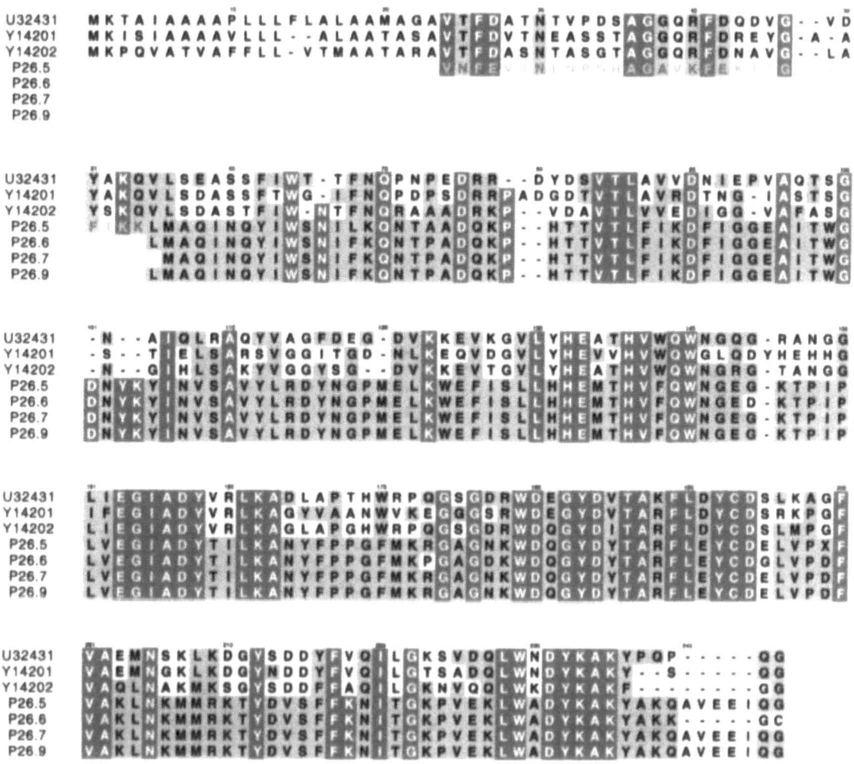

Fig. 3. Comparison of the different P26 peptide sequences (deduced from the cDNAs) with three homologous proteins, one from wheat (U32431) and two from barley (Y1420x). The proteins are named with their accession or clone number. The positions 1 to 24 correspond to the signal peptide. The sequence in light gray in P26.5 corresponds to the direct protein microsequencing of P26 extracted from pollen. Due to the choise of primers used for the amplification of the cDNA clones, the deduced peptidic sequences therefore start at residue 55/56. Dark gray boxes indicate strict conservations, light gray boxes show homologous residues

cDNAs encoding P26 were finally isolated following RT-PCR amplification of polyA RNAs from a total sunflower inflorescence. For reverse transcription we have used oligodT and random hexamers as primers while amplification was carried out with two sets of degenerated oligonucleotides (deduced from the P26 protein sequence) and oligo dT as primers. One set of oligonucleotides produced a fragment of the expected size with inflorescence RNAs while a control with leaf RNAs did not. Following cloning of the PCR products, fou clones were isolated and sequenced. They were found to correspond to three very similar cDNAs containing the coding sequence of P26. These cDNAs were not full-length due to the internal position of the primers used for amplification (Fig. 3).

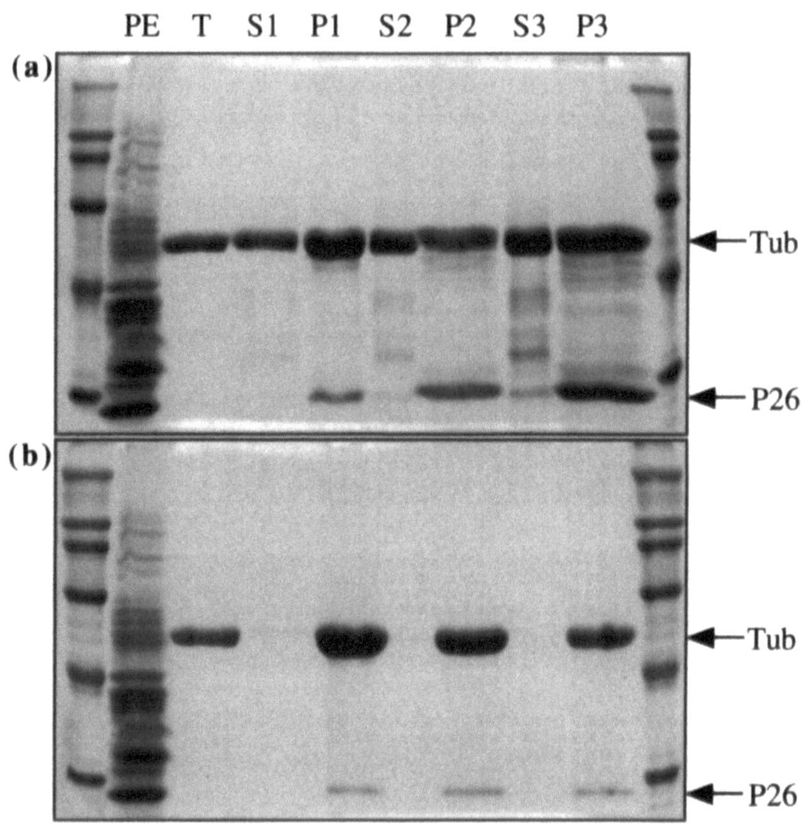

Fig. 4. Coprecipitation of microtubules and P26 in the absence (a) or presence (b) of taxol. Tubulins from pig brains were polymerized in appropriate conditions and then centrifuged: the microtubules are in the pellet and unpolymerized proteins remain in the supernatant. PE: pollen extract; T: polymerized tubulins without pollen extract; Sx: supernatant with pollen extract; Px pellet with pollen extract. Three increasing concentrations (numbered 1 to 3) of pollen extract were assayed . Without taxol, increasing pollen extract concentration leads to increased amount of P26 in the pellet. This effect is not clear if microtubules are fixed with taxol

A database search showed that P26 is very similar to three other plant proteins of unkown function (Görlach et al. 1996). These proteins contain an N-terminal signal peptide, suggesting that P26 has a signal peptide as well. A further analysis of the protein revealed a putative zinc-metalloprotease domain. Limited homologies also exist with sugar-associated enzymes.

12.3 Biochemical characterization of P26

Since no function is known neither for P26 nor for the releated plant proteins we decided to characterize this protein further as to its biochemical properties. The likely presence of a signal peptide opens the possibility of post-translational modifications such as glycosylation. Its putative presence in a large protein complex has also raised the probability of direct interactions with other protein partners.

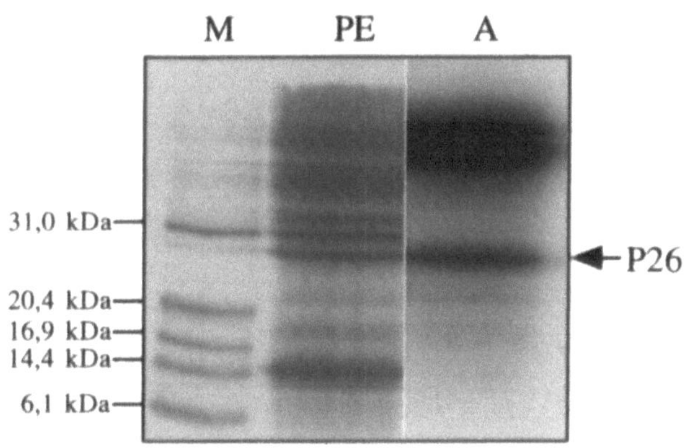

Fig. 5. In vitro self-phosphorylation of a pollen extract. M: molecular weight marker; PE: coomassie blue colored proteins separated on SDS-PAGE; A: autoradiogram revealing radiolabelled proteins. P26 is indicated with an arrow

12.3.1 P26 interacts with microtubules

Our first assay was directed by the fact that P26.4 (the internal chymotrypsic fragment of P26) exhibits a small sequence similar to the microtubule binding domain of TAU. When pig brain tubulins were in vitro polymerized at constant concentration in the presence of increasing amounts of total pollen protein extract P26 coprecipitates with the microtubules in a concentration dependent manner

(Fig. 4). This effect is not evident when microtubules were stabilized with taxol, suggesting that P26 integrates itself into polymerizing microtubules or that taxol occupies the P26/microtubule interaction site. To confirm this colocalization with microtubules in vitro, purified native P26 protein will be necessary.

12.3.2
P26 is phosphorylated by pollen extracts

The activity of MAPS (microtubule associated proteins) is regulated by phosphorylation events (Billingsley and Kincaid 1997). The apparent microtubule/P26 association raised the possibility that P26 is a new type of MAP. When in vitro phosphorylation was performed for one minute in a crude extract of pollen proteins in the presence of ^{32}P-ATP four strongly labeled proteins were detected following SDS-PAGE (Fig. 5). One of these proteins was P26. The use of monoclonal antibodies directed against phospho-amino acids suggests that tyrosine (and maybe serine) is phosphorylated.

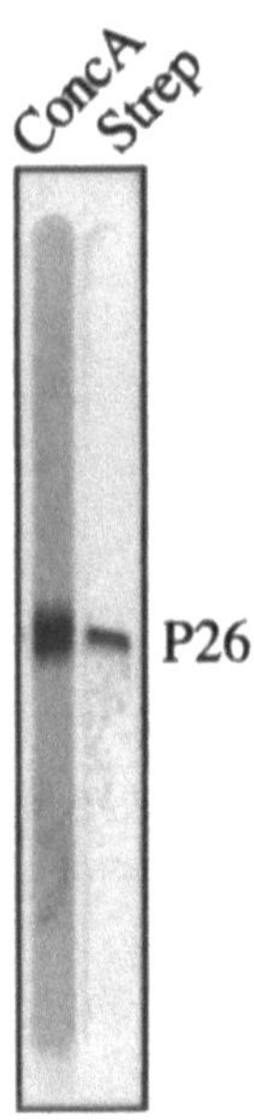

Fig. 6. Glycosylation detection assay on P26 Western blot using biotin/concanavalin A. The bound lectin was visualized by streptavidin/alkaline-phosphatase. ConcA: positive test with lectin treatment; Strep: streptavidin/alkaline-phosphatase without lectin treatment. The protein is detected in both tests demontrating glycosylation and biotinylation of P26

12.3.3
P26 is glycosylated and biotinylated

To see if P26 is glycosylated we first used four different biotinylated lectins that can be revealed with alkaline-phosphatase/streptavidin. Surprisingly we observed a very strong signal with P26 in the control lane incubated only with streptavidine, which suggests that the protein is naturally biotinylated. The signal was enhanced by concanavalin A, indicating that P26 is also glycosylated (Fig. 6). Additional evidence for glycosylation of the protein comes from affinity chromatographic studies showing that P26 binds to a concanavalin A Sepharose column. Microsequencing of the eluted protein showed that the bound protein was P26 and not a contaminant of similar molecular weight. Biotinylation of P26 was finally confirmed using a monoclonal antibody directed against biotin. (Fig. 7).

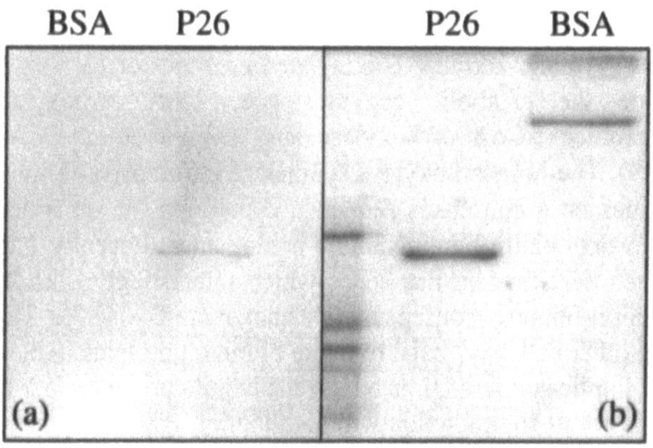

Fig. 7. Immunodetection of biotin on P26 using monoclonal antibodies coupled to alkaline-phosphatase. (a) Western blot treated with anti-biotin antibodies revealing P26; (b) corresponding stained gel in a symetrical view. This test confirms the presence of biotin in P26

From these studies we conclude that P26 contains several types of post-translational modifications: glycosylation, phosphorylation, and biotinylation. The sites for these modifications have not been identified yet.

12.4 Discussion

Preliminary experiments suggest that P26, a major protein of sunflower pollen, could be involved in an interaction with PLIM-1 in a large multiprotein complex, most likely the actin cytoskeleton. PLIM-1 is indeed related to the animal protein CRP, an actin cytoskeleton-associated protein. In the present study we have also observed a possible interaction between P26 and microtubules. These observations raise the question as to whether P26 could be a protein serving as a link between actin cytoskeleton and microtubules (Cai et al. 1997). Contact regions between actin cytoskeleton and microtubules have indeed been observed in a variety of cell types. Using protein microsequencing on SDS-PAGE-purified P26, we designed degenerated oligonucleotides that allowed us to isolate several different partial p26 cDNA clones. The deduced protein sequences revealed homologies with described ESTs of unknown function. A datalibrary search also revealed homology stretches with osid-related enzymes (hexose transporter, chitinases, hexose-glycine ligase). Biochemical studies of P26 have shown that the protein contains several types of post-translational modifications such as glycosylation, phosphorylation, and biotinylation. By analogy with highly similar proteins identified by database search P26 is likely to contain a signal peptide that is removed in the mature protein. This suggests that P26 transits via the ER and could be associated with transport vesicles. P26, as a biotinylated protein, has most likey an enzymatic activity. Biotinylated enzymes or carriers are described so far include the catabolic enzymes acetyl-CoA carboxylase, pyruvate carboxylase, propionyl-CoA carboxylase and methylcrotonyl-CoA carboxylase (Knowles 1989). The holocarboxylase synthetase participates in the biotinylation of these enzymes on a consensus sequence containing the dipeptide MK where lysine is the covalent biotin-binding amino-acid residue. Recently, a biotin bearing protein has been described in pea seeds which interestingly, like P26, does not exhibit the biotin-binding consensus (Delahaye et al. 1997). This raises the possibility that other pathways exist to ligate biotin on proteins in plants. As biotin is a coenzyme implicated in CO_2 or NH_2 transfers, its presence in a putative osids protein ligase is not in contradiction with its function. The phosphorylation of P26 could represent a way to regulate its activity.

At present it is yet difficult to unite all above-described features of P26 in a single functional concept. If the protein is associated with the cytoskeleton, its localization is expected to be cytoplasmic. The possible presence of a signal peptide however suggests that the protein is rather localized to an extracytoplasmic space, maybe even extracellular. This seems to be confirmed by the fact that P26 is glycosylated. Glycosylation however is not an exclusive feature of extracytoplasmic proteins since also cytoplasmic and nuclear proteins can be glycosylated as well (usually on Serine or Threonine residues).

To understand the role of P26 in pollen development a number of key questions have to be answered. Questions such as: What is/are the stage(s) is pollen development in which P26 is active? How is the development of these stages affected in plants not expressing the protein? Also, should the binding data with PLIM-1 and microtubules have a biological meaning P26 must be shown to co-localize with these proteins in the pollen grain. The presence of biotin in P26 suggests that the protein carries an enzymatic function: what then is the precise nature of this enzymatic activity? What is the role of phosphorylation in P26? Does it regulate the enzymatic activity of the protein or does it control another activity or possibly even its localization ? What is the nature of the glycosylated residue(s) ? What role does glycosylation play in the function of the protein ?

Some of these questions may be answered quickly, others obviously need a lot of work and patience. But the simple fact that P26 is extremely abundant in sunflower pollen grains lets us wonder what this protein is really doing there.

References

Baltz R, Evrard JL, Domon C, Steinmetz A (1992) A LIM motif is present in a pollen-specific protein. Plant Cell 4: 1465-1466

Beckerle MC (1997) Zyxin: zinc fingers at sites of cell adhesion. Bioassays 19(11): 949-957

Billingsley ML, Kincaid RL (1997) Regulated phosphorylation and dephosphorylation of tau protein: effects on microtubule interaction, intracellular trafficking and neurodegeneration. Biochem J 323: 577-591

Cai G, Moscatelli A, Cresti M (1997) Cytoskeletal organization and pollen tube growth. Trends Plant Sci 2(3): 86-91

Delahaye L, Duval M, Viguier D, Yaxley J, Job D (1997) Cloning and expression of the pea gene encoding SBP65, a seed-specific biotinylated protein. Plant Mol Biol 35(5): 605-621

Görlach J, Volrath S, Knauf-Beiter G, Hengy G, Beckhove U, Kogel KH, Oostendorp M, Staub T, Ward E, Kessmann H, Ryals J (1996) Benzothiadazole, a novel class of inducers of systemic acquired resistance, activates gene expression and disease resistance in wheat. Plant Cell 8: 629-643

Knowles JR (1989) The mechanism of biotin dependent enzymes. Ann Rev Biochem : 195-222

Rudd JJ, Franklin FCH, Franklintong VE (1997) Ca2+ independent phosphorylation of a 68 kDA pollen protein is stimulated by the self-incompatibility response in Papaver rhoeas. Plant J 12(3): 507-514

13 Functional studies of PLIM-1, a pollen-specific protein

Eliasson Å, Gass N and Steinmetz* A

13.1 Introduction

Pollen germination and pollen tube growth are unique biological processes which - in response to signals produced by the secretory and transmitting tissue cells of the stigma and style respectively - involve an active assembly of cytoskeletal structures leading to profound morphological modifications of the vegetative cell. The formation of the pollen tube is central to the sexual reproduction of higher plants as it allows transport of the male gametes (sperm cells) through the maternal tissues to the embryo sac containing the egg cell.

The molecular mechanisms governing early and late events of pollen tube formation are not well understood, largely because little is known about the biological signals as well as the proteins involved. The use of inhibitors interfering with the assembly of microtubules (colchicine) and actin microfilaments (cytochalasin D) has shown that pollen tube growth is controlled primarily by actin polymerization (for a review see Mascarenhas 1993). It is therefore expected that proteins participating in the assembly of the actin cytoskeleton are also required for pollen tube growth. It is also likely that formation of the unique structure of the pollen tube involves an additional set of proteins specifically expressed in the mature and germinating pollen grain. These proteins should include membrane-associated receptors, structural cytoskeletal proteins, diffusable signaling proteins, as well as pollen-specific transcription factors. The aim of our studies is to identify some of these proteins and to elucidate their role in pollen germination and pollen tube growth. A better understanding of these fundamental processes in plant reproduction is likely to have profound implications in plant biotechnology, since it might allow for instance a targeted interference with the reproductive process at the compatibility level.

The full understanding of the function of a protein requires knowledge at three

* for correspondence: Institut de Biologie Moléculaire des Plantes du CNRS; 12, rue du Général Zimmer; 67084 Strasbourg Cedex; France

different levels, namely: 1/ the physiological level, answering the question as to what is the phenotype of the organism or the cell not expressing the protein or expressing an abnormal (mutant) protein; 2/ the biochemical level, specifying the pathway in which the protein operates and defining the position of the protein in this pathway; 3/ the molecular level, identifying the functional domains of the protein, and the nature of its direct interacting partners.

It is reasonable to assume that mature pollen grains already contain transcripts for a number of the components involved in the germination and tube growth process. We have therefore used the differential cDNA library screening approach combined with Northern hybridization experiments to isolate cDNA clones encoding novel, pollen-specific proteins from sunflower (Herdenberger et al. 1990). One of these proteins, originally called SF3 and renamed PLIM-1, is characterized by the presence of two double zinc finger structures identified as LIM domains (Baltz et al. 1992a; 1992b). LIM domains, which have been mainly described in animal proteins, have been shown to act as protein binding interfaces in a variety of proteins generally associated with multiprotein structures such as transcription or cytoskeletal complexes (for a review see Dawid et al. 1998) Their acronym is derived from the initials of the three proteins in which this motif was first identified: LIN-11 from *Caenorhabditis elegans* (Freydt et al. 1990), ISL-1 from *Rattus norvegicus* (Karlsson et al. 1990), and MEC-3 from *Caenorhabditis elegans* (Way and Chalfie 1988). The protein PLIM-1 from sunflower was found to be encoded by a multigene family, and its mRNA appears to be very abundant in mature pollen grains (Baltz et al. 1992a, unpublished data). Recent experiments also showed that PLIM-1 is part of a multiprotein complex (Baltz et al. in preparation).

In the present paper we describe the various approaches we are using to identify the function of the protein PLIM-1. They include immunolocalization studies, expression in transgenic plants of PLIM-1-GFP fusion proteins under the control of the *plim-1* promoter, overexpression of native and mutant proteins in transgenic plants, expression of antibody derivatives (scFv) to selectively inactivate the protein in the developing pollen grains, and the two hybrid approach to isolate and identify PLIM-1 interactors. While some of these experiments have been concluded, others are still in progress and will be discussed as well in the following section.

13.2 Results and Discussion

Generally a protein functions in a specific cellular structure or organelle and thus it is of primary importance to determine its cellular location. To this end a couple of experimental approaches are presently available: 1/ immunolocalization on fixed (dead) tissues using specific antibodies and 2/ direct immunofluorescence experiments on live cells or tissues expressing the protein carrying a naturally fluorescing tag such as GFP (Green Fluorescent Protein).

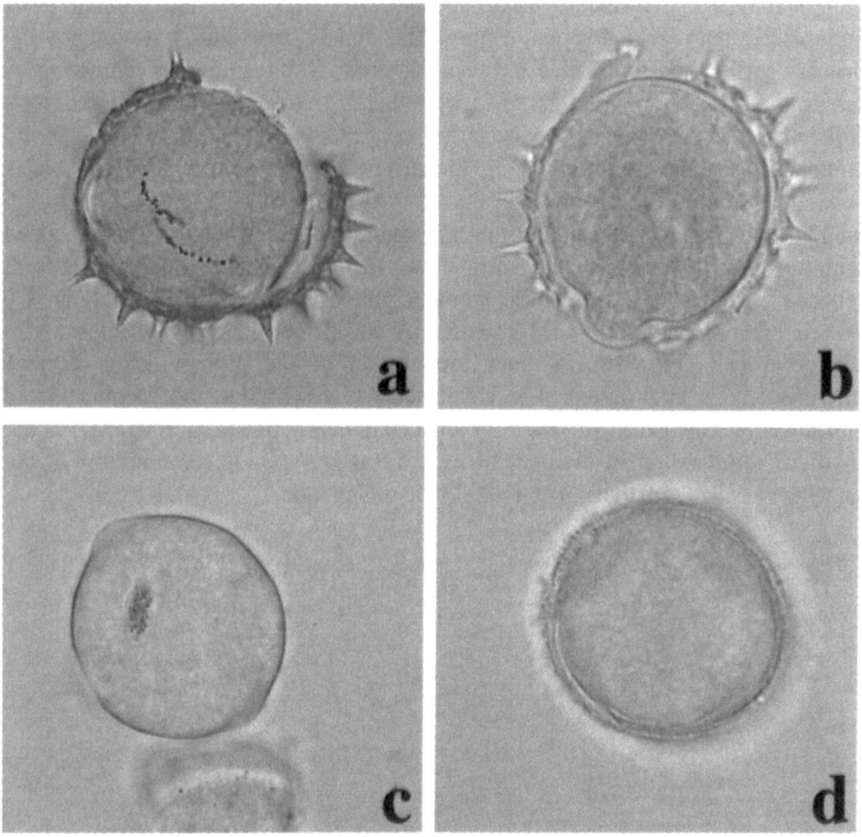

Fig. 1. Immunolocalization of PLIM-1 on sunflower and tobacco pollen. Glutaraldehyde fixed and cryocut pollen grains were treated with PLIM-1 polyclonal antibodies and then revealed with goat anti-rabbit secondary antibody coupled to gold.particles a) sunflower pollen treated with PLIM-1 antibodies; b) sunflower pollen treated with preimmune serum; c) tobacco pollen treated with PLIM-1 antibodies; d) tobacco pollen treated with preimmune serum

13.2.1
Immunolocalization studies of PLIM-1

To localize the .PLIM-1 protein in pollen grains we have used polyclonal antibodies directed against the full length PLIM-1 protein. Earlier work had shown that these antibodies recognize a single protein of 30 kDa (PLIM-1) in protein extracts from mature sunflower pollen (Baltz et al. submitted). To carry out the immunolocalization experiments mature pollen grains were fixed in 3.2% glutaraldehyde as described by Doris and Steer (1996), cryocut into 9 µm sections and placed onto gelatin-coated slides. The slides were then treated with the primary antiPLIM-1 antibodies and secondary anti-rabbit antibodies coupled to 11 nm gold particles. The gold particles were then visualized by optical microscopy following silver enhancement. Negative controls were the pre-immune serum and/or the secondary antibody alone. In sunflower pollen grains PLIM-1 localizes to the two sperm cells as shown in Fig. 1 a. At this point it is not possible to tell whether the protein is located within the sperm cells or whether it is associated with a vegetative cell structure surrounding them. In mature pollen grains from tobacco - which differ from the sunflower pollen in that they are bicellular - a protein reacting with anti-PLIM-1 antibodies was detected closely associated with the germ cell (Fig. 1c). A finer localization of the protein using electron microspcopy is in progress. The localization of PLIM-1 to the sperm cells of sunflower and the germinative cell of tobacco indicates that a priori PLIM-1 is not involved the division of the generative cell, but possibly in the transport of the generative cell/sperm cells to and within the pollen tube.

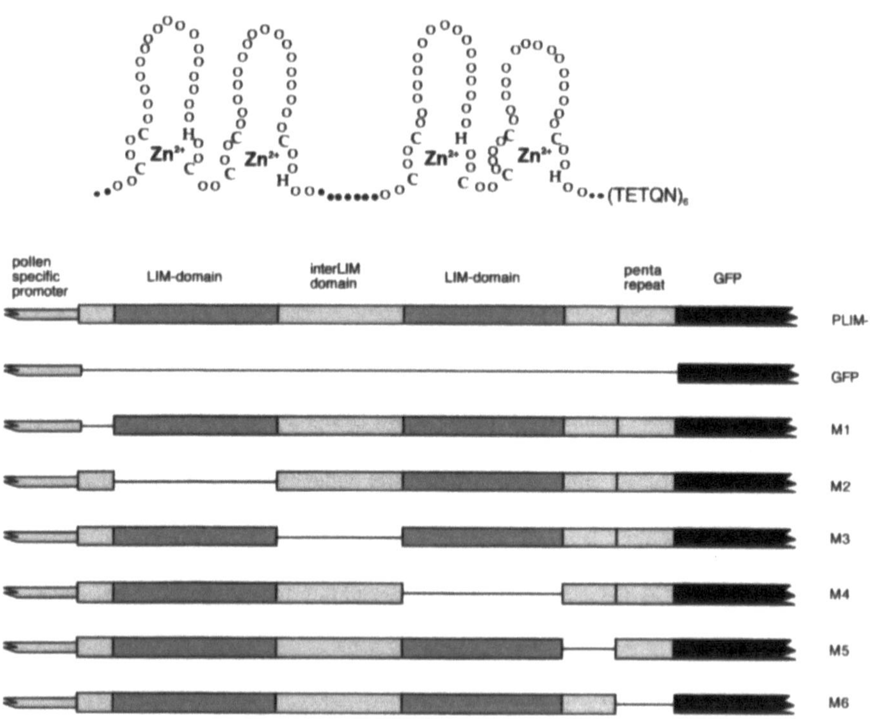

Fig. 2. Structure of PLIM-1 and deletion mutants. The PLIM-1-GFP fusion constructs inserted into pBIN19 and finally transformed into tobacco. The various peptide domains are indicated. A thin line represents a deletion

13.2.2 Use of a GFP fusion protein in localization and domain identification studies

The Green Fluorescent Protein (GFP) is a widely used marker protein which, when fused to a test protein serves as a fluorescent tag allowing observation - under blue light - of the protein in live cells and tissues (Chalfie et al. 1994). We have tried to use this marker 1/ to confirm our immunolocalization studies (see above) on pollen from transformed tobacco plants and 2/ to identify the domains of the protein involved in the association with the germ cell. Six deletion mutants C-terminally fused to GFP have been produced by mutagenesis (Fig. 2), placed under the control of the *plim-1* promoter, and introduced into the plant transformation vector pBIN19. Following transformation of tobacco using the leaf disk method, transformants were selected and plants were grown up to anthesis.

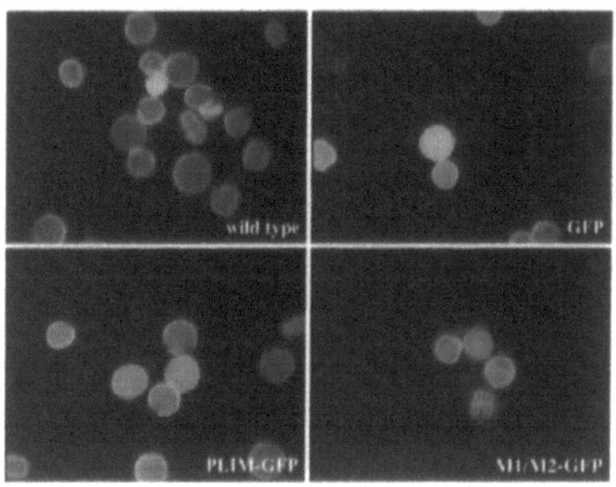

Fig. 3. Fluorescence microscopy of transgenic tobacco pollen. Fluorescence of GFP in transgenic tobacco pollen grains viewed in an optical microscope. The images are of the constructions shown in Fig. 2. Mutants M1 and M2 appeared the same, and mutants M3-M6 have not been tested. The pictures show the F_0 generation, thus segregation of the chromosomes during meiosis can lead to a mix of insert-containing (fluorescent) and insert-less (non-fluorescent) pollen grains

Non-fixed, entire, mature pollen grains of the transgenic plants (the F_0 generation only) were analyzed for green fluorescence in a fluorescence microscope. The fluorescence was compared to that of non-transformed tobacco pollen grains (Fig. 3). Our present analysis only includes the native protein (fused to GFP) and two of the mutants (M1 and M2) as well as GFP alone as a control. As can be seen from Fig. 3, the results obtained using this approach do not confirm the localization of PLIM-1 to a germ cell-associated structure. The fusion protein remains rather diffuse in the vegetative cell (Fig. 3c), as does GFP alone (Fig. 3b). The same diffuse localization was also observed for the mutants M1 and M2 (Fig. 3d). The fluorescence within and not just around the vegetative nucleus was confirmed with a confocal microscope for the PLIM-1 fusion proteins. The confocal imaging also indicated that the PLIM-1 fusion protein was not in the germinative cell (data not shown).

We can conclude from these data that the *plim-1* promoter from sunflower pollen is active in tobacco pollen and that GFP can be correctly expressed and detected by fluorescence microscopy in pollen grains. These data also show that PLIM-1 fused to the N-terminus of GFP loses its capacity to bind to the germ cell in tobacco pollen. This can be explained by a structure of the fusion protein in which one or several domains of the protein are masked by GFP and therefore becomes unavailable for interaction with their natural binding partners. It is also possible that freely diffusing fluorescing fusion proteins mask the correctly localized proteins.

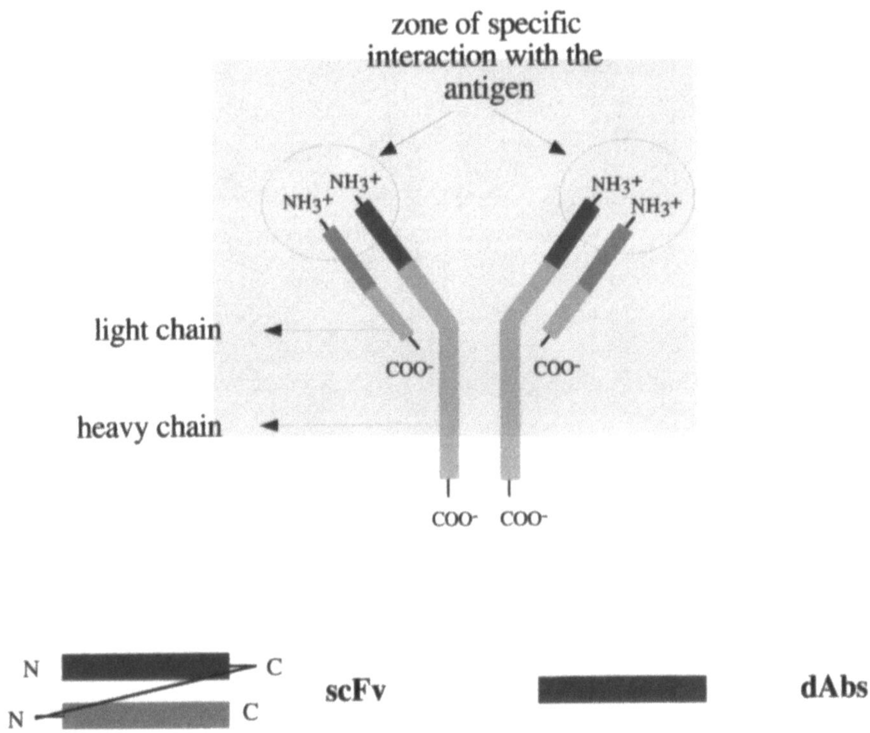

Fig. 4. Schematic structure of an antibody, an scFv and a dAb. Light grey bars show the constant regions of the heavy and light chains, dark grey bars the variable region of the light chain and black bars the variable region of the heavy chain. In an scFv the C-terminus of the variable region of the heavy chain is linked to the N-terminus of the variable region of the light chain via a flexible peptide linker. A dAb consists exclusively of the variable region of the heavy chain

13.2.3 Inactivation studies of PLIM-1 in vivo using monoclonal antibody derivatives

One way to study the physiological function of a protein is to inactivate/delete the gene by homologous recombination (Miao et al. 1995). Although this technique is commonly used in animals to create knock-outs, it is difficult to carry out in plants due to the low recombination frequency; furthermore it can be used only in the case of single copy genes. A second, frequently used approach involves inactivation of the corresponding mRNAs by antisense RNA expression. Since this method is based on the formation of double stranded RNA stretches between the messenger and antisense RNAs this may lead to artifactual, undesired inactivations of related mRNA species or unrelated mRNA species sharing a minimum of common sequence stretches. The observed phenotype can therefore result from the inactivation of not one specific, but several related mRNA species.

In our studies to elucidate the physiological function of PLIM-1 we have chosen an alternative approach which consists in the expression of antibody derivatives directed against PLIM-1. Monoclonal antibodies recognize a single epitope of a protein. The mRNAs expressing the light and heavy chains of a specific antibody can be isolated from selected hybridoma cell lines, reverse-transcribed into cDNA and cloned into a plasmid vector. The variable regions of the antibodies are required for the specificity of interaction with the antigen (Fig. 4). By combining the variable region of the heavy chain (VH) with that of the light chain (VL) via a linker peptide, antibody derivatives (scFvs = single chain variable fragments) with affinities similar to those of the parent antibody molecules can be obtained (Bird et al. 1988). Alternatively single domain antibodies (dAbs) consisting exclusively of the variable region of the heavy chain have been described to maintain high specificity and affinity for its epitope (Ward et al. 1989). An advantage of this technology is that the specificity of monoclonal antibodies can be tested *in vitro* on protein extracts prior to the introduction of their coding sequences into plants. Therefore only those antibodies that exclusively interact with the protein of interest can be selected for scFv or dAb production. The rationale behind this approach is that expression of an scFv or a dAb directed against a given protein under the control of an appropriate promoter will lead to its binding in vivo to the corresponding epitope on the protein and therefore to the inactivation of this domain. If this domain is functionally important, its inactivation will result in the production of transgenic plants presenting a specific phenotype.

To use this approach a panel of monoclonal antibodies were raised against PLIM-1 and their binding properties were analysed by enzyme linked immunosorbant assay (ELISA). Of these, four monoclonal antibodies were selected and their epitopes identified on PLIM-1 using different deletion mutants (Fig. 5). While construction of scFvs is still in progress, the bacterially-expressed and affinity-purified single domain antibodies (dAbs) were tested on western blots carrying pollen proteins. Although an interaction with a 30 kDa protein (corresponding to PLIM-1) was observed, additional bands appeared, indicating that in our case the dAbs are not specific enough (data not shown). We hope to restore the original specificity of these antibodies with the constructions of the scFvs .

The antibody approach has been used successfully to inactivate plant hormones (Owen et al. 1992). In spite of its rather heavy design it may be a most valuable alternative to study the function of a protein at the physiological level.

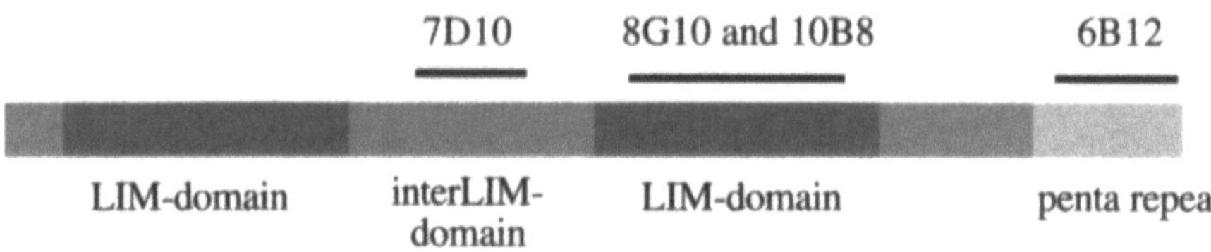

Fig. 5. Epitope map of PLIM-1. The names of the monoclonal antibodies recognizing the various epitopes are indicated above the thin bars

13.2.4
Identification of PLIM-1 interactors

The immunolocalization data indicate that PLIM-1 is not a freely diffusable protein but that it is associated with other cellular structures in mature pollen grains (sperm cells in sunflower and germ cell in tobacco). This association occurs via interactions with specific protein partners. The identification of these proteins would lead to a better understanding of the functional role of the protein in the pollen grain. Based on results from other groups working on LIM-domain proteins we know that LIM-domain proteins are often involved in protein-protein interactions (Schmeichel and Beckerle 1994). Previously, blot overlay experiments and native gels had shown that PLIM-1 indeed interacts with other proteins in the pollen grain, but it was not possible to isolate sufficient amounts of these interacting proteins in order to identify them (Baltz et al. in preparation). Knowing that PLIM-1 is part of a multimolecular complex and therefore has to bind to at least one other protein of this complex we decided to use the yeast-two-hybrid system to isolate directly interacting partner proteins at the mature pollen stage.

The yeast-two-hybrid screen was performed with PLIM-1 as the bait protein on an expression library of mature sunflower pollen. We used the Stratagene Hybrizap 2.1 kit, with which it was possible to obtain a large and thus representative pollen library due to the use of phagemid vectors. However, we did not obtain a sufficiently high transformation efficiency to isolate enough positive clones. Of the twenty-eight clones that were isolated all showed an unspecific interaction with PLIM-1 when compared to the empty bait vector.

In spite of this failure to isolate PLIM-1 interactors it appears to us that the yeast two hybrid approach is yet the most appropriate to use in the case of the pollen: pollen grains are indeed simple organisms comprising basically one transcriptionally and translationally active cell, which should minimize the number of false positives, however we need to optimize the conditions for yeast transformation 10-100 fold.

13.3 Conclusions

Our studies aimed at identifying the function of pollen LIM protein PLIM-1 from sunflower have revealed that PLIM-1 is associated with the two gametes in sunflower and the germ cell in tobacco. When expressed in tobacco pollen as a fusion protein with GFP under the control of the *plim-1* promoter the pollen grains exhibit diffuse fluorescence of the vegetative cell, but no accumulation in specific structures could be observed, suggesting that GFP interferes with the interaction of PLIM-1 with its normal protein partner. The physiological function of PLIM-1 is being studied using the antibody expression approach to specifically inactivate PLIM-1 (and not a related protein). The yeast two hybrid system is presently used to isolate PLIM-1 interactors.

Acknowledgements

We are indebted to M. Kohnen and F. Hentges for kindly providing the monoclonal PLIM-1 antibodies and to J-L. Evrard for excellent help in cryocutting of pollen grains. Å. Eliasson was a recipient of a Wennergren Center Foundation postdoctoral fellowship.

References

Baltz R, Domon C, Pillay DTN, Steinmetz A (1992) Characterization of a pollen-specific cDNA from sunflower encoding a zinc finger protein. Plant J 2: 713-721

Baltz R, Evrard JL, Domon C, Steinmetz A (1992) A LIM motif is present in a pollen-specific protein. Plant Cell 4: 1465-1466

Bird RE, Hardman KD, Jacobson JW, Johnson S, Kaufman BM, Lee SM, Lee T, Pope SH, Riordan GS, Whitlow M (1988) Single-chain antigen-binding proteins. Science 242: 423-426

Chalfie M, Tu Y, Euskirchen G, Ward WW, Prasher DC (1994) Green Fluorescent Protein as a marker for gene expression. Science 263: 802-805

Dawid IB, Breen JJ, Toyama R (1998) LIM domains: multiple roles as adapters and functional modifiers in protein interactions. Trends Genet 14: 156-162

Doris FP, Steer MW (1996) Effects of fixatives and permeabilisation buffers on pollen tubes: Implications for localisation of actin microfilaments using phalloidin staining. Protoplasma 195: 25-36

Freydt G, Kim SK, Horvitz HR (1990) Novel cysteine-rich motif and homeodomain in the product of the *Caenorhabditis elegans* cell lineage gene *lin-11*. Nature 344: 876-879

Herdenberger F, Evrard JL, Kuntz M, Tessier LH, Klein A, Steinmetz A, Pillay DTN (1990) Isolation of flower-specific cDNA clones from sunflower (*Helianthus annuus* L). Plant Sci 69: 111-122

Karlsson O, Thor S, Norberg T, Ohlsson H, Edlund T (1990) Insulin gene enhancer binding protein Isl-1 is a member of a novel class of proteins containing both a homeo- and a Cys-His domain. Nature 344: 879-882

Mascarenhas JP (1993) Molecular mechanisms of pollen tube growth and differentiation. Plant Cell 5: 1303-1314

Miao ZH, Lam E (1995) Targeted disruption of the TGA3 locus in *Arabidopsis thaliana*. Plant J 7: 359-365

Owen M, Gandecha A, Cockburn B, Whitelam G (1992) Synthesis of a functional antiphytochrome single-chain Fv protein in transgenic tobacco. Bio/technology 10: 790-794

Schmeichel KL, Beckerle MC (1994) The LIM domain is a modular protein-binding interface. Cell 79: 211-219

Ward ES, Güssow D, Griffiths AD, Jones PT, Winter G (1989) Binding affinities of a repertoire of single immunoglobulin variable domains secreted from *Escherichia coli*. Nature 341: 544-546

Way JC, Chalfie M (1988) *mec-3*, a homeobox-containing gene that specifies differentiation of the touch receptor neurons in *C. elegans*. Cell 54: 5-16

14 Pollen coat signals and ovule penetration in *Gasteria verrucosa* (Mill.) H. Duval

Willemse* MTM

14.1 Introduction

In his interesting chapter concerning "Homologies and Phylogeny" Michel Favre-Duchartre (1984) considers the "Pollen reception by female organ". A short text expresses the pollination and progamic phase of gymnosperms and angiosperms. About the differences in the acceptation and germination of pollen between an open carpellar construction and a closed one in case of: "angiocarpy", Favre-Duchartre notes:" This phylogenetic tendency to the protection of female gametophytes by (1) nucellus, (2) ovular integument(s),(3) carpels, and (4) inferior ovary (produced either by caulinar proliferation or, more often, by concrescence of staminal and perianth bases with the surface over the ovary) has been compared by Mangenot (1973, p 53) to Russian dolls (my, Favre-Ducharte, translation):" the female organ of angiosperms calls to mind the Russian dolls (those series of wooden dolls, the larger ones containing smaller and smaller ones): inside the gynoecium are found ovules; inside each ovule one discovers a megasporangium-nucellus, and within the latter a female prothallium". Mangenot suggests that this process may be further promoted if the carpels themselves are completely protected."

This description characterises the common idea of angiospermy and to cite Favre-Duchartre once more: "the prevention of self-fertilization represents an important advantage afforded by angiospermy", this justifies the view of a row of barriers which has to be to overcome by pollen and pollen tube to promote cross-fertilisation.

The heterogeneous pollen tube pathway from stigma to micropyle is a prepared interface to realise the interaction between male and female gametophytes leading to fertilisation. However, in gymnosperms and angiosperms there are different types of pathways, such as the pollination droplet, the wet or dry stigma, elongated

* for correspondence: Laboratory of Plant Cytology and Morphology; Agricultural University; Arboretumlaan 4; 6703 BD Wageningen; The Netherlands

integuments, the hollow or solid stylar canal and the tenui- or crassinucellar ovular types. Seldom there are real mechanical barriers such as in *Triticum sp* with a very narrow path for one pollen tube to enter the placental locule (Vishnyakaya and Willemse 1994).

Pollination includes a signal of activation and recognition preparing the pistil. From pollen arrival the germination of compatible or even incompatible pollen can start. In fact, there is no protection preventing self-fertilisation of the female gametophyte but there is an interaction system between the gametophytes from pollination to fertilisation.

The pistil with the heterogeneous pollen tube pathway and interface is a product of the sporophyte. In gymnosperms this interaction system of the sporophyll, with pollination droplet between the long integument and a crassinucellate nucellus seems of a more simple construction compared with the pistil of the angiosperms.

Differences and similarities between gymnosperms and angiosperms exist in the way of acceptation of pollen and of pollen tube growth. Next to structural and functional differences, environmental conditions such as the way of pollination, the high variety in the number of crossing barriers, the cost of flowering and the duration of the generation time, should also be considered in functional comparative studies.

14.2 Pollen-pistil interface and interaction: the interface

The pistil receives the pollen on its stigma and recognises the type of pollen. The stigma and style guide the pollen tube and supply nutrients up to the entrance of the ovarian locule. On the stigma or in the style the incompatible pollen tube may stop growing. From the entrance to the ovarian locule the pollen tubes run free or along the carpelar epidermis or grow in the locular fluid towards a micropyle which should be found and penetrated. Some micropyles clearly show an attractant present in the micropylar canal. In some plants in the embryo sac a late acting incompatibility can be expressed.

Stigma, style and ovary offer a heterogeneous pollen tube pathway. During pollen tube growth there is a continuing and dynamic interaction between the pollen before and after germination and the pistil. The intermediate site of this interaction is expressed as a dynamic interface that can have a different and transient character in various parts of the pistil. The interface is characterised by products outside the cells of the pollen and the pistil and is the domain of the apoplast. In this interface interactions such as signalling, acceptance, rejection, nutrition, protection and exchange of products take place.

On the stigma the interface is expressed for the pollen as the pollen coat around and between the exine and the pollen wall intine with substances that can extrude. The interface of the stigma exists of the stigmatic exudate, such as a fluid or pellicula, a proteinaceous layer, and the cell wall of the stigmatic cell. The contact between pollen and stigma results in an exchange of signals and products in the mixing or contacting interfaces.

In the style the interface concerns the pollen tube and pollen tube tip. The pollen tube tip exudate interacts with its surroundings. The pollen tube wall functions as a possible protective molecular sieve, commonly supplied with callose. The stylar interface is a liquid or are cell walls, both containing various compositions of carbohydrates, proteins and lipids. During the passage of the pollen tube exchange of products takes place, mainly for signalling and nutrition.

In the ovary locule and the micropyle the interface consists of a fluid or the upper epidermal wall layer of the endocarp. Here, as in the style, also an exchange between pollen tube tip exudate and epidermal cell exudate takes place. At the micropyle there is a micropylar exudate to guide the pollen tube to the entrance of the ovule. Once in the ovary, the pollen tubes get their nutrients and are divided over the ovules and directed to the micropyles.

The pollen tube pathway is heterogeneous and in the interface there is a dynamic interplay between pollen and pistil.

In the next paragraphs some aspects of pollination signals and the penetration of the pollen tube in the micropyle will be presented, mainly focussing on the pollen-pistil interaction in *Gasteria.*

14.3
The pollen tube pathway in *Gasteria*

As a sign of the receptive period of the *Gasteria* pistil a droplet of exudate appears on the stigma. Pollen grains touch and will sink in the stigmatic droplet, mixing the pollen coat substances in the stigmatic droplet. Thereafter pollen grains germinate and the pollen tubes penetrate the solid part that is present in the tip of the hollow style. When the pollen tubes reach the open stylar canal, the pollen tube pathway follows the fluid in the stylar canal and the placental fluid in the ovary locule. In the stylar fluid the pollen tube wall has nearly any callose. In the placental fluid this callose reappears in the wall. The final goal of the pollen tube is the ovular micropyle in which the micropylar exudate is present (Franssen-Verheijen and Willemse 1993).

Gasteria needs cross-pollination for seed set. Sears (1937) discovered the ovular incompatibility system in *Gasteria.* It includes that self as well as cross pollen grains germinate on the stigma, pass the fluid stylar canal and placental fluid and penetrate into the micropyle. Because of the incompatibility reaction in the ovule and late response of the incompatibility reaction, this type of incompatibility is considered as ovular gametophytic incompatibility or as late acting incompatibility (Seavey and Bawa 1986).

The ovular incompatibility system of *Gasteria* is genetically based on two or more loci (Brandham and Owens 1977; Naaborg and Willemse 1992). Compared with the normal pollen tube growth of cross pollen, the pollen tube has a retarded tube growth rate after self pollination. It contains a higher number of callose plugs and it has less callose in its wall (Willemse and Franssen-Verheijen 1988). These phenomena indicate similar reactions as common for a gametophytic self-incompatibility system (Nettancourt 1977).

14.3.1 Pollen coat signals

Signal molecules present in the pollen coat mix with substances on the stigmatic surface, as occurs during sporophytic incompatibility (Brownlee 1994). The stigmatic exudate of *Gasteria* has a weak hydrophobic nature, this means that lipophilic substances are present. During pollination the lipophilic components of the pollen coat of *Gasteria* mix with the stigmatic droplet. However, around the exine of immersed pollen grains some droplets of the pollen coat can be observed (Figs.1 a-b) The hydrophilic pollen coat components, such as proteins, mix with the hydrophilic part of the stigmatic exudate. In vitro another mixing of pollen coat substances occurs when a hydrophylic germination medium is used. A lipophilic part of the pollen coat covers the surface of the germination medium. As in vivo on the pollen exine a number of pollen coat droplets can be observed (Figs. 1 c-d). The hydrophilic part mix within the medium.

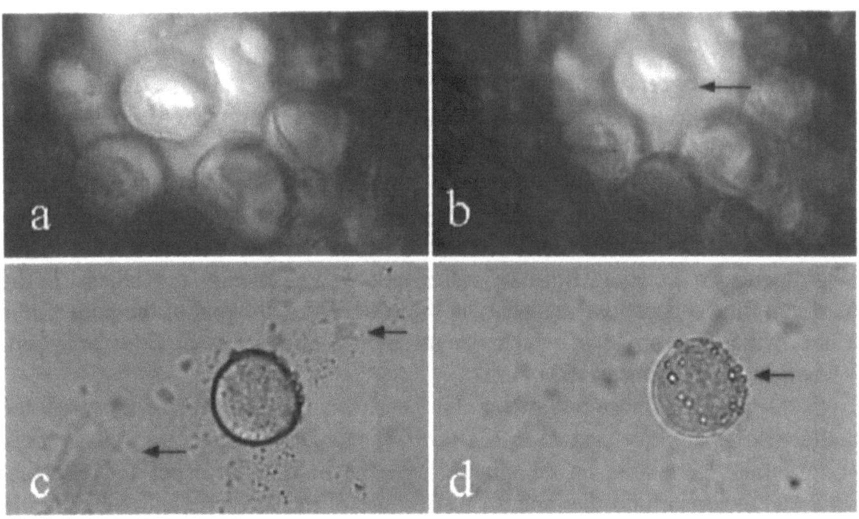

Fig. 1. Pollen grains in the stigmatic fluid observed on the outer level (a) and deeper in the fluid (b). Note the scarce droplets of pollen coating on the exine (arrow). Pollen grains in germination medium with outflow on the surface of hydrophobic pollen coat material (c, arrows) and some droplets of pollen coating on the exine (d, arrow). X 30

Dependent of the nature of lipophilic and hydrophilic substances in the pollen coat and the composition of the stigmatic droplet, in wet stigmas a selective mixing of pollen coat substances with the stigmatic droplet can occur.

Pollen coat substances are of sporophytic origin arising from the tapetal cells. But they also contain remnants of the locular fluid and other products synthesised by the tapetal cells and later on finally the remnants of degenerated tapetal cells. In fact, a mixture of all types of products from the locule and tapetal cells can be present. Pollen coat substances are considered to contain the signals for the early recognition between self or cross pollen in case of sporophytic incompatibility, but also other signals are present.

In *Gasteria,* after pollination with normal coated or washed cross pollen the normal cross pollen were able to set fruit but without the pollen coat the yield was strongly reduced (Willemse 1996). This means that the pollen coat contains signals to enhance fruit set after cross-pollination. In previous experiments with *Gasteria* it was found that cross-pollination increases the water volume in the placental fluid (Willemse and Franssen-Verheijen 1992). Self and cross-pollination and the addition of a small droplet of hydrophilic components from the pollen coat on the stigma increased the pistil and pistil fluid weights (Willemse 1996). This signal of water uptake by the pistil is present in the hydrophilic fraction of the pollen coat containing the proteins. After cross-pollination the fluid pollen tube pathway becomes more diluted, which promotes the cross pollen tube growth. The pollen tubes of self-pollen have a retarded pollen tube growth but do reach the ovules. Therefore, the pollen coat signal can be characterised as an activation signal to prepare and complete the fluid pollen tube pathway. The signal substances are of sporophytic origin. On the other hand, also the recognition signal, that is of gametophytic origin in *Gasteria,* is expressed in the two types of pollen tube growth. Since the normal fruit set and rate of water uptake in the pistil is connected with the compatible reaction, it is not clear how these signals should be distinguished.

14.3.2
Pollen coat glycoproteins

The role of glycoproteins as signalling substances during incompatibility is well known (Dickinson et al 1992). In the hydrophilic part of the pollen coat, collected in the germination medium, several proteins and a group of glycoproteins are found (Willemse and Vletter 1995). Also after extraction with cyclohexane, a lipid solvent, pollen coat proteins can be detected in incompatible plants and compatibele mutants. After this extraction the composition of the proteins present differs between the normal incompatible "cross" pollen and compatible "self" pollen. In comparison with the normal "cross pollen", the "self pollen" have three protein spots more and the spots representing the group of glycoproteins are different. This pattern characterises a normal incompatible plant.

The differences within the group of glycoproteins may indicate that these glycoproteins are involved in the pollen signal as the recognition reaction

(Willemse 1998). After 2Dgel electrophoresis of a pollen mixture in the germination medium the group of proteins consists of a row of four spots in the first dimension and of three in the second dimension (Fig. 2a). Preliminary results of a cyclohexane extraction of the pollen coat show that the pollen of the compatible plant, called "cross pollen", have a single glycoprotein series of four spots (Fig. 2c). In contrast, the pollen of the normal incompatible plant, called "self pollen", has a double series of three spots (Fig. 2b).

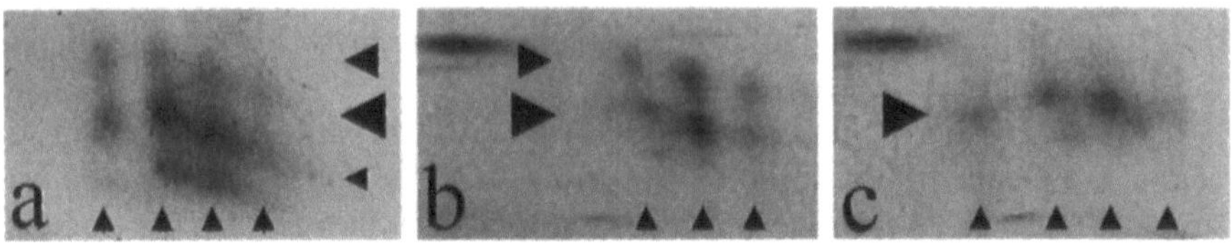

Fig. 2. Glycoprotein pattern of pollen coat after electrophoresis in two dimensions. The group of glycoproteins present in the medium has horizontal four spots in the first (small arrowheads) in the first dimension and vertical three spots (small, medium and large arrowhead), in the second dimension (a). After cyclohexane extraction the glycoprotein of one incompatible plant consist of three spots and two spots respectively (b) and of four spots in only one row in the mutant compatible plant

The suggestion that they express the two or more loci of the incompatibility genes by the group of glycoproteins is attractive to keep in mind.

14.4 Pollen tube penetration in the ovule

Pollen coat substances activate the pistil after landing on the stigma. Some types of activation after pollination are known, such as enhanced protein synthesis in the *Petunia* ovules (Deurenberg 1977), or ethylene production of the pistil (Hoekstra and Weges 1986). The ovular penetration by pollen tubes needs the directed growth towards the micropyle and is a result of the attraction of the pollen tubes. In *Gasteria* the micropylar exudate consists of a part of the squeezed filiform apparatus and contains carbohydrates and proteins. The proteins seem to be involved in the attraction (Franssen-Verheyen and Willemse 1993; Willemse et al 1995).

The in-vitro micropylar penetration technique used in *Gasteria* enables pollen tube growth into the micropyle of isolated ovuleson an adapted agar medium (Willemse et al 1995). With this in-vitro pollination system the ovule activation could be demonstrated as a result of cross-pollination. The ovular activation was reflected by a high percentage of pollen tube penetration in the micropyle of isolated ovules after 3 hours of pollination (Willemse 1996).

14.4.1 Pollen tube micropylar penetration in vitro

The in vitro penetration system offers the possibility to show the process of pollen tube penetration. The way of penetration of one tube or more tubes can be observed by the use of video microscopy. Some events of pollen tube behaviour in vitro were registered: 1/ the penetration of one pollen tube into the micropyle; 2/ the penetration of more tubes into a micropyle (multiple penetration) and 3/ the micropylar refusal of pollen tubes (Willemse 1999). Pollen tube refusal near the micropyle results in a/ passing of the pollen tube along the micropyle or b/ turning of the pollen tube from the micropyle or c/ arrest or stop of the pollen tube including the swelling of the pollen tube tip in front of the micropyle.

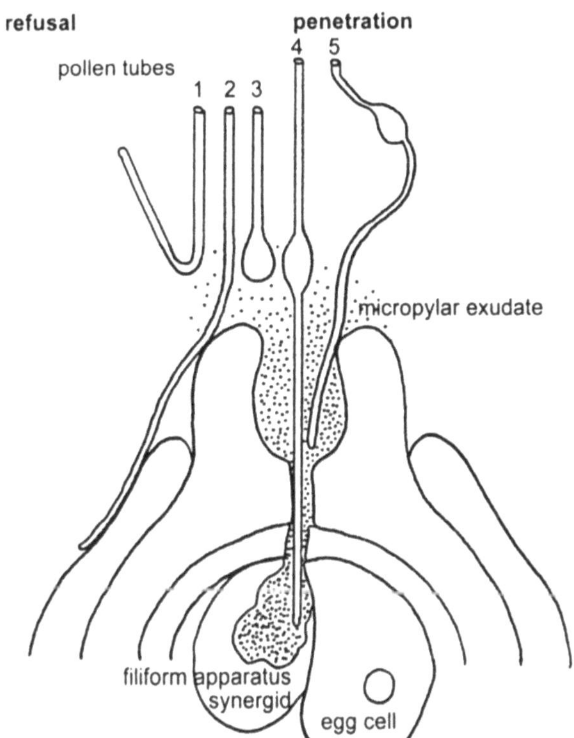

Fig. 3. Schematic representations of the pollen tube penetration patterns in a micropyle as observed in vitro with pollen tube refusal (1-3) as well as acceptance (4,5). 1. Reorientation of pollen tube near the micropyle. 2. The passing of pollen tubes along the micropyle and ovule. 3. The stop of pollen tubes by swelling of the tip. 4. Penetration of the pollen tube after stop and recovery with commonly the reorientation of the pollen tube followed by a retarded growth. 5. The arrest or stop, recovery and reorientation of the pollen tube followed by an expansion of the pollen tube expressed as loops outside the micropyle

The process of pollen tube penetration in vitro can be divided in two parts. First the pollen tubes arrive and stop and the pollen tube tip swells. This behaviour can be influenced by the presence of the micropylar exudate and its flow in the environment, probably causing a change of pollen tube osmotic pressure.

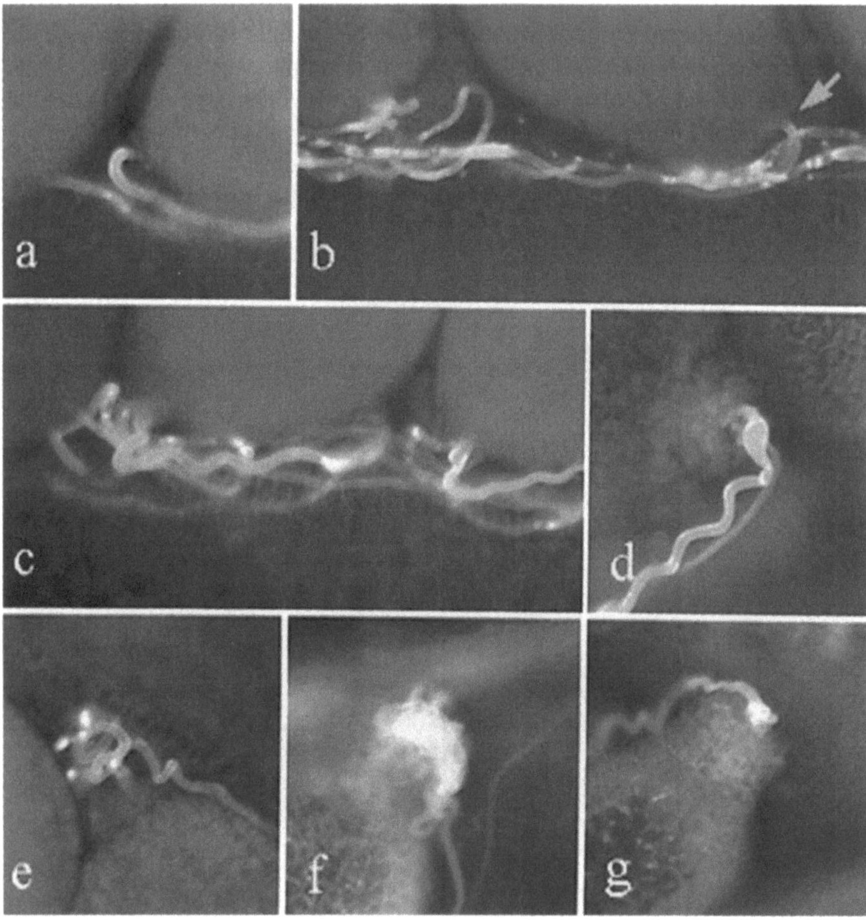

Fig. 4. Different patterns of pollen tube penetration in vivo represented after staining with aniline blue. a: straight pollen tube penetration; b and c: multiple pollen tube penetration; d: pollen tube marked by a swelling before penetration, note the twisting pollen tube; e: pollen tube showing the formation of simple loops; f: complex loops of pollen tubes above the micropyle; g: twisting pollen tube. X 100

Thereafter, the pollen tube growth restarts from the swollen tip and the tube orientates itself in the direction of the micropyle due to the presence of the attractant that influences and orientates the pollen tube growth. After the penetration in the micropyle the pollen tube continues a retarded growth. The pollen tube penetrates and follows the path of the micropylar exudate into one of the synergids. The pollen tube tip penetrates the funnel shaped micropyle, reaches the nucellus, passes the nucellar epidermis cell wall and then will enter one of the synergids. Another possibility is that the pollen tube still elongates but the main part of this pollen tube growth is expressed in the formation of long loops outside the micropyle. These possibilities of pollen tube behaviour, observed near and in the micropyle during in vitro pollination, are summarised in Fig. 3.

14.4.2
Pollen tube micropylar penetration in vivo

With aniline blue stain and UV microscopy, the callose in the pollen wall could be used to mark the pollen tube penetration in an ovary in vivo. After careful dissection, avoiding the distortion of the original position of the pollen tubes, the different patterns of pollen tube penetration could be observed. The penetration process of pollen tubes in the micropyle is directed towards the micropyle, this pattern could be found easily and in high frequency (Figs. 4a-b, arrow). Also the pollen tubes involved in multiple penetration can have such a straight penetration (Figs.4b-c, arrows). The often observed swelling of the pollen tube tip as found in vitro occurred also in vivo in a lower frequency (Figs. 4d). Also, the loops of the pollen tubes outside the micropyle were observed. The patterns of looping were simple (Fig 4e) or very complicated (Fig 4f). Torsion of pollen tubes near the micropyle were observed more frequently in vivo (Figs. 4d,e,f). Passing of the micropyle happened also in vivo, but the refusal of pollen tubes like the turning off from the micropyle or the swelling and stop in front of the micropyle was not observed in vivo.

From these observations it could be concluded that pollen tube penetration in vivo seldom involved the swelling and stop of the pollen tube before penetration as occurred in vitro. One of the interfering conditions in vivo is the surrounding placental fluid.

14.5
Conclusion

Signals during the progamic phase, from pollination to penetration of the pollen tube into the micropyle, create interactions between pollen and pistil in the interface of the pistil. Signal substances present in the pollen coat, in the pollen tube pathway and in the micropyle lead to recognition, activation and attraction of pollen and pollen tubes. During the progamic phase between pollen and pistil complex interactions occur.

The pollen tube pathway in *Gasteria* is heterogeneous but compatible as well as incompatible pollen grow and penetrate the micropyle. The final incompatibility reaction takes place in the ovule but from the onset of pollen tube germination signs of gametophytic incompatibility are present.

The pollen coat signal involves pistil activation as well as pollen recognition, this signal is of sporophytic origin. Glycoproteins are probably signal molecules partly for recognition and partly activation.

The pollen tube attractant in the micropyle influences the growth and orientation of the pollen tube. The behaviour of pollen tube penetration in vivo differs from that in vitro.

Acknowledgements

The author thanks Dr JHN Schel for critical reading of the manuscript and Mr AB Haasdijk for printing the figures.

References

Brandham PE, Owens SJ (1977) The genetic control of self-incompatibility in the genus *Gasteria* (Liliaceae). Heredity 40:155-169

Brownlee C (1994) Signal transduction during fertilization in algae and vascular plants. New Phytol 127:399-418

Deurenberg JJM (1977) Differentiated protein synthesis with polysomes from *Petunia hybrida* ovaries before fertilisation. Planta 133:201-206

Dickinson HG, Crabbe MJC, Gaude T (1992) Sporophytic self-incompatibility systems: S gene products. Int Rev Cytol 140: 525-592

Favre-Duchartre M (1984) Homologies and phylogeny. In Johri BM (ed) Embryology of angiosperms. Springer Verlag Berlin, Heidelberg, New York, Tokyo pp 697-734

Franssen-Verheijen MAW, Willemse MTM (1993) Micropylar exudate in *Gasteria* (Aloaceae) and its possible function in pollen tube growth. Am J Bot: 80: 253-262

Hoekstra FA, Weges R (1976) Lack of control by early pistillate ethylene of the accelerated wilting in *Petunia hybrida* flowers. Plant Physiol 80: 403-408

Mangenot G (1973) Données elementaires sur l'angiospermie. Ann Univ Abidjan 6: 1-245

Naaborg AT, Willemse MTM (1992) The ovular incompatibitlity system in *Gasteria verrucosa* (Mill.) H. Duval. Euphytica 57: 231-240

Nettancourt D de (1977) Incompatibility in Angiosperms. Springer Verlag, Berlin, Heidelberg, New York

Sears ER (1937) Cytological phenomena connected with self-incompatibility in flowering plants. Genetics 22: 130-171

Seavey SR, Bawa KS (1976) Late-acting self-incompatibility in Angiosperms. Bot Rev 52: 195-219

Vishnyakaya MA, Willemse MTM (1994) Pollen-pistil interaction in Wheat. Acta Bot Neerl 43: 51-64

Willemse MTM (1996) Progamic phase and fertilisation in *Gasteria verrucosa* (Mill.) H. Duval: pollination signals. Sex Plant Reprod 9: 347-352

Willemse MTM (1999) Pollen coat signals with respect to pistil activation and ovule penetration in *Gasteria verrucosa* (Mill.) H. Duval. In Cresti M, Cai C (eds). Springer Verlag, Heidelberg, pp 145-156

Willemse MTM, Franssen-Verheijen MAW (1988) Pollen tube growth and its pathway in *Gasteria verrucosa* (Mill.) H. Duval. Phytomorphol 38: 127-132

Willemse MTM, Franssen-Verheijen MAW (1992) Pollen tube growth in the ovary of *Gasteria*. In Batygina TB (ed) Embryology and seed production. Academy of sciences of Russia, St. Petersburg. Proceedings of the XI international symposium, Nauka Leningrad, pp 613-614

Willemse MTM, Plyush TA, Reinders MC (1995) In vitro micropylar penetration of the pollen tube in the ovule of *Gasteria verrucosa* (Mill.) H. Duval and *Lilium longiflorum* Thunb: conditions, attraction and application. Plant Sci 108: 201-208

Willemse MTM, Vletter A (1995) Appearance and interaction of pollen and pistil pathway proteins in *Gasteria verrucosa* (Mill.) H.Duval. Sex Plant Reprod: 8: 161-167

15 Pollen tubes: cellular organization and control of growth

Derksen*[1] J, van Amstel[2] ANM, Rutten[3] ALM, Knuiman[1] B, Li[4] YQ and Pierson[1] ES

15.1 Introduction

Pollen tubes are the specialized cells in seed plants that carry the sperm cells towards the ovules where double fertilization occurs. Pollen tubes, together with other cell types such as root hairs are exclusively tip growing cells, where expansion and wall secretion take place at the tip only (Derksen and Emons 1990). The growth rate of pollen tubes depends on the species and growth conditions, for example in pine the growth rate is less than one μm per hour (de Win et al. 1996) while in maize it is up to 160 μm per hour (Mascarenhas 1993). Pollen tubes also show similarity to haustoria cells as they grow in the alien tissues of the ovary, which provide nutrition (Johri 1992). Pollen tubes have been studied extensively, but mostly on a few taxa which easily grow in vitro, e.g. for Angiosperms *Lilium*, *Nicotiana, Tradescantia* and *Petunia* (Steer and Steer 1989; Derksen and Emons 1990; Mascarenhas 1993; Pierson and Cresti 1992; Derksen et al. 1995a; Derksen 1996; Taylor and Hepler et al. 1997; Li et al. 1997) and for Gymnosperms *Pinus* (de Win et al. 1996) and *Picea* (Lazzaro 1996). A number of properties have been thought to be of relevance for tip growth but comparison between species shows dissimilarities as well as similarities. Presently we will address the factors that determine growth by reviewing the cellular organization of pollen tubes.

*[1] Department of Experimental Botany; Catholic University of Nijmegen, Toernooiveld 1; 6525 ED Nijmegen; The Netherlands

[2] Department of Biology, The Plant Laboratory; The University of York; York YO1 5YW; United Kingdom

[3] Present address: Structural Biology Programme; EMBL, Meyerhofstrasse 1; 69012 Heidelberg; Germany

[4] Department of Biological Sciences and Biotechnology; Tsinghua University; 100084 Beijing; China

15.2 Organelle distribution

Rosen and coworkers (1964) and Sassen (1964) were the first to show the accumulation of secretory vesicles (SV) in the tip area where the site of growth is. In tobacco (Derksen et al. 1995b; Fig.1) and probably in most species (Derksen et al. 1995a), a single class of SV is found at the tip. In addition, smooth endoplasmic reticulum (SER) accumulates at the tip of lily (Lancelle and Hepler 1992), tobacco (Derksen et al. 1995b) and pine pollen tubes (de Win et al. 1996), as seen in electron microscopic preparations after rapid cryo-fixation, but also in earlier reports using chemical fixatives (Steer and Steer 1989). The accumulations probably correspond to the clear cap or hyaline zone observed in living cells in the light microscope (Iwanami 1959). As shown by Rutten and Knuiman (1993), Brefeldin A, a highly specific inhibitor of the anterograde protein transport (ER→SV) successively causes the disappearance of dictyosome lamellar stacks, the trans-Golgi network and the SV (Fig. 2), and finally stops growth. In pine pollen tubes that grow slowly a typical massive SV accumulation at the tip is absent (de Win et al. 1996). Thus, the supply of SV to the wall is one of the ruling factors determining the speed of growth.

The distribution of a number of other organelles may largely differ between various species, most pronounced differences are found between Angiosperms (e.g. tobacco, Derksen et al. 1995b; Fig 1) and pine pollen tubes (de Win et al. 1996; Fig. 3). These observations have led to the conclusion that the -uneven-distribution of these organelles does not relate to tip growth but may relate to growth rates (Derksen et al. 1995a). On the other hand, differences in cellular organization of pollen tubes may reflect species-specific adaptations to their natural environments, i.e. the ovaries and cones.

Fig. 1. Median longitudinal section through the tip of a cryo-fixed and freeze-substituted tobacco pollen tube cultured *in vitro* in a sucrose borate medium (Derksen et al. 1995b). Secretory vesicles (SV) are accumulated in the very tip. m: mitochondria, d: dictyosomes, r: rough endoplasmic reticulum, I: electron dense primary wall secreted in the tip, II: electron translucent outer part of the primary wall behind the tip. **Fig. 2.** Median longitudinal section through the tip of a cryo-fixed and freeze-substituted tobacco pollen tube treated for 1 h with 1μg/ml Brefeldin A after two hours of culture under standard conditions in sucrose-borate medium. The tip is depleted of secretory vesicles and almost completely filled with smooth endoplasmic reticulum. **Fig. 3.** Longitudinal section through a cryo-fixed and freeze-substituted pine pollen tube cultured *in vitro* for two days in a sucrose borate medium (de Win et al. 1996). In the tips mitochondria (arrow heads) are found near the cell membrane. V: vacuoles in the tip, a: amyloplasts. The very tip is not seen due to the curvature of the tubes. **Fig. 4.** Dry-cleaved preparation (Derksen et al. 1984) of a tobacco pollen tube. After dry cleaving, the inner side of the membrane with the cortical cytoskeleton and attached organelles can be seen in the electron microscope. Microtubules (arrow heads) are often accompanied by microfilaments, er: endoplasmic reticulum. Arrows: coated pits (reticulate) and coated vesicles (circular and dark)

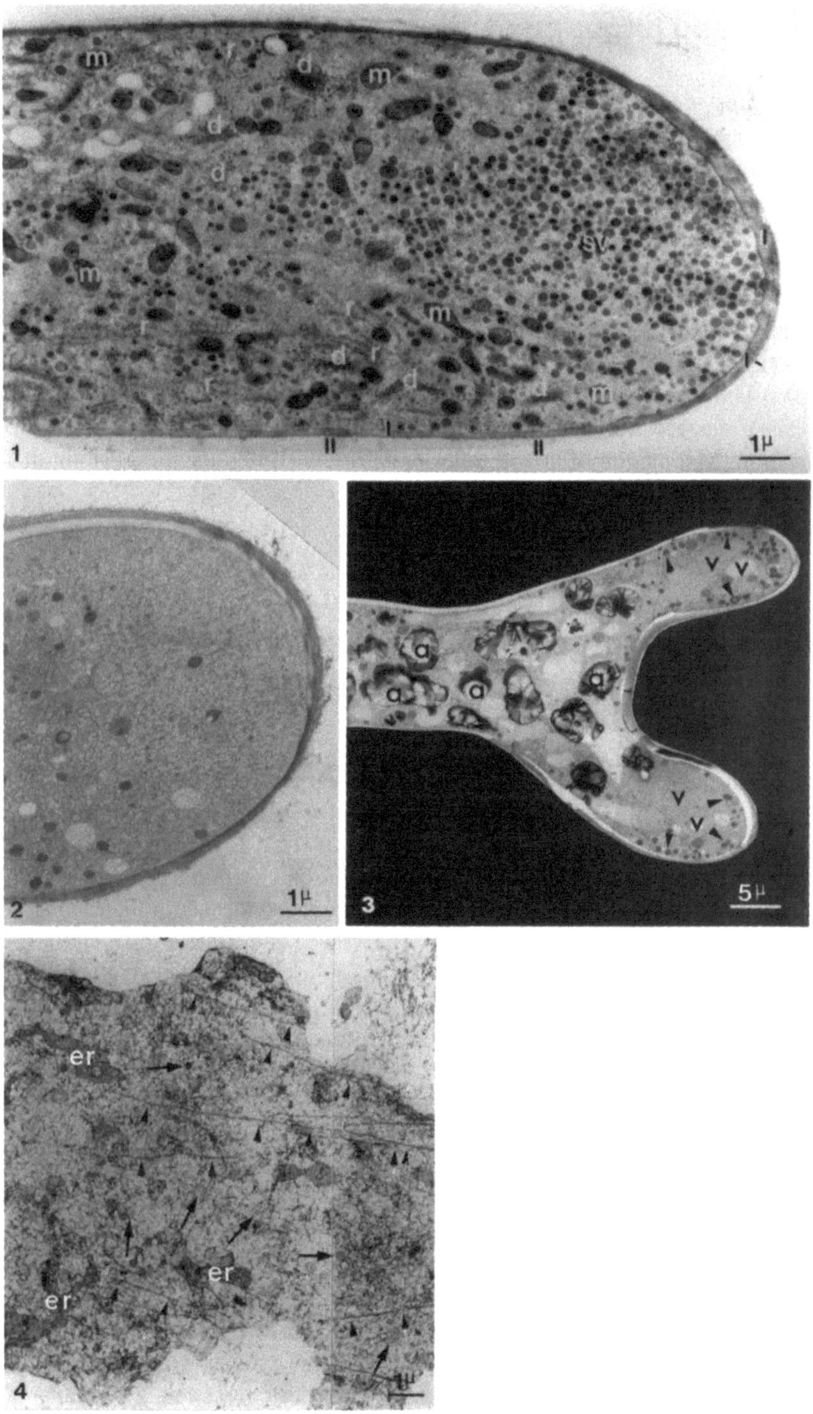
1
m
d
r
sv
1μ
2
1μ
3
a
5μ
4
er
1μ

15.3
Growth and turgor pressure

Turgor pressure is the driving force in cell expansion. However, from the observations of Brefeldin A (Rutten and Knuiman 1993) and Cytochalasin D (Picton and Steer 1981) inhibited pollen tubes, it has become clear that normal turgor pressure cannot force the tube to expand without insertion of new wall material. The role of turgor pressure was recently assessed by Benkert and coworkers (1997) who showed that pollen tubes accurately regulate their turgor pressure between 0.1-0.4 Mpa. Artificial increase in turgor pressure does not increase growth rate, but causes the tubes to burst near double pressure values. In addition, the osmolality of the medium did not correlate with growth rates or tube lengths. Thus, it may be concluded that pollen tubes regulate their turgor pressure to values needed for tube expansion, though turgor does not control tube elongation. Modulation of growth rates therefore must rely upon regulation of SV fusion rates and perhaps wall extensibility.

15.4
Cytoskeleton and cytoplasmic streaming

The orientation of microtubules (MT) and actin filaments (AF) is mainly axial in pollen tubes. MT are mainly cortical (Fig. 4) and often are accompanied by AF with which they appear physically connected (Derksen et al. 1985; Pierson et al. 1986; Lancelle et al. 1987; Raudaskoski et al. 1987). AF are present throughout the cytoplasm (Parthasarathy et al. 1985; Pierson et al. 1986; Pierson 1988). MT have been shown to regulate the position of the nucleus and generative cell in tobacco pollen tubes (Åström et al. 1995). In younger tobacco pollen tubes, MT degradation has little effect on growth (e.g. Derksen and Traas 1984), but in older vacuolized tubes the cytoplasmic organization is lost, the number of SV in the tip decreases and finally growth stops (Joos et al. 1994). In tobacco pollen tubes Cytochalasin treatments, known to affect AF function, cause MT to rearrange in swirls and to loose their axial orientation (Derksen and Traas 1984). In sub-protoplasts of tobacco pollen tubes where the connections between MT and AF are disrupt, AF still occur in polar arrays, but MT show typical swirls or concentric arrays and the cytoplasm is irregularly distributed (Rutten and Derksen 1990; 1992). Together, these observations have led to the idea that MT function as an anchor for AF to maintain cytoplasmic organization, especially in older, vacuolized pollen tubes where the cytoplasm is less dense (Derksen et al. 1995a). In the tip of growing tobacco pollen tubes (Cai et al. 1993; Del Casino et al. 1993; Derksen et al. 1995a) MT traverse the cytoplasm, but it remains uncertain whether they have a direct function in pollen tube growth. Together with myosins on the surface of the organelles, AF generate cytoplasmic streaming (Shimmen and Yakota 1994; Tang et al. 1989; Miller et al. 1995; 1996). Detailed analyses of motion patterns (Pierson et al. 1990; de Win et al. 1997; 1998; de Win et al. 1999) provide information on the underlying AF. The movements of the organelles are highly individual; bulk flow and steric hindrance occur only marginally. Remarkably, the SV at the tip show Brownian motion only (Pierson et al. 1990), probably because of the absence of AF in that area (Miller et al. 1996). Motion is much less vigorous in the slowly growing pine pollen tubes (de Win et al. 1996). Inhibition of acto-myosin driven motion by cytochalasins results in retention of SV around the dictyosomes and as a consequence inhibition of growth, even though the dictyosomes remain normally dispersed in the tube (Picton and Steer 1981). These observations show that AF and cytoplasmic streaming are essential for the supply of SV to the growing tip. A number of MT and AF-associated proteins, such as motor, bundling and regulatory proteins have been detected in pollen tubes (Derksen et al. 1995a; Li et al. 1997), the presence of cytoskeletal membrane-bound proteins, i.e. spectrins and integrins, is solely deduced from immunological cross-reactions with animal homologues. Their possible relevance for pollen tube growth has been recently evaluated (Li et al. 1997).

15.5 Exocytosis, endocytosis and calcium

With a sufficient supply of SV to ensure growth, the site of fusion becomes of crucial importance to the future shape of the cell. Obviously, the site of fusion must be limited to a small area in the tip of the growing tubes. However, only little direct information on both docking and fusion of SV in plant cells is available. A number of proteins are known to be involved in docking and fusion of SV in animal cells, but in plants only the presence of annexins, which promote vesicle fusion in a Ca^{2+}-dependent way, has been firmly established (Battey and Blackbourn 1994; Battey et al. 1996). Accordingly, growth and SV fusion in pollen tubes also depend on relatively high Ca^{2+} concentrations (Picton and Steer 1983; Steer 1988b; Steer and Steer 1989). Measurements using the Ca^{2+} sensitive vibrating probe, fluorescent Ca^{2+} dyes and inhibitors of the Ca^{2+} channels show high, micromolar, tip-focused concentrations caused by an influx in the tip and uptake in a sink behind the tip (Hepler et al. 1994; Feijó et al. 1995; Derksen et al. 1995a; Holdaway-Clarke et al. 1997; Li et al. 1997). In addition, a correlation between Ca^{2+} flux and growth has been observed (Pierson et al. 1994, 1996, Malhó et al. 1995, Holdeway-Clarke et al. 1997). The sink for Ca^{2+} is assumed to be the SER (Evans et al. 1991; Holdaway-Clarke et al. 1997), which is always present in large quantities in and behind the tip. Like the accumulations of SV, the abundance of SER at the tip (see above) is believed to result from the absence of actin filaments in that area, which in turn is thought to be caused by the high concentration of Ca^{2+} (Pierson et al. 1994; Miller et al. 1996). A crucial problem, however, remains. How are the secretory events initiated and maintained exclusively at the tip ? Several possibilities exist. First, the membrane composition in the tip may differ considerably from that in the tube, either by a membrane-bound skeleton similar to that occurring in animal cells (Derksen et al. 1995a; Li et al. 1997), or by selective retrieval of membrane material by endocytosis via coated pits (Steer 1988a; Steer and Steer 1989) in the area behind the growing tip

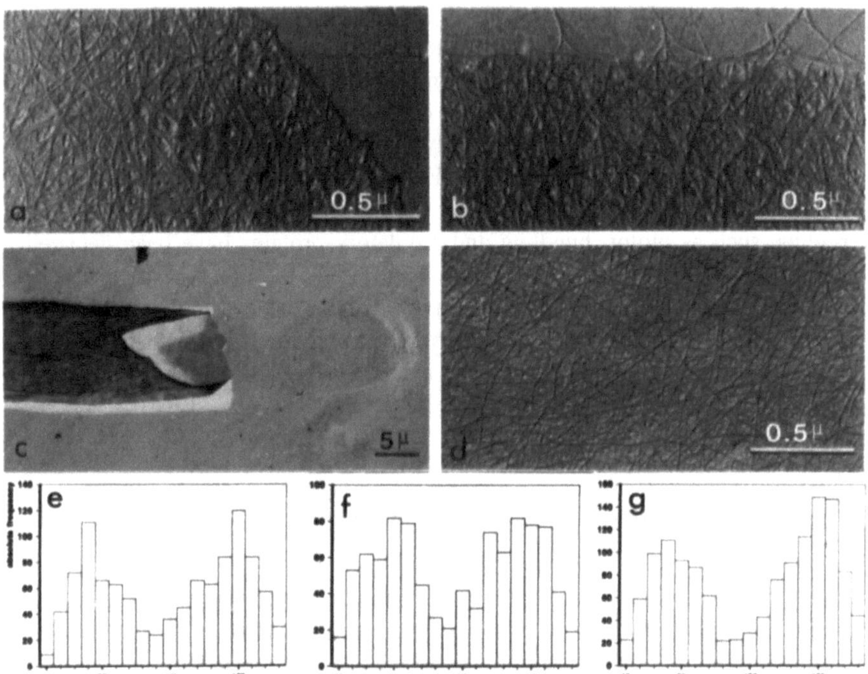

Fig. 5. Texture of tobacco pollen tubes after extraction with H_2O_2/acetic acid (1:1) overnight (a and b) and alkaline extraction (8 hrs) with pure ethanolamine (c and d). After extensive washing the specimen were mounted on poly-L-lysine coated formvar-carbon coated copper grids, dehydrated, critical-point dried and cleaved with adhesive tape (van Amstel 1995). Cleaving provides inside wall fragments as well as fractures through the wall. Preparation were shadow-casted with platinum at an angle of 45^0. The orientation of the microfibrils was analysed from micrographs as shown in a,b and d). The orientation of the cellulose microfibrils (CMF) was measured using a Kontron videoplan with a standard programme. The distribution of the CMF orientions with respect to the long axis of the tube is shown for the tip, 0-2.5 μm (c), behind the tip, 2.5-10 μm (d) and in the tube at 10-15 μm (e). In all cases the CMF are oriented in a Z-helix with the highest abundance around 45^0. Alkaline and H_2O_2/acetic acid (1:1) extractions show similar results

(Derksen et al. 1995a;b; Blackbourn and Jackson 1996; Fig. 4). Second, the activity of the -putative- Ca^{2+}channels may be modulated in various ways, such as stretch activation (Pierson et al. 1994; 1996), voltage dependency (Malhó et al. 1995; 1996), and the activity of Ca^{2+} binding proteins such as calmodulin and membrane associated ATPases (Tirlapur et al. 1994; Obermeyer and Weisenseel

1991). Notably, turgor, which is probably involved in stretch activation, is effected by a Ca^{2+} dependent K^{+} channel (Obermeyer and Kolb 1993). All these mechanisms are not mutually exclusive and allow the construction of feedback loops that may control tip growth.

15.6 Cell wall

Generally, the very tip is considered to be the weakest part of the pollen tube as it is the preferential site of disruption. To maintain the cylindrical morphology of the pollen tube, expansion must follow a distinct pattern governed by tip to base changes in the elastic/plastic properties of the wall while the force is determined by the, non-vectorial, turgor pressure in the tube (Derksen 1996). The high expansion rate and the weak wall structure in the very tip provide the environment for stretch-activation of Ca^{2+}channels.

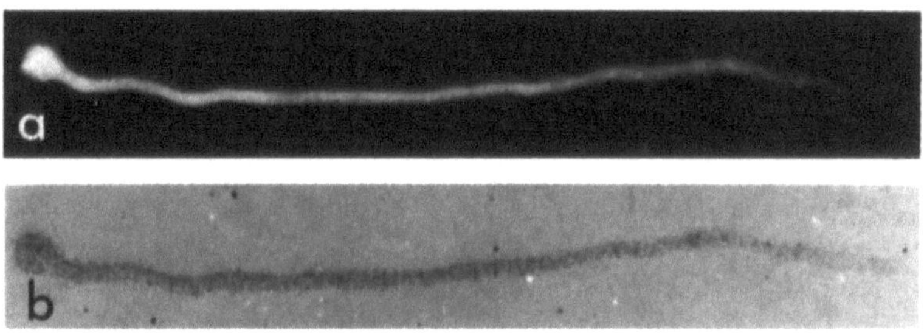

Fig. 6. Calcofluor white staining (a) and ^{3}H-UDP-glucose incorporation (b) in the wall of the same pollen tube. Extracted pollen tubes were stained with calcofluor white as described by van Amstel (1995). The ^{3}H-UDP-glucose (1μCi/10 mg pollen) was added 1.5 h after germination. Samples were taken at various intervals. To determine the incorporation in the CMF, samples were extracted overnight with H_2O_2/acetic acid (1:1) and after extensive washing, mounted on Vectabond-treated object slides. Preparations were covered with Kodak stripping film and exposed for at least 1 week (van Amstel 1995). Both autoradiograph and Calcofluor white staining show a clear label that increases from tip to grain. In the tip, label is almost absent

Cellulose microfibrils (CMF), the main stress resistant elements in plant cell walls, seem unlikely to create a tip to base gradient in stress resistance. CMF have been claimed to be absent from the tip of petunia pollen tubes (Mühlethaler and Linskens 1956). In both petunia and lily, Sassen (1964) and Dashek (1966) reported at least few CMF with orientations around 45^{0} with respect to the tube axis, which is the orientation providing the least resistance. Similar observations have been made for tobacco (Fig. 5) and pine (Derksen et al. submitted). As shown in Fig. 6, the density of H^3 labeled UDP-glucose in CMF-ghosts is extremely low in the tip. Using cellobiohydrolase in combination with gold-labeled antibodies, Ferguson and coworkers (1998) showed that little or no cellulose is present in the primary wall of tobacco. The pure cellulosic (β1-4) nature of the CMF-ghosts, however, is disputed (Herth et al. 1974; Steer and Steer 1989). From a rheological point of view the low abundance the observed fibrils is more significant than their composition.

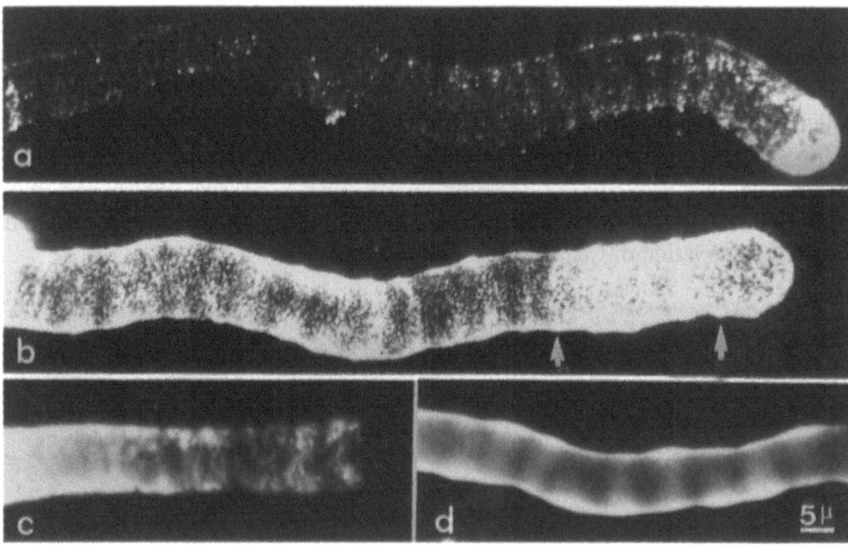

Fig. 7. Pectin and callose staining of the wall. Immunolabeling of esterified pectin was carried out with JIM 7 (a), JIM 5 was used to identify acidic pectin (b). In both cases FITC-labelled sheep anti-rat was used as a second antibody (Li et al. 1994). Living tobacco pollen tubes were stained with decolorized Aniline Blue (c and d) as described by van Amstel (1995). Esterified pectin is almost exclusively present in the tip, though remnants in a banded pattern are visible in the tube (a). Acidic pectin is found in the tip and in the tube (b). Behind the tip, at first the staining becomes more intense (in between arrows), but further from the tip the staining is less intense and has a banded look (b). With aniline blue, no label can be detected in the tip (arrow head) (c). The label with Aniline Blue in the tube at first shows spot- to linear shaped deposits that are arranged in bands (c). In older parts of the tube the staining becomes intense and smooth (c), though sometimes the banded aspect is preserved (d)

Using monoclonal antibodies, esterified and acidic pectins have been shown in the primary wall of pollen tubes (Fig. 7; Li et al. 1992; 1994). Probably a shift in composition occurs from mainly esterified pectins in the tip to de-esterified, acidic pectins in the older outer wall parts in the tube (Li et al. 1994). Such a shift is thought to be caused by the activity of methyl/acetyl-esterases secreted with the

SV in the tip (Li et al. 1994). In their egg box, Ca^{2+} linked, configuration, acidic pectins are very rigid and stress resistant (van Cutsem and Messiaen 1994). Their gradual appearance can provide the necessary increase in tip to base wall rigidity. Acidic pectins in the tube occur in banded patterns that are supposed to result from a pulsating growth, with alternating periods of wall thickening and rapid tip expansion (Plyushch et al. 1995; Pierson et al. 1995; Li et al. 1996).

A gradual tip to base change in wall ultrastructure is present in tobacco (Derksen et al. 1995b), petunia (Derksen et al. 1999) and pine (Derksen et al. submitted). In angiosperm pollen tubes the wall deposited in the tip is re-enforced by a secondary wall of a mainly callosic nature that is exclusively present in the tube (Kroh and Knuiman 1982; Steer and Steer 1989; Schlüpmann 1994; Ferguson et al. 1998). As no callose lining or secondary wall occurs in pine pollen tubes (Derksen et al. submitted), a function of this callose lining in confining secretory events to the tip seems improbable. The formation of a secondary wall may relate to the relative weakness of the primary wall that may easily corrode (Derksen et al. 1995a) and that at least in tobacco is nearly absent in pollen tubes grown in the style. The function of the callosic wall is thought to isolate the pollen tube from the genetically different stylar environment (Gorsky-Brylass 1986). This assumption is supported by the relation between wall structure and stylar environment (Li et al. 1995), and the absence of callosic linings or secondary walls in pine pollen tubes that grow in the open spaces of the female cones (Derksen et al. submitted). However, the permeability of callose depends on the chemical environment (Stone and Clarke 1992) and for example uptake in the cytoplasm of the S-alleles products in *Nicotiana alata* may occur via the callosic layer (Gray et al. 1991). On the other hand, the route of uptake has not yet been identified and active uptake may well occur in the coated pit-rich zone behind the tip (Derksen et al. 1995b). The thickness of the callosic layer gradually increases towards the base of the pollen tube, as shown for petunia (Herrero and Dickinson 1981), tobacco and lily (van Amstel 1994). Callose first occurs in small spot- or linearly shaped deposits (fig. 7c), but in the older parts of the tube callose has a more smooth look, although often banded patterns are observed in both lily and tobacco (van Amstel 1994; Fig. 7d). The staining patterns with Sirofluor may reveal local differences in callose content as well as differences in callose structure (for complexity of callose structure: Evans et al. 1984; Stone and Clarke 1992). The banded patterns of callose do not co-align with the banded patterns seen after pectin staining (Li et al. 1992; 1994; 1995; Fig 7).

15.7 Conclusions

Tip growth in pollen tubes requires the presence of sufficient turgor pressure and production and transportation of sufficient SV to the tip, by which the Ca^{2+}dependent fusion rate determines the growth rate. The question to the confinement of the fusion site to the tip, however is paramount in tip growth.

Various mechanisms exist that are not mutually exclusive, but present observations show that the structural and mechanical properties of the wall after its secretion provide a fitting environment for stretch-activation of the Ca^{2+} channels in the tip. In addition, positive feed back loops, selective endocytosis behind the tip and a membrane-bound cytoskeleton may reïnforce or stabilize the system.

References

Åström H, Sorri O, Raudakoski M (1995) Role of microtubules in the movement of the vegetative nucleus and generative cell in tobacco pollen tubes. Sex Plant Reprod 8: 61-69

Battey N H, Blackbourn HD (1994) The control of exocytosis in plant cells. New Phytol 125: 307-338

Battey NH, James NC, Greenland AJ (1996) C-DNA isolation and gene expression of the maize annexins p33 and p35. Plant Physiol 112: 1391-1396

Benkert R, Obermeyer G, Bentrup FW (1997) The turgor pressure of growing lily pollen tubes. Protoplasma 198: 1-8

Blackbourn HD, Jackson AP (1996) Plant clathrin heavy chain: sequence analysis and restricted localisation in growing pollen tubes. J Cell Sci 109: 777-786

Cai G, Bartalesi A, Del Casino C, Moscatelli A, Tiezzi A, Cresti, M (1993) The kinesin-immunoreactive homologue from *Nicotiana tabacum* pollen tube: biochemical properties and sub-cellular localization. Planta 191: 496-506

Dashek WV (1966) The lily pollen tube: aspects of fine chemistry and nurtrition in relation to fine structure. Ph D Thesis. Marquette University. Milwaukee, Wisconsin. USA

Del Casino C, Li YQ, Moscatelli A, Scali M, Tiezzi A, Cresti M (1993) Distribution of microtubules during the growth of tobacco pollen tubes. Biol Cell 79: 125-132

Derksen J Emons AMC (1990) Microtubules in tip growth systems. In Heath IB (ed) Tip growth in plant and fungal cells. Academic Press Inc. San Diego, California, pp 147-181

Derksen J, Pierson ES, Traas JA (1985) Microtubules in vegetative and generative cells of pollen tubes. Eur J Cell Biol 38: 142-148

Derksen J, Traas JA (1984) Growh of tobacco pollen tubes *in vitro*; effects of drugs interfering with the cytoskeleton. In (Willemse MTM, van Went J.L. (eds). Sexual reproduction in seed plants, ferns and mosses. Proc. 8th Int Symp on Sex Reprod in seed plants, ferns and mosses. Pudoc, Wageningen, pp 64-67

Derksen J, Rutten T, van Amstel T, de Win A, Doris F, Steer M (1995a) Regulation of pollen tube growth. Acta Bot Neerl 44: 93-119

Derksen J, Rutten T, Lichtscheidl IK, de Win AHN, Pierson ES, Rongen G (1995b) Quantitative analysis of the distribution of organelles in tobacco pollen tubes: implications for exocytosis and endocytosis. Protoplasma 188: 267-276

Derksen J (1996) Pollen tubes: a model system for plant cell growth. Bot Acta 109: 341-345

Derksen J, van Wezel R, Knuiman B, Ylstra B, van Tunen AJ (1999) Pollen tubes of flavonol deficient *Petunia* show striking alterations in wall structure leading to disruption. Planta, in press

De Win AHN, Knuiman B, Pierson ES, Geurts H, Kengen HMP, Derksen J (1996) Development and cellular organization of *Pinus sylvestris* pollen tubes. Sex Plant Reprod 9: 93-101

De Win AHN, Worring M, Derksen J, Pierson ES (1997) Classification of organelle trajectories using region-based curve analysis. Cytometry 29: 136-146

De Win AHN, Pierson ES, Timmer C, Lichtscheidl IK, Derksen J (1998) Interactive computer-assisted position acquisition procedure designed for the analysis of organelle movement in pollen tubes. Cytometry 32: 263-267

De Win AHN, Pierson ES, Derksen J (1999) Rational analyses of organelle trajectories in tobacco pollen tubes reveal characteristics of the actomyosin cytoskeleton. Biophys J, in press

Evans DE, Briars SA, Williams LE (1991) Active calcium transport by plant cell membranes. J Exp Bot 42: 285-303

Evans NA, Hoyne PA, Stone BA (1984) Characteristics and specificity of the interaction of a fluorochrome from aniline blue (Sirofluor) with polysaccharides. Carbohydr Polym 4: 215-230

Feijó JA, Malhó R, Obermeyer G (1995) Ion dynamics and its possible role during *in vitro* pollen germination and tube growth. Protoplasma 187: 155-167

Ferguson C, Teeri TT, Siika-aho M, Read SM, Bacic A (1998). Location of cellulose and callose in pollen tubes and grains of *Nicotiana tabacum*. Planta 206: 452-460

Gorsky-Brylass A (1986) The "callose"stage of the generative cells in pollen grains. Grana 10: 21-30

Gray JE, McClure BA, Bonig I, Anderson, MA, Clarke AE (1991) Action of the style product of the self-incompatibility gene of *Nicotiana alata* (S-RNase) on in vitro-grown pollen tubes. Plant Cell 3: 271-283

Hepler PK, Miller DD, Pierson ES, Callaham DA (1994) Calcium and pollen tube growth. In (Stephenson AG, Kao D-H (eds) Pollen-pistil interactions. Am Soc Plant Physiol, pp 111-123

Herrero M, Dickinson HG (1981) Pollen tube development in *Petunia hybrida* following compatible and incompatible intraspecific matings. J Cell Sci 47: 365-383

Herth W, Franke WW, Bittiger H, Kuppel A, Keilich G (1974). Alkali-resistent fibrils of β-1-3 and β-1-4 glucans: structural polysaccharides in the pollen tube wall of *Lilium longiflorum*. Cytobiologie 9: 344-367

Holdaway-Clarke TL, Feijó JA, Hackett GR, Kunkel JG, Hepler PK (1997) Pollen tube growth and the intracellular cytosolic calcium gradient oscillate in phase while extracellular calcium is delayed. Plant Cell 9: 1999-2110

Iwanami, Y. (1959) Physiological studies on pollen. Yokohama Municipal University 116 (C-34, Biol. 13), 1-137

Johri BM (1992) Haustorial role of pollen tubes. Ann Bot 70: 471-475

Joos U, van Aken J, Kristen U (1994) Microtubules are involved in maintaining the cellular polarity in pollen tubes of *Nicotiana sylvestris*. Protoplasma 179: 5-15

Kroh M, Knuiman B (1982) Ultrastructure of cell wall and plugs of tobacco pollen tubes after chemical extraction of polysaccharides. Planta 154: 241-250

Lancelle SA, Cresti M, Hepler PK (1987) Ultrastructure of the cytoskeleton in freeze-substituted pollen tubes of *Nicotiana*. Protoplasma 140: 141-150

Lancelle SA, Hepler PK (1992) Ultrastructure of freeze-substituted pollen tubes of *Lilium longiflorum*. Protoplasma 167: 215-230

Lazzaro MD (1996) The actin microfilament network within eleongating pollen tubes of the gymnosperm *Picea abies* (Norway spruce). Protoplasma 194: 186-194

Li YQ, Brunn L, Pierson ES, Cresti M (1992) Periodic deposition of arabinogalactan epitopes in the cell wall of pollen tubes of *Nicotiana tabacum*. Planta 188: 532-538

Li YQ, Chen F, Linskens HF, Cresti M (1994) Distribution of unesterified and esterified pectins in cell walls of pollen tubes. Sex Plant Reprod 7: 145-152

Li YQ, Chen F, Faleri C, Ciampolini F, Linskens HF, Cresti M (1995) Presumed phylogenetic basis of the correlation of pectin deposition pattern in pollen tube walls and the stylar structure of angiosperms. Proc Kon Ned Acad Wet Amsterdam 98: 39-44

Li YQ, Zhang HQ, Pierson ES, Huang FY, Linskens HF, Hepler PK, Cresti M (1996) Enforced growth-rate fluctuation causes pectin ring formation in the cell wall of *Lilium longiflorum* pollen tubes. Planta 200: 41-49

Li YQ, Moscatelli A, Cai G, Cresti M (1997) Functional interactions among cytoskeleton, membranes and cell wall in the pollen tube of flowering plants. Int Rev Cytol 176: 133-199

Malhó R, Read ND, Trewavas AJ, Pais MS (1995) Calcium channel activity during pollen tube growth and reorientation. Plant Cell 7: 1173-1184

Malhó R, Trewavas AJ (1995) Localized apical increases of cytosolic free calcium control pollen tube orientation. Plant Cell 8: 1935-1949

Mühlethaler K, Linskens HF (1956) Elektronenmikroskopische Aufnahmen von Pollenschläuchen. Experientia 15: 253-255

Mascarenhas JP (1993) Molecular mechanisms of pollen tube growth and differentiation. Plant Cell 5: 1303-1314

Miller DD, Scordilis SP, Hepler PK (1995) Identification and localization of three classes of myosins in pollen tubes of *Lilium longiflorum* and *Nicotiana alata*. J Cell Sci 108: 2549-2563

Miller DD, Lancelle SA, Hepler PK (1996) Actin filaments do not form a dense meshwork in *Lilium longiflorum* pollen tube tips. Protoplasma 195: 123-132

Obermeyer G, Kolb HA (1993) K^+ channels in the plasma membrane of lily pollen protoplasts. Bot Acta 106: 26-31

Obermeyer G, Weisenseel M (1991) Calcium channel blocker and calmodulin antagonists affect the gradient of free calcium ions in lily pollen tubes. Eur J Cell Biol 56: 319-327

Obermeyer G, Lützelschwab M, Heumann HG, Weisenseel W (1992) Immunolocalization of H^+-ATPases in the plasma membrane of pollen grains and pollen tubes of *Lilium longiflorum*. Protoplasma 171: 55-63

Parthasarathy MV, Perdue TD, Witztum A, Alvernaz J (1985) Actin network as a normal component of the cytoskeleton in many vascular plant cells. Am J Bot 72: 1318-1323

Picton JM, Steer MW (1981) Determination of secretory vesicle production rate in by dictyosomes in pollen tubes of *Tradescantia* using Cytochalasin D. J Cell Sci. 49: 261-272

Picton JM, Steer MW (1983) Evidence for the role of Ca^{++}ions in the tip extension in pollen tubes. Protoplasma 115: 11-17

Pierson ES. (1988) Rhodamine-phalloidin staining of F-actin in pollen after dimethyl-sulphoxide permeabilization: a comparison with the conventional formaldehyde preparation. Sex Plant Reprod 1: 83-87

Pierson ES, Cresti (1992) Cytoskeleton and cytoplasmic organization of pollen and pollen tubes. Int Rev Cytol 140: 73-125

Pierson ES, Derksen J, Traas JA (1986) Organization of microfilaments and microtubules in pollen tubes grown *in vitro* or *in vivo* in various angiosperms. Eur J Cell Biol 41: 14-18

Pierson ES, Lichtscheidl IK, Derksen J (1990) Structure and behaviour of organelles in living pollen tubes of *Lilium longiflorum*. J Exp Bot 41: 1461-1468

Pierson ES, Miller DD, Callaham DA, Shipley AM, Rivers BA, Cresti M, Hepler PK (1994) Pollen tube growth is coupled to the extracellular calcium ion influx and the intracellular calcium gradient: effect of BAPTA-type buffer and hypertonic media. Plant Cell 6: 1815-1828

Pierson ES, Li YQ, Zhang HQ, Willemse MTM, Linskens HF, Cresti M (1995) Pulsatory growth of pollen tubes: investigation of a possible relationship with the periodic distribution of cell wall components. Acta Bot Neerl 44: 121-128

Pierson ES, Miller DD, Callaham DA, Van Aken J, Hackett G, Hepler PK (1996) Tip-localized calcium entry fluctuates during pollen tube growth. Dev Biol 174: 160-173

Plyushch TA, Willemse MTM, Franssen-Verheijen MAW, Reinders MC (1995) Structural aspects of pollen tube growth and micropylar penetration in vitro in *Gasteria verrucosa* (Mill) H. Duval and *Lilium longiflorum* Thunb. Protoplasma 187: 13-21

Raudaskoski M, Åström H, Perttila K, Virtanen I, Louhelainen J (1987) Role of the microtubule skeleton in pollen tubes: an immunocytochemical and ultrastructural approach. Biol Cell 61: 177-188

Rosen, W.G. (1964) Chemotropism and fine structure of pollen tubes. In Linskens HF (ed) Pollen Physiology and Fertilization. North-Holland Publish Company, Amsterdam, pp 159-166

Rutten TLM, Derksen J (1990) Organization of actin filaments in regenerating and outgrowing subprotoplasts from pollen tubes of *Nicotiana tabacum*. Planta 180: 471-479

Rutten TLM, Derksen J (1992) Microtubules in pollen tube subprotoplasts: organization during protoplast formation and protoplast outgrowth. Protoplasma 167: 231-237

Rutten TLM, Knuiman B (1993) Brefeldin A effects on tobacco pollen tubes. Eur J Cell Biol 61: 247-255

Sassen MMA (1964) Fine structure of Petunia pollen grain and pollen tube. Acta Bot Neerl 13: 175-181

Schlüpmann H, Bacic A, Read SM (1994) Uridine diphosphate glucose metabolism and callose synthesis in cultured pollen tubes of *Nicotiana alata* Link et Otto. Plant Physiol 105: 659-670

Shimmen T, Yakota E (1994) Physiological and biochemical aspects of cytoplasmic streaming. Int Rev Cytol 155: 97-139

Steer MW (1988a) Plasma membrane turnover in plant cells. J Exp Bot 39: 987-996

Steer MW (1988b) The role of calcium in endocytosis and exocytosis in plant cells. Physiol Plant 72: 213-220

Steer MW, Steer JM (1989) Pollen tube tip growth. New Phytol 111: 323-358

Taylor LP, Hepler PK (1997) Pollen germination and pollen tube growth. Ann Rev Plant Physiol Plant Mol Biol 48: 461-491

Tang X, Hepler PK, Scordilis S.P (1989) Immunochemical and immunocytochemical identification of a myosin heavy chain polypeptide in *Nicotiana* pollen tubes. J Cell Sci 92: 569-574

Tirlapur UK, Scali M, Moscatelli A, Del Casino C, Cai G, Tiezzi A, Cresti M (1994) Confocal image analysis of spatial variation in immunocytochemically identified

calmodulin during pollen hydration, germination and pollen tube tip growth in *Nicotiana tabacum* L. Zygote 2: 63-68
Van Amstel (1995) Construction of Plant Cell Walls. Ph D Thesis. Catholic University Nijmegen. The Netherlands
Van Cutsem P, Messiaen J (1994) Biological effets of pectic fragments in plant cells. Acta Bot Neerl 43: 231-245

16 In vitro selection for heat-tolerance in Lilies

Chi[1] HS, Straathof[2] TP, Löffler[2] HJM and Van Tuyl[2]* JM

16.1 Introduction

The lily (*Lilium* L.) is one of the most important crops in the flower bulb industry. Lilies are used worldwide as cut flowers, as pot plants or garden plants. With an area of about 3900 hectares (CBS 1996), The Netherlands is the world leading country in the production of lily bulbs, followed by the United States and Japan. In The Netherlands approximately one billion bulbs are produced annually and 65% of the bulbs is exported (CBS 1996), mainly for cut flower production. The remaining 35% is used for year round flower production in The Netherlands. Most of the flowers produced in The Netherlands are exported, with an economic value of approximately 300 million guilders (CBS 1996).

The genus *Lilium* comprises about 85 species, classified into seven sections (De Jong 1974). More than half of the species originate from Asia. The ancestors of the modern lilies such as the Asiatic and Oriental hybrid lilies originate mostly form China and Japan (Beattie and White 1993). However, since most of the lilies are bred and cultivated in The Netherlands, the modern commercial cultivars are only suitable for cultivation in a moderate climate. Only few commercial cultivars are adapted to tropical or subtropical areas. Therefore, the subtropical lily industry is hampered by poor growth of the bulbs and a low quality of the cut flowers. Widening of the genetic basis to introduce heat-tolerance from wild species into the commerical lily assortment is therefore very important in subtropical lily breeding programmes. Three interesting species originate from tropical or subtropical countries, i.e. *L. neilgherense* from south of India, *L. formosanum* from Taiwan and *L. longiflorum* from south of Japan and Taiwan (Woodcock and Stearn 1950; Shii 1983). The culture characteristics of these species are vigour, earliness, heat-tolerance and suitability for year-round forcing (Chin et al. 1996).

To breed efficiently for heat-tolerant cultivars, male gametophytic selection

[1] National Chiayi Institute of Technology; Department of Horticulture; Taiwan

[2]* for correspondence: DLO-Centre for Plant Breeding and Reproduction Research (CPRO-DLO); PO Box 16; 6700 AA Wageningen; The Netherlands

(MGS) could be an important tool. The MGS is based on the assumption that the selective trait is controlled by genes expressed in both the gametophytic and the sporophytic phase (Hormaza and Herrero 1992; 1996). A gametophytic-sporophytic genetic overlap (GSGO) of 60-80% has been demonstrated in many plant species (Ottaviano and Mulcahy 1989). Because of haploidy and the very large size of pollen populations, MGS is expected to be extremely efficient. Three steps are necessary to develop an in vitro pollen selection system for heat-tolerance. First, genetic variation in heat-tolerance must be detected in the species (or genus). Beside a plant test, a bioassay for determining the heat-tolerance of a genotype at pollen level could hereby be useful. Second, a selection system must be developed. Selection pressure can be applied to the pollen before pollination (Frova et al. 1995) or during the pollen tube growth in the style and during fertilization (Petolino et al. 1990). Third, selected and control seedlings must be tested for heat-tolerance and selection response must be found. If the selection system is carried out in vitro, an in vitro assay for heat-tolerance can be efficient.

In this study we investigated the possibilities of a bioassay for determination of heat-tolerance of pollen from different *Lilium* species. Furthermore, an in vitro pollination system (Van Tuyl et al. 1991) was used to select for heat-tolerant seedlings. The possibilities for an in vitro pollen selection system for heat-tolerance in lilies is discussed.

16.2 Materials and methods

16.2.1 Development of a bioassay for heat-tolerance of pollen

Bulbs of one accession of *L. formosanum*, one accession of *L. neilgherrense* and 14 accessions of *L. longiflorum* (Table 1) were planted in pots in a CPRO-DLO greenhouse in April and flowered in June and July. Fresh pollen was collected from plants and used immediately. Pollen was germinated on a modified BK-medium (Brewbaker and Kwack 1963). As a modification, the component H_3BO_3 was 3-times as high as in the BK-medium and supplemented with 10% of sucrose and solidified with 0.5% phytagel. To stimulate germination, a thin linear row of pollen was brought on the pollen germination media in the middle of a 9 cm petri-dish. Petri-dishes were placed in climate rooms at four different temperatures, i.e. 20, 25, 30 and 35°C. The experiment was carried out in 5 repeats.

After 24 hours, the length of pollen tubes was measured under the stereomicroscope using millimeter-paper under the petri-dish. The length of the pollen tubes was measured from the edge of the pollen row. Furthermore, the germination percentage was estimated in classes of 10 percent. All data were subjected to analysis of variance.

16.2.2
In vitro pollination at different temperatures

Bulbs of the *L. longiflorum* cultivars White American, Indian Summer and Gelria were planted in pots in a CPRO-DLO greenhouse in June and flowered in September. One day before anthesis, the flower buds of the *L. longiflorum* cultivars White American and Gelria were collected from the greenhouse and transferred to the laboratory. The flower buds were placed on water in a laminar flow cabinet at a temperature of 22-25°C till bloom. The aseptic pollen was collected and used for in vitro pollination. The viability of aseptic pollen was determined by an in vitro germination assay as described in paragraph 16.2.1.

Table 1. *Lilium* accessions used in the heat-tolerance experiments

Code	CPRO-number	Origin / cultivar	Species	origin related to climate zone*
1	940272	native in Taiwan	*L. formosanum*	3
2	86105	native in India	*L. neilgherrense*	3
3	940317	Avita	*L. longiflorum*	2-3
4	771017	Indian Summer	*L. longiflorum*	1
5	960100	native in Japan and Taiwan	*L. longiflorum*	3
6	970002	Snow Queen	*L. longiflorum*	2
7	970001	Gelria	*L. longiflorum*	1
8	970008	White American	*L. longiflorum*	1
9	83139	Hinemoto	*L. longiflorum*	2-3
10	83293	White Diamond	*L. longiflorum*	2-3
11	78903	Nellie White	*L. longiflorum*	2
12	88257	Romero	*L. longiflorum*	1
13	81284.2	Blancivetta	*L. longiflorum*	2
14	85234	Osnat	*L. longiflorum*	2
15	960155	White Fox	*L. longiflorum*	3
16	960158	Lorina	*L. longiflorum*	1-2

*1 = grown in cold climates; 2 = grown in moderate climates; 3 = grown in warm climates (Van Tuyl, personal communication).

Flower buds with pedicle of the *L. longiflorum* cultivar Indian Summer were cut from the plants in the greenhouse two to four days before anthesis. The complete flowers were sterilized in 70% ethanol for 3 min, and commercial bleach containing 2% NaOCl for 30 min, and subsequently rinsed three times, i.e. 5, 10 and 20 min, in sterile distilled water. After rinsing, petals and anthers were dissected with a scalpel and the pistil, style, ovary and pedicle were placed vertically in a 20 cm test tube partly filled with medium. Tubes were closed with cotton plugs and covered by a double layer of aluminium foil. Tubes were placed in a climate room with a light intensity of 12 $W.m^{-2}$, a photoperiod of 16 hrs and a temperature of 25°C. A MS-medium (Murashige and Skoog 1962) supplemented with 8% of sucrose, solidified with 0.4% phytagel and a pH of 6.0 was used to culture the flower bud.

When exudate production indicated stigmatic receptivity, the pistils were pollinated by aseptic pollen in the flow cabinet. After in vitro pollination, the test tubes were placed in climate rooms with three different temperatures, i.e. 25, 30 and 35°C. Five days after pollination, all test tubes were placed at 25°C. Pollen tube growth in the style was observed using aniline blue fluorescence (Kho and Baer 1968). Ovule culture was applied to the seed pods which were ripe between 70-90 days after pollination. The ovules were incubated on a MS-medium supplemented with 5% sucrose and 0.1 mg/l NAA, solified with 0.4% phytagel and a pH of 5.8. The ovules were incubated in the dark. Emerged plantlets were transferred to a medium containing half strength of MS salts, 5% sucrose and 0.4% phytagel at a pH of 5.8 (Van Tuyl et al. 1991).

16.3 Results

16.3.1 Development of a bioassay for heat-tolerance of pollen

The analysis of variance of the pollen tube length measurements (Table 2) and the pollen germination measurements (Table 3) show a highly significant temperature and genotype effect. Also a smaller but significant interaction between cultivar and temperature is present, which means that not all the accessions react identically at different temperatures.

Table 2. Analysis of variance of the pollen tube length

source	df	ss	ms	F	F pr
temperature	3	314.0	104.7	326.8	<0.001
genotype	15	527.5	35.2	109.8	<0.001
temp * geno	45	255.3	5.7	17.7	<0.001
residual	256	82.0	0.3		
total	319	1178.8			

Table 3. Analysis of variance of the pollen germination experiment

source	df	ss	ms	F	F pr
temperature	3	77574.7	5171.6	222.5	<0.001
genotype	15	38873.4	12957.8	557.5	<0.001
temp * geno	45	44481.6	988.5	42.5	<0.001
residual	256	5950.0	23.2		
total	319	166879.7			

Df, degree of freedom; ss, sum of square deviation; ms, mean square; F, F value

In Table 4, the optimum temperature for pollen tube growth, length of the pollen tubes at 35°C, optimum temperature for pollen germination and germination percentage at 35°C is presented for all 16 accessions. The optimum temperature for pollen tube growth is for most accessions at 30°C. Only the *L. longiflorum* cultivar White American has a much lower optimum of 20°C. It was observed that 80 to 90% of the pollen tubes of a certain accession at a certain temperature had the same length. A large variation in pollen tube growth at 35°C was found between the accessions ranging from 1.0 of *L. longiflorum* cultivar Blancivetta to 6.1 mm of *L. neilgherrense*. The wild accesion of *L. formosanum*, *L. neilgherrense* and *L. longiflorum* which originate from warm climate zones (Table 1) showed long pollen tubes at 35°C. At 40°C, the pollen tube length of all the accesions was almost zero (data not shown). The optimum temperature for pollen germination was between 25 and 30°C for most accesions. The *L. longiflorum* cultivar White American had again the lowest optimum (20°C). Also variation in pollen germination at 35°C was found between the accesions ranging from 10% of *L. longiflorum* Nellie White to 96% of *L. longiflorum* Lorina. No relation between pollen germination at high temperature and the origin related to climate zone (Table 1) could be detected.

Table 4. Results of the pollen tube growth and pollen germination experiment

code	pollen tube growth		pollen germination	
	optimum temperature	length (mm) at 35°C	optimum temperature	%of germination at 35°C
1	30 °C	4.0	25-30 °C	80
2	30 °C	6.1	25-30 °C	88
3	30 °C	5.0	25-30 °C	86
4	25 °C	3.0	25-30 °C	98
5	30 °C	4.4	25-30 °C	88
6	30 °C	2.3	30 °C	56
7	30 °C	3.0	20-25 °C	68
8	20 °C	1.1	20 °C	68
9	30 °C	2.0	25 °C	40
10	30 °C	2.0	30 °C	98
11	30 °C	1.5	20-30 °C	10
12	30 °C	2.0	30 °C	70
13	30 °C	1.0	30 °C	12
14	30 °C	1.5	30 °C	28
15	25 °C	2.0	30 °C	88
16	25 °C	2.0	30 °C	96
average		2.7		67.1

16.3.2 In vitro pollination at different temperatures

Pollen tube growth in the style was studied. The style of *L. longiflorum* Indian Summer is in general 10 cm long. Most of the pollen tubes of *L. longiflorum* Gelria had a length of 4 cm four days after in vitro pollination and incubation at

35°C. At the same time and temperature, an average of 30 to 50 pollen tubes of White American reached the ovary. Furthermore in this combination, about 10-15 pollen tubes were found in one-third to half of the ovary one week after in vitro pollination.

Table 5. Results of the in vitro pollination experiment at different temperatures

cross	IS (4) x WA (8)				IS (4) x Ge (7)			
temperature	# ovary	# seeds	seeds/ovary	relative	# ovary	# seeds	seeds/ ovary	relative
25°C	13	1615	124	100	12	1140	95	100
30°C	15	1306	87	70	14	364	26	27
35°C	11	9	0.8	0.6	18	0	-	-

The results of in vitro pollination of two cross combinations at 25, 30 and 35°C are presented in Table 5. In total, 83 ovaries were in vitro pollinated. At 35°C, all the embryos were rescued by ovule culture. All the embryos rescued at 35°C were found in the upper half of the ovary. The number of seeds per ovary in the combination of the *L. longiflorum* cultivars Indian Summer x White American were higher than in the combination of Indian Summer x Gelria at all three temperatures. Especially at 35°C, the combination of Indian Summer x White American resulted in 9 seeds whereas the combination of Indian Summer x Gelria did not result in any seeds. The 9 seeds from the 35°C treatment germinated 2-4 weeks after in vitro culture of the mature seeds.

16.4 Discussion

For pollen germination as well as for the pollen tube length, highly significant temperature and genotype effects were found in the bioassay. Pollen germination and pollen tube growth was lower at 35°C than at 25 to 30°C for all accessions. But at 35°C large differences between the genotypes were still detected. Three accesions native from warm climates, i.e. *L. formosanum*, *L. neilgherrense* and *L. longiflorum*, and the *L. longiflorum* cultivar Avita had the longest pollen tubes at 35°C. A correlation between the origin related to the climate zone and the pollen tube length at 35°C of the accessions could, however, not be confirmed. A plant test to establish the heat-tolerance of the accessions studied at pollen level will probably provide more information.

Not only differences between genotypes are a condition for a succesful pollen selection system, also segregation of pollen within an accession is required. However, at a certain temperature, 80 to 90% of the pollen within an accession had almost the same length.

Two cross combinations were made in which the pollen tube growth was exposed to three temperatures. In both cross combinations, the mother plant was *L. longiflorum* Indian Summer. The origin of this cultivar is related to a cold climate. Five days after pollination, the ovaries incubated at 30 and 35°C were placed at 25°C to prevent premature yellowing of the seed pot. Still embryos from the 35°C treatment had to be rescued by ovule culture to prevent abortion.

Also both fathers, i.e. *L. longiflorum* White American and Gelria originate from cold climates. White American has a low optimum temperature for pollen tube growth compared to Gelria. This resulted at 35°C in a three times longer pollen tube of Gelria compared to White American. At 35°C the pollen germination percentage of both cultivars is the same. On the basis of these results it was expected that more seeds should be obtained from Gelria than of White American at the 35°C treatment. However, microscopic studies of the pollen tube growth in the style gave opposite results. The pollen tube growth of Gelria was retarded at 35°C, while the pollen tubes of White American reached the ovary. Furthermore, only embryos were obtained at 35°C with White American as father.

To establish if MGS can be useful to select for heat-tolerance in lilies the seedlings obtained at 35°C will be compared in a plant test with seedlings obtained at 25°C. Furthermore, new crosses combined with a high temperature treatment during pollen tube growth will be made with heat-tolerant accessions as father.

References

Beattie DJ, White JW (1993) *Lilium* - hybrid and species. In: The physiology of flower bulbs (De Hertogh A, Le Nard M, eds) Elsevier Amsterdam, pp 423-342

Brewbaker JL, Kwack BH (1963) The essential role of calcium ion in pollen germination and pollen tube growth. Am J Bot 50(9): 859-865

CBS (1996) Tuinbouwcijfers. Centraal Bureau voor de Statistiek, pp 29-74.

Chin SW, Kuo CE, Liou JJ, Shii CT (1996) Dominant expression of vigor and heat tolerance of *Lilium longiflorum* germplasm in distant hybridization with Asiatic and Oriental lilies. Acta Hort 430 (II): 495-502

De Jong PC 1974. Some notes on the evolution of lilies. North Am Lily Yearbook 27: 23-28

Frova C, Portaluppi P, Villa M, Sari Gorla M (1995) Sporophytic and gametophytic components of thermotolerance affected by pollen selection. J Hered 86: 50-54

Hormaza JI, Herrero M (1992) Pollen selection. Theor Appl Genet 83: 663-672

Hormaza JI, Herrero M (1996) Male gametophytic selection as a plant breeding tool. Sci Hort 65: 321-333

Murashige T, Skoog F (1962) A revised medium for rapid growth and bioassay with tobacco tissue cultures. Physiol Plant 15: 473-497

Ottaviano E, Mulcahy DL (1989) Genetics of angiosperm pollen. Adv Genet 26: 1-64

Petolino JF, Cowen NM, Thompson A, Mitchell JC (1990) Gamete selection for heat-stress tolerance in maize. J Plant Physiol 136: 219-224

Shii CT (1983) The distribution and variation of *Lilium formosanum* and *L. longiflorum* in Taiwan. Lily Yearbook North Am Lily Soc 36: 48-51

Van Tuyl JM, Van Dien MP, Van Creij MGM, Van Kleinwee TCM, Franken J, Bino RJ (1991) Application of in vitro pollination, ovary culture, ovule culture and embryo

rescue for overcoming incongruity barriers in interspecific *Lilium* crosses. Plant Sci 74: 115-126
Woodcock HBD, Stearn WT (1950) Lilies of the world. Country Life Ltd., London, pp 127-128

17 Monitoring sporophytic development of individual microspores of barley (*Hordeum vulgare* L.)

Kumlehn* J and Lörz H

17.1 Introduction

Microspores can be switched from their normal gametophytic fate of pollen formation to initiate sporophytic development leading to embryo formation by exposure to stress. Microspore-derived embryogenesis is of scientific interest in various respects: it constitutes a powerful model system to investigate induction of embryogenesis and early embryonic development. In contrast to zygotic embryogenesis, all developmental stages are freely accessable at a large scale for observation, molecular analysis and manipulation techniques. The possibility to produce doubled-haploid plants makes this system very useful in plant breeding programmes. Through a single step from one generation to another, individuals carrying defect alleles are systematically eliminated, whereas traits of interest can be fixed homozygously in the progeny. Moreover, microspore cultures can be coupled with genetic transformation techniques.

During earlier work on microspore embryogenesis, several questions arose which remained to be unambiguously answered so far: (1) Which characteristic cellular events are coupled with the initiation of microspore-derived embryogenesis ? (2) By which early-visible markers can sporophytically developing individuals be identified ? (3) What is the time course of the developmental events ? (4) How can these events be influenced by particular conditions ?

Embryogenic barley microspores have been described in mass cultures (Sunderland et al. 1979; Olsen 1991; Hoekstra et al. 1993) or in individual microspore cultures derived from pretreated anthers (Bolik and Koop 1991), but the current state of these studies is far from being satisfactory. In our opinion,

* for correspondence: Centre for Applied Plant Molecular Biology (AMP II); University of Hamburg; Ohnhorststrasse 18; 22609 Hamburg; Germany

evidence for developmental patterns can only be obtained by pursuing individuals throughout their development from an isolated microspore to embryo formation. Since the initiation of the sporophytic pathway takes place during exposure to stress, the process under investigation should encompass the period of inductive treatment. Identification of embryogenic microspores is desirable for proper characterization and evaluation of microspore cultures, for identification of suitable cells for manipulation techniques such as microinjection, and for efficient individual selection of specific cell types which might be sampled subsequently for detailed molecular analyses.

17.2
Technical prerequisits for a monitoring system

Barley is today's best suited species for microspore-derived embryogenesis amongst cereals. There are various well-established protocols used worldwide (reviewed by Jähne and Lörz 1995). To establish a monitoring system for individual development encompassing the whole process from sporophytic initiation to embryo formation, some modification of the culture procedure was necessary.

17.2.1
Including the period of stress exposure to the procedure of cell observation

For sporophytic induction of cereal microspores, it was, so far, necessary to expose whole spikes or at least dissected anthers to certain pretreatments, e.g. reduced availability of water, low or high temperature and/or exogenuously supplied auxin. Attempts to ensure viability and sporophytic development of isolated barley microspores without pretreatment inside the anthers had failed (Bolik and Koop 1991; Mordhorst et al. 1993; Hoekstra et al. 1996). We have now established culture conditions under which microspores can be switched to sporophytic development in an already isolated state which makes observation of the cellular processes during this period feasible. Isolated barley microspores can be efficiently induced to embryogenesis in melibiose solution (for details see below). The treatment can be conducted at temperatures between 4 and 32 °C, with a respective exposure time of 14 to 2 days.

17.2.2
Individual cell handling

For the establishment of a monitoring system, a method for handling of individual cells or small cell groups is desirable. In this respect we profit from foregoing research on selection and transfer of isolated somatic and germ cells, in vitro fertilization and culture of zygotes: for these purposes, a sophisticated system with glass capillaries interfaced to a computer-controlled micropump has been developed (Koop and Schweiger 1985; Kranz et al. 1991; Kumlehn et al. 1997). This system is ideally suitable for handling isolated microspores (Fig. 1).

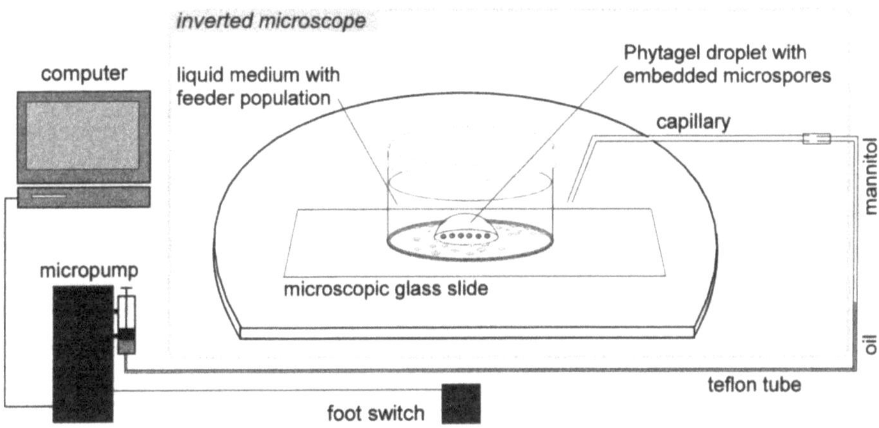

Fig. 1. Schematic representation of handling and culture of individual barley microspores

17.2.3 Immobilization and culture of isolated microspores

By using the capillary system, isolated microspores can be individually selected from a primary population, drawn up in a small volume of medium and released into a liquid droplet of mannitol supplemented with phytagel positioned inside a glass ring which had been mounted directly onto a microscopic glass slide (Fig. 1). Upon cell transfer, this droplet was solidified by adding a minute amount of $CaCl_2$ solution. Subsequently, the solidified gel was surrounded by liquid medium which later on was exchanged and/or supplemented with a feeder population according to the experimental design (Fig. 1).

17.2.4 Documentation of the developmental process

The groups of immobilized microspores were photographed regularly in order to monitor the individual development. For high-magnification microscopy, a small distance between objective and specimen is necessary. By using an inverted microscope, detailed recordings were feasible, since the cells of interest had been immobilized directly on a thin microscopic glass slide. Computer-aided cell tracking was reported for identification of embryogenic individuals in somatic cell cultures (Golds et al. 1992; Toonen et al. 1997). Although manual adjustment of the images is more time-consuming, we preferred this approach, because subcellular details of interest can be focused individually.

17.3 Individual development of isolated microspores

The immobilized microspores were exposed for 2 days to restricted nutrient supply (0.4 M melibiose, 1 mM $CaCl_2$, 2 mM MES) combined with a temperature of 32° C. Subsequently, this starvation medium was replaced by a suitable nutrient medium (e.g. after Mordhorst and Lörz 1993) which in addition contained a proper population density of sporophytically induced microspores. Compared to culture in liquid medium, the proportion of surviving and developing microspores was markedly reduced when immobilized in a gel. This is assumed to be due on the one hand to additional stresses, e.g. relatively long exposure to isolation medium before immobilization and repetitive changes of temperature and osmotic conditions during the transfer procedure, and on the other hand to mechanically restricted exchange of population signals through the gel matrix during culture. As a result, it was possible to monitor only a few individuals during their sporophytic development in this first attempt, and a further optimization of the method is desirable.

The observed microspores showed a huge variety of developmental features. A fairly typical example of sporophytic development is shown in Fig. 2. During the pretreatment, the microspore expands continuously and, after a certain period of pretreatment, the cytoplasm begins to increase its volume. At the same time, the strongly vacuolated nucleolus indicates intense transcription activity of the cell. After transfer to nutrient medium, the cytoplasm increasingly occupies the interior of the exine, a process which characteristically commences opposite to the pore. As shown in Fig. 2d, nuclear divisions occur before the whole interior of the exine is occupied by dense cytoplasm, whereas cell wall formation cannnot be clearly recognized. Consequently, genetic transformation should be conducted prior to this developmental stage in order to avoid chimerism. In contrast, Bolik and Koop (1991) identified cytoplasm-rich barley microspores lacking large vacuoles as embryogenic, and suggested those structures to be used for transformation experiments. The subsequent development of the monitored structure shown in Fig. 2 is characterized by further expansion through repeated cell divisions resulting in a burst of the exine and the formation of a proembryogenic mass.

Cell expansion of barley microspores was assumed to be coupled with initiation of embryogenic development (Bolik and Koop 1991; Hoekstra et al. 1992). Through this study, it was possible to trace back embryogenic development also to microspores which did not follow this rule, that is, some microspores commenced cell division resulting in embryogenic development without preceding microspore expansion.

Formation of star-like structures that was found to be characteristic for successful tobacco and wheat microspore cultures (reviewed in Touraev et al. 1997), also appears upon inductive treatment in barley. These structures show a centrally located nucleus inside a small volume of cytoplasm, whereas centripetally oriented cytoplasmic strands penetrate the highly vacuolated cell periphery. Although few cell divisions were observed in some of these structures, we did not detect any example where embryogenic development proceeded. These findings are in accordance with observations reported by Bolik and Koop (1991). Some star-like structures accumulated starch granules before degeneration.

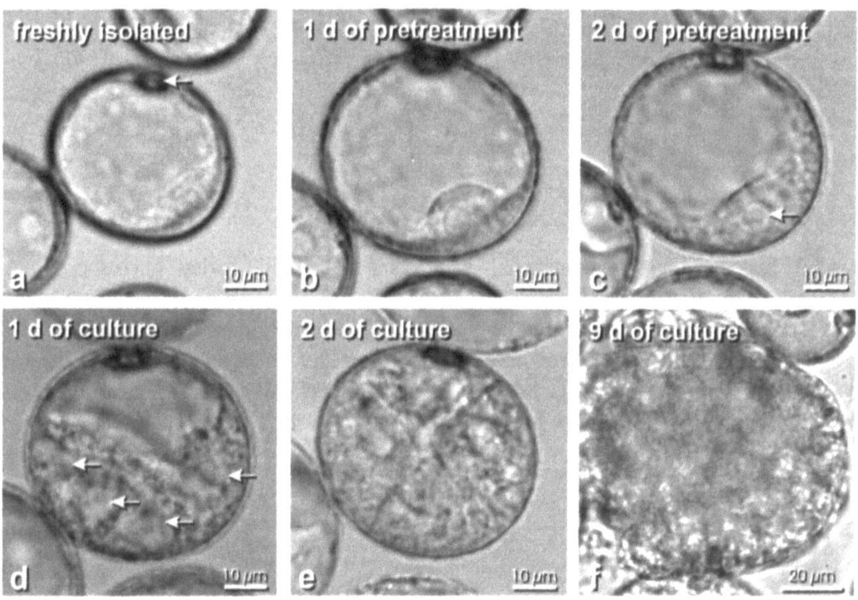

Fig. 2. Sporophytic development of an individual barley microspore after immobilization on a microscopic glass slide. (a) Freshly isolated late uninucleate microspore; an arrow indicates the microspore porus. (b) The same microspore after 1 day of pretreatment in melibiose solution at 32 °C; note the expansion of the cell. (c) The same microspore after 2 days of pretreatment; the volume of the cytoplasm commences to increase; the arrow indicates the strongly vacuolated nucleolus. (d) The same structure after 1 day of culture in nutrient medium; the cytoplasm has occupied about half of the interior of the exine; the arrows indicate several weakly recognizable nuclei. (e) The same structure after 2 days of culture; note the symmetric compartmentation of the structure by strong cell walls. (f) The same structure after 9 days of culture; the expanding embryogenic structure has burst the exine; note the reduced magnification compared with the preceding images

Among the huge variety of developmental patterns, we observed another interesting example in which two nucleoli were recognized in a freshly isolated microspore at the uninucleate stage. As was observed in isolated egg cells and cultivated zygotes of maize and wheat (Kranz et al. 1995; Kumlehn et al. 1998), the number of nucleoli correlates in general with the overall ploidy level of the egg cell or the zygote and does not change upon entering G2 phase. If this holds also true for the microspore shown in Fig. 3, it constitutes sporophytic development either from an unreduced or from a spontanuously doubled-haploid microspore.

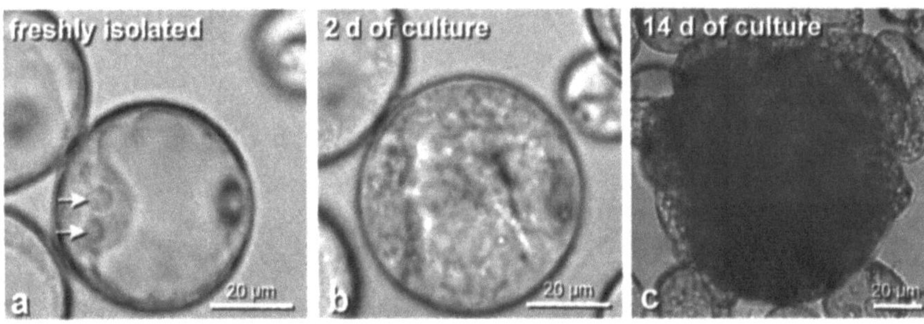

Fig. 3 Sporophytic development of a presumably diploid barley microspore after immobilization on a microscopic glass slide. (a) Freshly isolated late uninucleate microspore; an arrow indicates the two nucleoli. (b) Multicellular structure after 2 days of pretreatment and 2 days of culture in nutrient medium. (c) Embryogenic structure derived from the same microspore with burst exine after 2 weeks of culture; note the reduced magnification compared with the preceding images

17.4 Future prospects

The results presented have to be regarded merely as a preliminary step towards a detailed analysis of sporophytic development as a dynamic process. The method needs further refinement to obtain higher efficiency which is a prerequisite for statistic processing of the observations. This way, it might be possible to correlate certain cellular events to the initiation of sporophytic development, and thereby to get evidence on key events leading to microspore-derived embryogenesis. Detailed microscopic analyses, e.g. by vital staining techniques or in situ hybridization examining any desired developmental stage of specimen monitored before, should result in additional insights into the developmental mechanisms. Information gained on structural kinetics might be used subsequently as a framework for detailed analyses on how embryogenesis is governed molecular-genetically. This can be approached by employing techniques recently developed for the use of samples comprising merely a few or even single plant cells (Dresselhaus et al. 1994; Richert et al. 1995). The identification and selection of embryogenic microspores will allow us to precisely examine molecular processes of microspore-derived embryogenesis at distinct developmental stages by using samples encompassing individuals of only one targeted cell type.

Concerning the optimization of microspore cultures for plant breeding purposes, further studies should reveal the developmental window of pollen formation during which a switch to the sporophytic pathway is possible at all, and at which developmental stage it is most suitable with respect to efficient embryogenesis, genetic transformation, spontaneous or induced genome doubling, or green plant formation.

Acknowledgments

This study was supported by the European Commission DG XII Biotechnology Programme (contract BIO4-CT96-0275). We thank Dr. Margret Sauter for critical reading of the manuscript.

References

Bolik M, Koop HU (1991) Identification of embryogenic microspores of barley (*Hordeum vulgare* L.) by individual selection and culture and their potential for transformation by microinjection. Protoplasma 162: 61-68

Dresselhaus T, Lörz H, Kranz E (1994) Representative cDNA libraries from few plant cells. Plant J 5: 605-610

Hoekstra S, van Zijdeveld MH, Louwerse JD, Heidekamp F, van der Mark F (1992) Anther and microspore culture of *Hordeum vulgare* L. cv. Igri. Plant Sci 86: 89-96

Hoekstra S, van Zijdeveld MH, Heidekamp F, van der Mark F (1993) Microspore culture of *Hordeum vulgare* L.: the influence of density and osmolality. Plant Cell Rep 12: 661-665

Hoekstra S, van Bergen S, van Brouwershaven R, Schilperoort RA, Heidekamp F (1996) The interaction of 2,4-D application and mannitol pretreatment in anther and microspore culture of *Hordeum vulgare* L. cv. Igri. J Plant Physiol 148: 696-700

Golds TJ, Babczinsky J, Rauscher G, Koop HU (1992) Computer-controlled tracking of single cell development in *Nicotiana tabacum* L. and *Hordeum vulgare* L. protoplasts embedded in agarose/alginate films. J Plant Physiol 140: 582-587

Jähne A, Lörz H (1995) Cereal microspore culture. Plant Sci 109: 1-12

Koop HU, Schweiger HG (1985) Regeneration of plants from individually cultivated protoplasts using an improved microculture system. J Plant Physiol 121: 245-257

Kranz E, Bautor J, Lörz H (1991) In vitro fertilization of single isolated gametes of maize mediated by electrofusion. Sex Plant Reprod 4: 12-16

Kranz E, von Wiegen P, Lörz H (1995) Early cytological events after induction of cell division in egg cells and zygote development following in vitro fertilization with angiosperm gametes. Plant J 8: 9-23

Kumlehn J, Brettschneider R, Lörz H, Kranz E (1997) Zygote implantation to cultured ovules leads to direct embryogenesis and plant regeneration of wheat. Plant J 6: 1473-1479

Kumlehn J, Lörz H, Kranz E (1998) Differentiation of isolated wheat zygotes into embryos and normal plants. Planta 205: 327-333

Mordhorst AP, Lörz H (1993) Embryogenesis and development of isolated barley (*Hordeum vulgare* L.) microspores are influenced by the amount and composition of nitrogen sources in culture media. J Plant Physiol 142: 485-492

Olsen FL (1991) Isolation and cultivation of embryogenic microspores from barley (*Hordeum vulgare* L.). Hereditas 115: 255-266

Richert J, Kranz E, Lörz H, Dresselhaus T (1996) A reverse transcriptase-polymerase chain reaction assay for gene expression studies at the single cell level. Plant Sci 114: 93-99

Sunderland N, Roberts M, Evans LJ, Wildon DC (1979) Multicellular pollen formation in cultured barley anthers. J Exp Bot 30: 1133-1144

Toonen MAJ, Verhees JA, Schmidt EDL, van Kammen A, de Vries SC (1997) AtLTP1 luciferase expression during carrot somatic embryogenesis. Plant J 12: 1213-1221

Touraev A, Vicente O, Heberle-Bors E (1997) Initiation of microspore embryogenesis by stress. Trends Plant Sci 2: 297-302

18 Effects of ABA during the pretreatment of barley anthers on androgenesis of *Hordeum vulgare* L. cultivars Igri and Digger

van Bergen S, Kottenhagen MJ, van der Meulen RM, Wang* M

18.1 Introduction

For induction of barley doubled haploids via androgenesis, at least three requirements are necessary. First, the developmental stage of the microspores should be at the verge of mitosis (Hoekstra et al. 1992). Second, the right pretreatment of the anthers is needed in order to switch the gametophytic pathway into a sporophytic development of the microspores. And third, the right hormone in the culture medium of microspores which are isolated from pretreated anthers, is necessary for induction of plant production. The pretreatment for barley can be a cold treatment of spikes (Huang and Sunderland 1982), or anthers can be pretreated on a mannitol solution (Robert-Oehlschlager and Dunwell 1990). The latter pretreatment enhances doubled-haploid production in barley; the microspore viability is increased and a higher efficiency of plant production is achieved (Hoekstra et al. 1994). Knowledge about the biological processes involved in the effects of pretreatment might help in the design of a genotype independent protocol for androgenesis.

Little is known about the processes that are affected by the pretreatment. Heberle-Bors (1989) suggests that during pretreatment, starvation due to the absence of useful saccharides, is the basis for the occurrence of certain degradation processes in pollen grains, resulting in the induction of pollen embryogenesis. Especially in tobacco, the starvation effect is pronounced if pretreatment is performed in nitrogen-free medium with mannitol. Pretreatment of anthers from barley, resulting in plant production, can be performed using similar conditions as for pollen pre-culture of tobacco, i.e. on mannitol dissolved in what is called a pre-medium. Mannitol is thought to create osmotic stress. The presence of the sugars mannitol during pretreatment and maltose during culture, are

* for correspondence: Center for Phytotechnology RUL/TNO; TNO Department of Plant Biotechnology; Wassenaarseweg 64; 2333 AL Leiden; The Netherlands

important to have proper osmotic pressure for microspore-derived barley plant production (Hoekstra et al. 1993).

Davies and Jones (1991) suggested that, in response to certain stress treatments like osmotic pressure and cold, ABA is produced. It has been reported that pretreatment of tobacco anthers by mannitol caused an increase in endogenous ABA level (Imamura and Harada 1980). In addition, relatively high endogenous ABA levels are correlated with high embryogenic potential of *Pennisetum* and carrot cells (Rajasekaram et al. 1987a,b; Kiyosue et al. 1992). In this paper, a comparative study on regeneration efficiency between cultivars Igri and Digger was carried out. A significant correlation between stress-induced ABA content and regeneration efficiency was found. Addition of ABA and different ABA analogues in pre-medium containing no mannitol showed stimulation on regeneration efficiency. An important role for ABA during pretreatment is suggested in this study.

18.2 Material and methods

18.2.1 Material

Donor plants of *Hordeum vulgare* L. cvs. Igri and Digger were grown in a phytotron under conditions described previously (Hoekstra et al. 1992). Monoclonal antibody to free (+)ABA was purchased from Idetek, Inc. (San Bruno, Calif., USA). Rabbit anti-mouse alkaline-phosphatase conjugate, (+) ABA and bovine calf serum (grade suitable for enzyme-linked immunosorbent assay; ELISA) and Fluoresceine Di-Acetate (FDA)were obtained from Sigma Chemical Company (St. Louis, MO. USA).

18.2.2 Induction of androgenesis

Growth of the donor plants *Hordeum vulgare* L. cv Igri, selection of material (anthers containing late uninucleate microspores), mannitol pretreatment of anthers (in pretreatment solution containing 0.37 M mannitol, 10^{-2} M $CaCl_2$, 10^{-3} M $MgSO_4.7H_2O$, 10^{-3} M KNO_3, $2x10^{-4}$ M KH_2PO_4, 10^{-6} M KI and 10^{-7} M $CuSO_4.5H_2O$, 440 $mOs.kg^{-1}$) and culture conditions were as described previously (Hoekstra et al. 1993) with the following modifications. Anthers (without Ficoll) or microspores were cultured in medium I at 350 $mOs.kg^{-1}$ (Hoekstra et al. 1992).

For microspore culture, anthers were pre-treated on mannitol solution for a period of 4 days and microspores in these anthers were isolated as described by Hoekstra et al. (1993). After isolation, 1 x 10^4 embryogenic microspores were plated in 1 ml medium I (Hoekstra et al. 1992) in a 3.5 cm Petri-dish. After 14 days culture, the amount of ELS was counted. For anther culture, 30 anthers (equivalent to 1/2 spike) of several spikes were randomly used per parameter to be tested. After 4 days the anthers were transferred to medium I (Hoekstra et al. 1993) without Ficoll at 350 $mOs.kg^{-1}$. After 14 days of culture, the amount of ELS was counted. For the effect of ABA/ABA analogues during pretreatment, anthers were incubated in different pretreatment solutions in the presence of ABA/ABA analogues (Sigma, St. Louis, MO). After a period of 4 days the anthers were transferred to medium I.

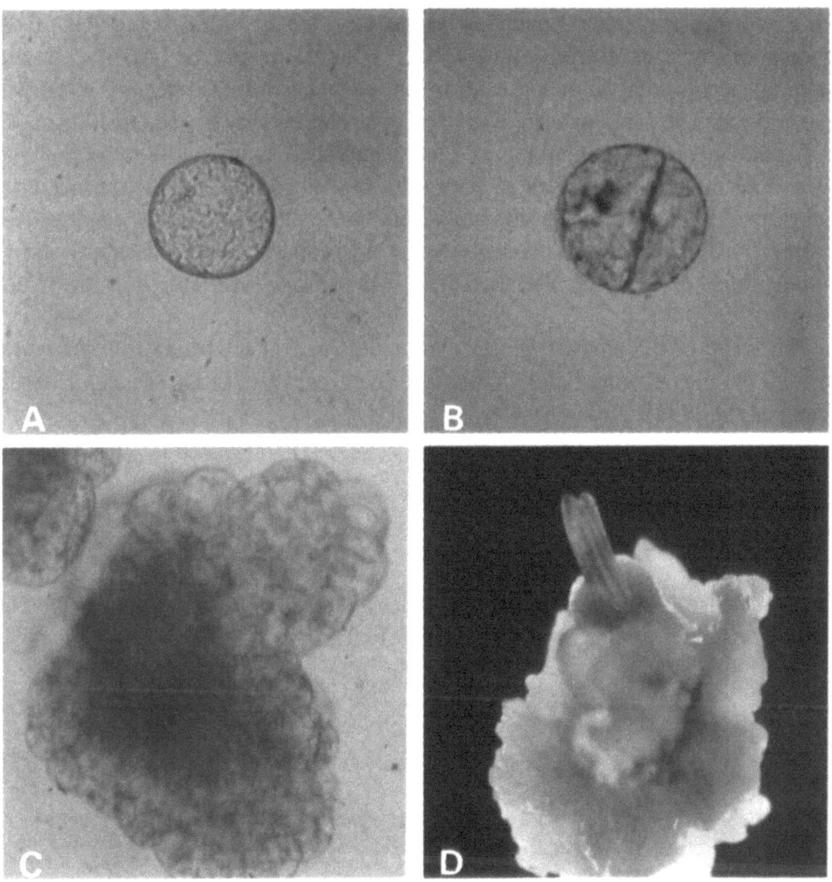

Fig. 1. Development of a microspore to a plant in barley cultivar Igri. A: Barley anthers cultivar Igri were pretreated in mannitol for 4 days as described in materials and methods. Then the microspores were isolated. The average size of microspore is about 50 µm. B: After 3 days of culture, microspore cell division can be observed. C: After 14 days of microspore culture, multicellular structures whit embryogenic potential were observed and these structures are called embryo like structure (ELS). D: After 21 days of culture, germination of embryos was observed

18.2.3 Viability measurement/FDA staining

The number of viable microspores during anther pretreatment was determined by squashing the anther tissue in 1 ml pretreatment solution. The squashed material was stained with about 20 μl (5 mg/mL in acetone) FDA.

18.2.4 ABA extraction and ELISA assay

Anthers were put into Eppendorf vials and frozen with liquid N_2. Pestle maceration of the frozen anthers was not necessary for ABA extraction (unpublished results, M Wang et al.). The contents of the Eppendorf vials were freeze dried. The samples were extracted at a rotation shaker (IKA) 150 rpm, 4°C with cold methanol (containing 100 mg/l butylated hydroxytoluene and 0.5 g/l citric acid monhydrate) twice each time in 0.5 ml for 24 hours. The vials were centrifuged at 8,000 *g* for 15 min at 4°C and the supernatants of the samples were pooled. The samples were concentrated in a Speed-vac system (Savant) and the volume adjusted to 25 μl with methanol. The samples were stored at -80°C until assay.

The assay of the amount of ABA in anthers and in pretreatment solution was carried out by enzyme-linked immunosorbent assay (ELISA) as described by Wang et al. (1995).

18.2.5 Experimental data

Mean values ± SD are presented unless stated otherwise. Significance of difference in mean values was tested with Students' *t*-test.

18.3 Results

18.3.1 Correlation between ABA content during pretreatment and regeneration efficiency

In barley (cv. Igri.) microspore regeneration experiments, anthers containing uninucleate microspores were used. After a pretreatment on mannitol, round shaped microspores with a red interference of the exine wall were found (Fig. 1A). After 3 days of culture, first divisions were visible (Fig. 1B). Between day 7 and

11 in culture, depending on the cultivar, the microspores broke out of their exine and small compact structures were formed, the so-called embryo-like structures (ELS). At day 14, most of dividing microspores were out of their exine (Fig. 1C). After about 21 days, germination of the embryos could be observed (Fig. 1D).

It is commonly known that a large variety in regeneration exists between different cultivars in various types of tissue culture systems. The number of ELS formation was used as a marker for regeneration efficiency. The regeneration efficiencys of the cultivars Igri (winter type), and Digger (spring type) were tested by counting the number of ELS on day 14 in culture of microspores of both cultivars Igri and Digger. Pretreatment with mannitol showed the highest regeneration efficiency for cultivar Igri (Table 1), while, without pretreatment, the lowest regeneration efficiency was found with cultivar Digger (Table 1).

Table 1. Correlations between regeneration efficiency, microspore viability and ABA contents

Samples	ELS/10^4 microspores	% viability	ABA pg/30 anthers
Igri pretreatment	1940 $\pm$ 80	58 $\pm$ 9.4	3680 $\pm$ 450
Igri control	190 $\pm$ 16*	24 $\pm$ 12.1*	1869 $\pm$ 270*
Digger pretreatment	300 $\pm$ 32*	22 $\pm$ 3.0*	2385 $\pm$ 200*
Digger control	100 $\pm$ 20*	13 $\pm$ 2.6*	1053 $\pm$ 150*

Barley anthers cultivars Igri and Digger were incubated under pretreatment (pre-medium containing 0,37 M mannitol) and non-pretreatment (directly in culture medium) conditions for 4 days. Sunsequently microspores from these anthers were isolated for determination of viability by FDA staining and further culture for the formation ELS. After 14 days of incubation, the number of ELS was counted. Data presented in this table are mean values $\pm$ S.D. of four independent experiments. * mean values significantly different ($P < 0.05$) from value obtained in sample "Igri pretreatment"

Previous studies on Igri showed that there is a positive relationship between the use of a mannitol pretreatment and regeneration efficiency (Hoekstra et al. 1993). We measured the viability of the microspores by staining the microspores with FDA and found that high regeneration efficiency caused by mannitol pretreatment was correlated with the percentages of viability of microspores isolated from anthers (Table 1). The ABA contents in both anthers and the pretreatment solution were measured after 4 days of pretreatment. For both Igri and Digger, it was found that ABA was synthesised in the anthers and diffuses out to the medium (data not shown). The ABA content in the anthers of cultivar Igri increased from day 0 from 450 pg/30 anthers to a level on day 4 of about 3700 pg/30 anthers (Table 1). For the cultivar Digger the ABA content in that anthers increased from a day 0 of 300 pg/30 anthers to a level on day 4 of about 2000 pg/30 anthers. The total content of ABA (medium plus anthers) on day 4 for Igri was about twice as high as for Digger. (Table 1). Without pretreatment, a lower ABA content in both Igri and Digger were found as compared to the pretreatment conditions.

18.3.2
Effects of ABA and ABA analogues during pretreatment

During mannitol pretreatment, *de novo* ABA synthesis was found in anthers for both cultivars Igri and Digger and without pretreatment, a low ABA content was measured in the anthers of both cultivars (Table 1). Therefore, we asked if addition of ABA under the control condition (no mannitol pretreatment) could give a stimulation on ELS formation. Anthers from Igri were incubated in pre-medium without mannitol but in the presence of 10^{-7}M ABA for 1, 2, 3 and 4 days and subsequently the microspores were isolated for culture. After 14 days of culture, the number of ELS formation was counted and data are presented in Table 2. It is clear that the addition of ABA was able to stimulate the formation of ELS and 1 day treatment with ABA gave the best stimulation (Table 2). However, the highest stimulation condition (10^{-7} M ABA for 1 day treatment, see Table 2) gave only about 413 ELS/10^4 microspores. This is still about 5 times lower than achieved with mannitol pretreatment (1940 ELS/10^4 microspores, see Table 1). This data suggests that either a higher ABA concentration is required or ABA alone is not able to give as much ELS formation as compared to mannitol treatment.

Table 2. Effects of ABA during pretreatment

Treatments	% ELS
Control	100 ± 10
10^{-7} M ABA for 1 day	$270 \pm 15^{**}$
10^{-7} M ABA for 2 days	$186 \pm 7^{*}$
10^{-7} M ABA for 3 days	$158 \pm 15^{*}$
10^{-7} M ABA for 4 days	$140 \pm 10^{*}$

Barley anthers cv. Igri were incubated in buffer without mannitol but with 10^{-7} M ABA for a period of 1, 2, 3 or 4 days and then microspores from these anthers were isolated for further culture for the formation of ELS. After 14 days of incubation, the number of ELS was counted. Data presented in this table are mean values ± S.D. of four independent experiments. * mean values significantly different ($P < 0.05$) from value obtained in control. ** mean value significantly different ($P < 0.05$) from values obtained in all the other samples

In Table 3, the results of experiments in which different ABA analogues were used during anthers in pre-medium buffer for 4 days are presented. It is clear that ABA-glucose ester showed the best effect on the induction of ELS and (+)-ABA induced about the same effect on ELS formation as (±)-ABA (Table 3).

18.4 Discussion

A stressful pretreatment is one of the requirements for doubled haploid production in barley. The difference in regeneration efficiency between cultivars Igri and Digger was correlated with both the ABA content in anthers during pretreatment and microspore viability after pretreatment (Tables 1 and 2). These results suggest that production of ABA caused by stressful treatment, may be essential for the viability of microspores and subsequently for the androgenesis. During pre-medium incubation of anthers (without addition of mannitol), addition of ABA showed significant stimulation on regeneration efficiency (the formation of ELS) (Tables 2 and 3). In our previous studies, addition of ABA during mannitol pretreatment of Igri anthers showed no stimulative effect on regeneration efficiency (Hoekstra et al. 1997). However, we also demonstrated that the presence of fluridone during the first 24 hours of pretreatment resulted in a large decrease of the number of plants obtained with mannitol pretreatment, suggesting that normally there is a need for *de novo* ABA synthesis during mannitol pretreatment (Hoekstra et al. 1997). So, an increase in endogenous ABA level is required for plant production. It is likely that under our experimental conditions (mannitol pretreatment), a stress-induced endogenous ABA level is sufficient to give a switch for microspores to enter the androgenesis. In addition, beside ABA, other factors apparently also play an essential role in induction of androgenesis.

Table 3. Effects of ABA analogues on ELS formation during pretreatment.

Samples	% ELS
(+)ABA	$65.2 \pm 30.0^{**}$
(±) ABA	100.0 ± 17.0
ABA methylester	$101.7 \pm 16.1^{**}$
ABA glucoseester	$161.4 \pm 32.0^{*}$

Barley anthers cv. Igri were incubated in buffer without mannitol but with 10^{-7} M (+) ABA or ABA analogues for a period 4 days and then microspores from these anthers were isolated for further culture for the formation ELS. After 14 days of incubation, the number of ELS was counted. Data presented in this table are mean values ± S.D. of four independent experiments. * mean value significantly different ($P < 0.05$) from value obtained in control (sample (±)ABA). ** mean values not significantly different ($P > 0.95$) from value obtained in control (sample (±) ABA)

With regard to a correlation between androgenetic capacity and ABA level, the literature appears to be rather controversial. Johansson et al. (1982) report that appropriate pretreatment of *Anemona canadensis* to induce androgenesis, reduced the endogenous ABA concentration at the end of pretreatment nearly 4 times, whereas Imamura and Harada (1980) demonstrate the presence of a peak in endogenous ABA level after 24 hours of mannitol pretreatment in tobacco anthers. We observed a fast increase in ABA content after 24 hours of mannitol pretreatment (data not show), and addition of ABA in pre-medium for the first 24 hours gave the highest stimulation effect on the formation of ELS (Table 2). Our data agree with the observations of Imamura and Harada (1980). Of course we should take in consideration the turn-over of ABA during the different periods of incubation. Imamura and Harada (1980) already proposed that a specific level for ABA was required for induction of androgenesis. These authors have shown in a time range experiment using 10^{-5} M ABA, that a 1 to 3 days period of pretreatment stimulate plant production in tobacco androgenesis. For wheat androgenesis, Hu et al. (1995) observe a stimulation by application of ABA (about 10^{-5} M) throughout the pretreatment. Kyo and Harada (1986) show that application of 5×10^{-6} M ABA in the second half of the pollen pretreatment stimulated embryo production in tobacco.

Reynolds and Crawford (1996) demonstrated that there was a direct and positive correlation with an increase of ABA and expression of an early cysteine-labeled metallothionein gene in developing pollen embryoids. The importance of endogenous ABA levels has been determined for morphogenic competence in *Pennisetum sp.* by Rajasekaran et al. (1987a). Furthermore, these authors show that application of ABA enhances somatic embryogenesis, and that inhibition of plant production is observed by addition of fluridone in the induction medium. The use of fluridone on donor plants even causes the loss of capacity for embryogenesis (Rajasekaran et al. 1987b).

The capacity of different ABA analogues to affect the formation of ELS during barley cultivar Igri anther pretreatment was studied. When the carboxyl group of the ABA molecule was modified by esterification with a methyl or a glucose group, the induction of ELS formation was different (Table 3). ABA-glucose ester gave a better induction on ELS formation as compared to ABA-methyl ester and ABA itself (Table 3). This observation was exactly opposite to what we have found for ABA-induced Rab gene expression in barley aleurone (Van der Meulen et al. 1993). The higher stimulative effect of ABA-glucose ester on ELS formation may be explained by the following possibilities: in the present experiments, the incubation time was 4 days. Due to this the difference in ABA/ABA-methyl ester and ABA-glucose ester might be due to a difference in ABA turn-over. For example ABA-glucose ester may be much less sensitive to ABA metabolitic enzymes, resulting in less degradation and higher stimulative effect though the molecule itself is less efficient as compared to ABA.

In the present study, a strong correlation between stress-induced ABA production and microspore viability and subsequently microspore regeneration efficiency was observed. It is possible that the high ABA level induced by stress protects microspores from death and thereby increases the number of candidates for further performance of cell division. Our near future studies will concentrate on the more detailed role of ABA.

References

Davies WJ, Jones HG (1991) Drought-induced changes in physiology and ABA. In: Abscisic Acid, Physiology and Biochemistry. Bio's scientific publishers limited Oxford, UK, pp 63-77

Heberle-Bors E (1989) Isolated pollen culture in tobacco: plant reproductive development in a nutshell. Sex Plant Reprod 2: 1-10

Hoekstra S, van Zijderveld MH, Louwerse JD, Heidekamp F, van der Mark F (1992) Anther and microspore culture of *Hordeum vulgare* L. cv. Igri. Plant Science 86: 89-96

Hoekstra S, van Zijderveld MH, Heidekamp F, van der Mark F (1993) Microspore culture of *Hordeum vulgare* L.: the influence of density and osmolarity. Plant Cell Reports 12: 661-665

Hoekstra S, van Zijderveld MH, van Bergen S, van der Mark F, Heidekamp F (1994) Genetic modification of barley for end use quality. *In* Henry RJ, Ronalds JA, eds, Improvement of cereal quality by genetic engineering. Plenum Press, pp 139-144

Hoekstra S, van Bergen S, van Brouwershaven RI, Schilperoort RA, Wang M (1997) Androgenesis in *Hordeum vulgare* L.: Effects of mannitol, calcium and abscisic acid on anther pretreatment. Plant Sci 126: 211-218

Hu TC, Ziauddin A, Simion E, Kasha KJ (1995) Isolated microspore culture of wheat (*Triticum aestivum* L.) in a defined media I. Effects of pretreatment, isolation methods and hormones. In Vitro Cell Dev Biol 31: 79-83

Huang B, Sunderland N (1982) Temperature-stress pretreatment in barley anther culture. Ann Bot 49: 77-88

Imamura I, Harada H (1980) Effects of abscisic acid and water stress on the embryo and plantlet production in anther culture of *Nicotiana tabacum* cv. Samsun. Z Pflanzen Physiol 100: 285-289

Johansson L, Andersson B, Eriksson T (1982) Improvement of anther culture technique: activated charcoal bound in agar medium in combination with liquid medium and elevated CO_2 concentration. Physiol Plant 54: 24-30.

Kiyosue T, Nakajima M, Yamaguchi I, Satoh S, Kamada H, Harada H (1992) Endogenous levels of abscisic acid in embryogenic cells, non-embryogenic cells and somatic embryos of carrot (*Daucus carota* L.). Biochem Physiol Pflanzen 188: 343-347

Kyo M, Harada H (1986) Control of the developmental pathway of tobacco pollen in vitro. Planta 168: 427-432

Rajasekaran K, Hein MB, Vasil IK (1987a) Endogenous abscisic acid and indole-3-acetic acid and somatic embryogenesis in cultured leaf explants of *Pennisetum purpureum* Schum. Plant Physiol 84: 47-51

Rajasekaran K, Hein MB, Davis GC, Carnes MG, Vasil IK (1987b) Endogenous growth regulators in leaves and tissue cultures of *Pennisetum purpureum* Schum. J Plant Physiol 130: 13-25

Reynolds TL, Crawford RL (1996) Changes in abundance of an abscisic acid-responsive, early cysteins-labeled metallothionein transcript during pollen embryogenesis in bread wheat. Plant Mol Biol 32: 823-829

Roberts-Oehlschlager SL, J.M. Dunwell JM (1990) Barley anther culture: Pretreatment on mannitol stimulates production of microspore-derived embryos. Plant Cell Tiss Org Cult 20: 235-240

Van der Meulen RM, Heidekamp F, Jastorff B, Horgan R, Wang M (1993) Effects of abscisic acid analogues on abscisic acid -induced gene expression in barley aleurone protoplasts: Relationship between structure and function of the abscisic acid molecule. J. Plant Growth Regul. 12: 13-19

Wang M, Heimovaara-Dijkstra S, Van Duijn B (1995) Modulation of germination of embryos isolated from dormant and nondormant barley grains by manipulation of endogenous abscisic acid. Planta 195: 586-592

19 Programmed cell death during androgenesis in *Hordeum vulgare* L.

Wang*[1] M, van Bergen[1] S, Lamers[2]GEM, Oppedijk[1] BJ and Schilperoort[3] RA

19.1 Introduction

For the production of (doubled) haploid plants from microspores, via gametophytic embryogenesis, a specific developmental stage (at the verge of mitosis) in combination with stressful pretreatment (starvation in combination with salt and osmotic stresses) of anthers and microspores generally is essential (Heberle-Bors 1989; Hoekstra et al. 1992; 1997). During this pretreatment, microspores change into cells that are called embryogenic microspores (Hoekstra et al. 1993). Little is known about the biological processes occurring during pretreatment, however, understanding these processes is not only of importance for breeding programs in agriculture but also of fundamental interest for the study of developmental switches. In barley, a few studies have been performed to investigate the effects of anther pretreatment on gametophytic embryogenesis. It was found that pretreatment with a salt solution containing mannitol is the best condition for plant production (Hoekstra et al. 1996). Davies and Jones (1991) suggest that, in response to certain stress treatments like osmotic pressure and cold, ABA is produced. Pretreatment of tobacco anthers by mannitol caused an increase in endogenous ABA level (Imamura and Harada 1980). Furthermore, relatively high endogenous ABA levels are correlated with high embryogenic potential of *Pennisetum* and carrot cells (Rajasekaran et al. 1987a; Rajasekaran et al. 1987b; Kiyosue et al. 1992; Reynolds and Crawford 1996). These observations suggest a possible role for ABA in androgenesis.

*[1] for correspondence: Center for Phytotechnology RUL/TNO; TNO Department of Plant Biotechnology; Wassenaarseweg 64; 2333 AL Leiden; The Netherlands

[2]Center for Phytotechnology RUL/TNO; EMCA unit; Institute of Molecular Plant Sciences; Leiden University; Wassenaarseweg 64; 2333 AL Leiden; The Netherlands

[3]Center for Phytotechnology RUL/TNO; Institute of Molecular Plant Sciences; Leiden University; Wassenaarseweg 64; 2333 AL Leiden; The Netherlands

Degradation of anther tissues is considered to be essential for induction of androgenesis, and under pretreatment conditions cell death in anther tissue is indeed observed (Sunderland et al. 1984).

Apoptosis is a form of programmed cell death with characteristic features such as a shrinking nucleus, condensed chromatin and fragmentation of DNA into ca 200 bp, which plays a major role in embryogenesis and development in animal cells (Wyllie 1980). In plants, it has been demonstrated that environmental stress (Ryerson and Heath 1996) as well as certain fungal infections can induce cell death with apoptotic characteristics (Wang et al. 1996a; Mittler and Lam 1996). Characteristic features of apoptosis, such as intra-nucleosomal cleavage of DNA leading to a ladder pattern of about 200 bp DNA fragments in gel electrophoresis, have been reported in developing tracheary cells (Chasan 1994), in development of the suspensory apparatus (Yeung and Meinke 1993), in tomato root cap cells (Wang et al. 1996a), barley aleurone cells (Wang et al. 1996b; 1998) and in diploid parthenogenesis and early somatic embryogenesis of Norway spruce (Havel and Durzan 1996). Plant hormones have a regulating impact on these processes as well (see review of Pennell and Lamb 1997). Recently we demonstrated that abscisic acid (ABA) inhibits DNA fragmentation in barley aleurone cells (Wang et al. 1996b).

In the present study, cellular and molecular events during pretreatment were studied. Cell death was found to be related to DNA fragmentation in both pretreatment and control conditions.

19.2 Materials and methods

19.2.1 Materials

Donor plants of *Hordeum vulgare* L. cv. Igri were grown in a phytotron under conditions described previously (Hoekstra et al. 1992). Monoclonal antibody to free (+)ABA was purchased from Idetek, Inc. (San Bruno, Calif., USA). Rabbit anti-mouse alkaline-phosphatase conjugate, (+)ABA and bovine serum albumin (grade suitable for enzyme-linked immunosorbent assay; ELISA) were obtained from Sigma Chem. Co. (St. Louis, Mo., USA).

19.2.2 Androgenesis

Selection of material and culture conditions of anthers were as described previously (Hoekstra et al. 1996) with a few modifications. In short, anthers containing late uninucleate microspores, were pretreated on mannitol with pre-medium for a period of 4 days and sampled at day 0, 1, 2, 3 and 4 for assay; at day 4 anthers were transferred to medium I (Hoekstra et al. 1992) without hormones. As a control, anthers were cultured on medium I without any pretreatment.

19.2.3 Cytological staining

Barley anthers were fixed in 2% glutaraldehyde in PBS (pH 7.3) at room temperature and prepared for TUNEL staining as described before (Wang et al. 1996b). The samples were examined with a fluorescence microscope (Leitz Diaplan). Micrographs were made using a Kodak Gold film (400 ASA).

For the viability test a FDA staining was used. Microspores were mixed with FDA solution to give a final concentration of FDA of about 0.1 μg/ml. After mounting on a slide and covering with coverslip, the samples were examined with fluorescence microscopy.

19.2.4 RNA, DNA isolations and ABA ELISA assay

Anthers were frozen in liquid nitrogen and ground to powder. Samples of anthers containing microspores were used for genomic DNA was isolated from the samples (Wang et al. 1996b).

For ABA assay, anthers were put into Eppendorff vials and frozen with liquid N_2. The contents of the Eppendorf vials were freeze dried. The samples were extracted at 150 rpm shaken, 4°C with cold methanol (containing 100 mg/l butylated hydroxytoluene and 0.5 g/l citric acid monohydrate) twice each time in 0.5 ml for 24 hours. The vials were centrifuged at 8,000 g for 15 min at 4°C and the supernatants of the samples were pooled. The samples were concentrated in a Speed-vac system (Savant) and the volume adjusted to 25 μl with methanol. The samples were stored at -80 C until assay. ABA content of anthers and pretreatment solution was determined by an enzyme-linked immuno-sorbent assay (ELISA) as described by Wang et al. (1995).

19.2.5 Experimental data

Mean values ± S.D. are presented unless stated otherwise. Significance of differences in mean values was tested with Student's *t*-test.

19.3 Results

19.3.1 DNA fragmentation in anther cells during pretreatment

When microspores were on the verge of mitosis, anthers were isolated from spikes of donor plants and pretreated with mannitol in a high calcium (20 mM) solution for 4 days (Hoekstra et al. 1992; 1997). These anthers, therefore, were under starvation, salt and osmotic stresses. After 4 days of pretreatment, microspores were isolated from pretreated anthers and subsequently incubated in culture

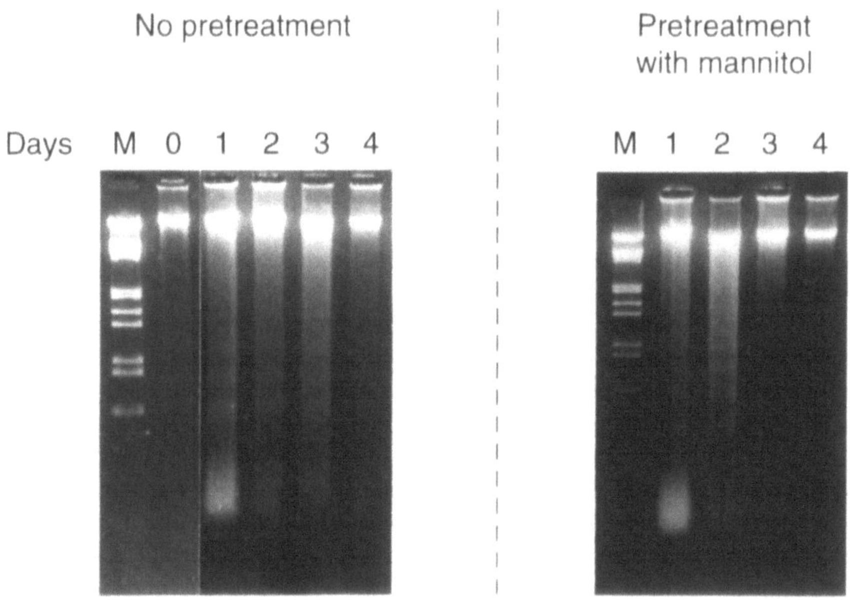

Fig. 1. DNA fragmentation in anthers with or without pretreatment. Genomic DNA (1 μg per lane), isolated from anthers incubated for various times, was loaded on a 1% (w/v) agarose in TBE buffer, stained with ethidium bromide and separated electrophoretically. The number above each lane shows the number of days of pretreatment. M: the DNA marker of lambda DNA digested with *HindII/EcoRI*. At least four independent experiments with similar results were carried out and one typical experiment is presented.

medium for induction of cell division and differentiation. During the pretreatment, the switch from the gametophytic to the sporophytic pathway takes place. As a control, anthers were directly incubated in the culture medium without above pretreatment for 4 days and then microspores were isolated from these anthers for further culture. Under control conditions, no or very low amount of plant production was observed (Hoekstra et al. 1996), while under pretreatment condition about 50 to 100 plants were produced from microspores isolated from 30 anthers.

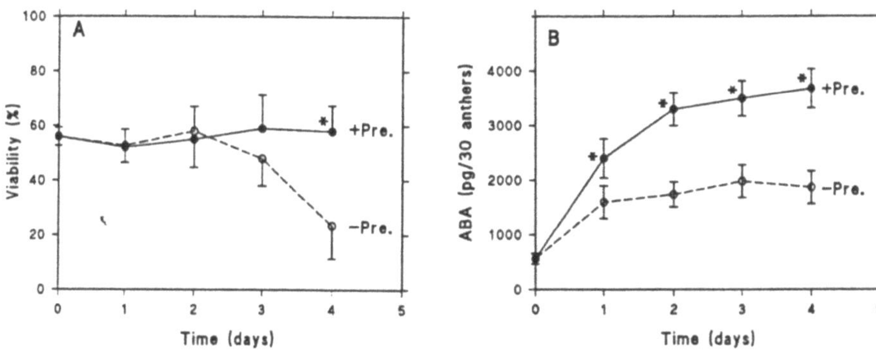

Fig. 2. Viabilities of microspores and total ABA content in anthers during pretreatment. A: FDA staining of microspores during pretreatment and without pretreatment. At each time point, 30 anthers were used for both viability test and ABA level assay. For ABA assay, the total extracts from both anther and incubation medium. The means ± SD of three independent experiments are presented in both A and B. * mean value significantly different ($P < 0.05$) from value obtained at the same time point from control (-Pre)

We were interested in the degradation (pattern) of DNA under pretreatment conditions. Gel electrophoresis of the DNA samples showed a clear 200 bp DNA "ladder" pattern under pretreatment conditions. Under these conditions, there were strong DNA fragmentations showing DNA oligomers of multiples of about 200 bp fragments after 1 and 2 days of incubation (Fig. 1). This indicated extensive intra-nucleosomal cleavage of DNA. Meanwhile, a smear of DNA degradation products was observed during incubation (Fig. 1). It was evident that the degree of DNA fragmentation during pretreatment first increased as compared to freshly isolated anthers and then decreased. Ater 4 days a situation with only one visible genomic DNA band remained. This genomic DNA band is probably DNA from surviving cells. When anthers were cultured without pretreatment, much weaker DNA fragmentation "ladders" could be observed at day 1 of incubation (Fig. 1). Apparently, the cells in anther tissue were dying independently of the pretreatment conditions. But pretreatment conditions induced stronger DNA fragmentation as compared to control conditions (without pretreatment). It is clear that anther tapetum and loculus cells are in a strong process of (programmed) cell death under pretreatment conditions.

19.3.2 The fate of anther cells and microspores during pretreatment

The viabilities of the microspores in the anther tissue with or without pretreatment were investigated. It was found that the viability of microspores in the anther was constant during pretreatment whilst a significant decrease in viability was observed without pretreatment (Fig. 2A). This decrease in viability occurred after 2 days of incubation (Fig. 2A). ABA content increased after one

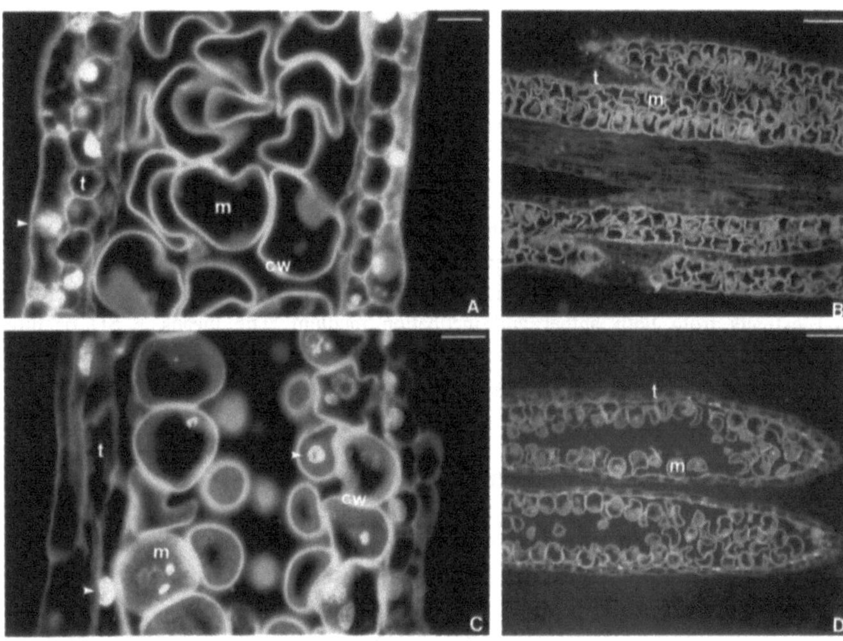

Fig. 3. TUNEL detection of DNA fragmentation in anthers under with and without pretreatment for 4 days. TUNEL staining of anthers with mannitol pretreatment (A,B) and anthers incubated in medium for 4 days without pretreatment (C,D). For each sample, 5 to 10 anthers were analyzed and representative examples are presented. cw: cell wall; m: microspore; t: tapetum cell, arrows: TUNEL stained nucleus. Bars in A and C are 150 μm. Bars in B and D are 30 μm

day of incubation under both pretreatment and control conditions (Fig. 2B). ABA continued to increase in pretreated anthers but not in untreated (control) anthers (Fig. 2B). After 4 days of pretreatment, total ABA content in the samples (anther tissue + microspores + incubation medium) was about twice that of the ABA content of non-pretreated (Fig. 2B). In order to confirm which cells in anthers were dying, we studied TUNEL staining in sections of the anther tissue. After 4 days of culture with (Figs. 3 A and B) or without (Figs. 3 C and D) pretreatment, a great loss of anther cells was observed and the arrangement of microspores in the anther was disturbed (Figs. 3A and C). It was clear that under both conditions, disappearance of tapetum cell layers and reduction of the number of loculus cells occurred (Fig. 3B and C). However, under pretreatment conditions a greater reduction of loculus cells took place than without pretreatment. This was in agreement with the DNA fragmentation data (Fig. 1). In summary, there were less microspores with TUNEL staining after pretreatment as compared to microspores without pretreatment (Fig. 3). In addition, under control conditions, there was also more cytosolic content in the microspores as compared to microspores with pretreatment (Fig. 3).

19.4 Discussion

Detection of the broken ends of DNA and the 200-bp "ladder" seen upon electrophoresis of DNA, has become a widely used experimental technique for showing that apoptosis is occurring (Wang et al. 1996b). The "programmed destruction" of a specific cell types in anthers reported by Goldberg et al. (1993) is likely due to apoptosis. During anther pretreatment, degradation of the tapetum probably occurred earlier and did not take more than 3 days which is in agreement with our data on the time course of DNA fragmentation with and without pretreatment in anthers (Figs. 1 and 3).

Upon harvest, the connection with the vascular tissue is interrupted and the anther is placed on a solution containing mannitol and calcium as major components. The exposure of anthers to this stress causes the synthesis of ABA. In the presence of high concentrations of mannitol and salt (e.g. calcium (pretreatment), much higher ABA levels were reached as compared to control. In anther tissues, cell death may be genetically controlled and independent of stress, but stress probably enhances this process (see Fig. 2). As the anther tissue is degrading, the microspores become gradually exposed to the pretreatment solution. A pretreatment-induced high level of ABA likely acts as an inhibitor of death of microspores. Moreover, the microspores enlarge under these osmotic circumstances and microspore-derived "ABA responsive" transcripts, corresponding to the specific ABA level, are expressed. The production of what is called embryogenic microspores, is established at the end of the pretreatment. The speculated roles for ABA during pretreatment can be supported by the following facts:

(A). Increased ABA level during pretreatment of anthers was correlated with constant viabilities of microspores in these anthers (Fig. 2), and a reduced number of microspores showing DNA fragmentation;

(B). A relatively low level of ABA in cultured anthers without pretreatment was correlated with an increased number of both dead microspores and microspores with fragmented DNA (Fig. 2).

(C). It has been demonstrated that ABA is able to inhibit both the DNA fragmentation and death of barley aleurone cells (Wang et al. 1996b) and the death of wheat aleurone cells (Kuo et al. 1996). In addition, it has been reported that ABA plays also an important role in induction of embryogenesis (Rajasekaran et al. 1987a; 1987b; Kiyosue et al. 1992).

After 4 days of incubation there was about a 2 times higher ABA content under the pretreatment condition than in control condition. A faster degradation of anther tissue was observed under pretreatment condition than under control condition. A constant high ABA content might be essential to prevent the death of microspores. It is likely that anther cells are dying of apoptosis during late development and pretreatment caused an increase in ABA content in microspores which subsequently inhibited the death of microspores. Under control conditions, two times lower ABA content was measured which might be the reason for the loss of viability of a large number of microspores.

The occurrence of regulated cell death processes in microspore regeneration may enable us to develop tools for improving androgenesis methods and to obtain a better understanding of the signals and molecular switches involved in development.

Acknowledgements

We thank B Van Duijn for critical reading of the manuscript and suggestions. We express our gratitude to P Hock for the lay-out of the figures.

References

Chasan R (1994) Tracing tracheary element development. Plant Cell 6: 917-919

Davies WJ, Jones HG (eds) (1991) In: Abscisic Acid Physiology and Biochemistry. Bio's scientific publishers limited.

Goldberg RB, Beals TP, Sanders PM (1993) Anther development: basic principles and practical applications. Plant Cell 5: 1217-1229

Havel L, Durzan DJ (1996) Apoptosis during diploid parthenogenesis and early somatic embryogenesis of norway spruce. Int J Plant Sci 157 : 8-16

Heberle-Bors E (1989) Isolated pollen culture in tobacco: plant reproductive development in a nutshell. Sex Plant Reprod 2: 1-10

Hoekstra S, Van Zijderveld MH, Louwerse JD, Heidekamp F, Van der Mark F (1992) Anther and microspore culture of *Hordeum vulgare* L. cv. Igri. Plant Sci 86: 89-96

Hoekstra S, Van Zijderveld MH, Heidekamp F, Van der Mark F (1993) Microspore culture of *Hordeum vulgare* L.: the influence of density and osmolality. Plant Cell Rep 12: 661-665

Hoekstra S, Van Bergen S, Van Brouwershaven IR, Schilperoort RA, Heidekamp F (1996) The interaction of 2,4-D application and mannitol pretreatment in anther and microspore culture of *Hordeum vulgare* L. cv. Igri. J Plant Physiol 148: 696-700

Hoekstra S, Van Brouwershaven IR, Van Bergen S, Schilperoort RA, Wang M (1997) The role of mannitol, calcium and ABA during pretreatment in barley androgenesis. Plant Sci 126: 211-218

Imamura J, Harada H (1980) Effects of abscisic acid and water stress on the embryo and plantlet production in anther culture of *Nicotiana tabacum* cv. Samsun. Z Pflanzen Physiol 100: 285-289

Kiyosue T, Nakajima M, Yamaguchi I, Satoh S, Kamada H, Harada H (1992) Endogenous levels of abscisic acid in embryogenic cells, non-embryogenic cells and somatic embryos of carrot (*Daucus carota* L.). Biochem Physiol Pflanzen 188: 343-347

Kuo A, Cappelluti S, Cervantes-Cervantes M, Rodriguez M, Bush DS (1996) Okadaic acid, a protein phosphatase inhibitor, blocks calcium changes, gene expression, and cell death induced by gibberellin in wheat aleurone cells. The Plant Cell 8: 259-269

Mittler R, Lam E (1996) Sacrifice in the face of foes: pathogen-induced programmed cell death in plants. Trends Microbiol 4: 10-15

Pennell RI, Lamb C (1997) Programmed cell death in plants. The Plant Cell 9: 1157-1168

Rajasekaran K, Hein MB, Vasil IK (1987a) Endogenous abscisic acid and indole-3-acetic acid and somatic embryogenesis in cultured leaf explants of *Pennisetum purpureum* Schum. Plant Physiol 84: 47-51

Rajasekaran K, Hein MB, Davis GC, Garnes MG, Vasil IK (1987b) Endogenous growth regulators in leaves and tissue cultures of *Pennisetum purpureum* Schum. J Plant Physiol 130: 13-25

Reynolds TL, Crawford RL (1996) Changes in abundance of an abscisic acid responsive, early cysteine labelled metallothionein gene during pollen embryogenesis in bread wheat (*Triticum aestivum*). Plant Mol Biol 32(5): 823-829

Ryerson DE, Heath MC (1996) Cleavage of nuclear DNA into oligonucleosomal fragments during cell death induced by fungal infection or by abiotic treatments. Plant Cell 8: 393-402

Sunderland N., Huang B, Hills GJ (1984) Disposition of pollen *in situ* and its relevance to anther/pollen culture. J Exp Bot 35: 521-530

Wang M, Heimovaara-Dijkstra S, van Duijn B (1995) Modulation of germination of embryos isolated from dormant and nondormant barley grains by manipulation of endogenous abscisic acid. Planta 195: 586-592

Wang H, Li J, Bostock RM, Gilchrist DG (1996a) Apoptosis: a functional paradigm for programmed plant cell death induced by a host-selective phytotoxin and invoked during development. Plant Cell 8: 375-391

Wang M, Oppedijk BJ, Lu X, Van Duijn B, Schilperoort RA (1996b) Apoptosis in barley aleurone during germination and its inhibition by abscisic acid. Plant Mol Biol 32: 1125-1134

Wang M, Oppedijk BJ, Caspers MPM, Lamers GEM, Boot, MJ, Geerlings DNG, Bakhuizen B, Meijer AH, Van Duijn, B (1998) Spatial and temporal regulation of DNA fragmentation in the aleurone of germinating barley. J Exp Bot 49:

Wyllie AH (1980) Cell death: The significance of apoptosis. Int Rev Cytol 68: 251-306

Yeung EC, Meinke DW (1993) Embryogenesis in angiosperms: development of the suspensory. Plant Cell 5: 1371-1378

20 Androgenesis and albinism in Poaceae: influence of genotype and carbohydrates

Caredda S and Clément* C

20.1 Introduction

Androgenesis consists in the in vitro regeneration of haploid plantlets from microspores which were initially programmed to develop into pollen grains. It has been succesfully developed and improved in Poaceae for the last three decades. Nowadays, androgenesis is carried out in routine in a few species and is further tentatively combined with genetic transformation in *Zea mays* (Jardinaud et al. 1995a; 1995b), *Triticum* (Loeb et al. 1994) and *Hordeum vulgare* (Kasha et al. 1995; Yao et al. 1997). However, albino plantlets are obtained following androgenesis with a fluctuating ratio between the number of green and albino androgenetic haploid plantlets (G/A ratio). This phenomenon exclusively occurs in Poaceae and decreases the potential yield of androgenesis (Jähne and Lörz 1995). Each species of this family is affected by albinism, especially *Hordeum vulgare* and *Oryza sativa*, which exhibit a G/A ratio lower than 1 under most culture conditions (Table 1). In *Hordeum vulgare*, *Triticum aestivum* and *Oryza sativa*, some genotypes even form 100 % of albino androgenetic plantlets (Chu et al. 1990; Orshinsky et al. 1990; Jähne et al. 1991; Xie et al. 1995). Albinism therefore represents an obstacle to the use of androgenesis in breeding programs that is only poorly documented and should be further investigated (Jähne and Lörz 1995).

The microspore enters the androgenetic process following two possible pathways: 1/ the microspore develops into a haploid callus from which haploid plantlets can be regenerated (Fujii 1970; Clapham 1971; 1973); 2/ the microspore directly develops into haploid embryo which further regenerates a haploid plantlet (Garrido et al. 1995). The latter process is the most commonly used today. It can be decomposed into 3 steps (Hoekstra et al. 1996; Touraev et al. 1996): 1/ the pretreatment destined to switch the pollen fate from the initial gametophytic program to the alternative sporophytic program; 2/ the induction phase, which

* for correspondence: Université de Reims Champagne Ardenne; Biologie et Physiologie Végétales; BP 1039; 51687 Reims cedex 2; France

Table 1. Occurence of albino plantlets in anther culture (AC) and microspore culture (MC). nc: not calculated

Species	Culture	G/A ratio	References
Triticum aestivum	AC	0-16	Orshinsky et al. 1990
	AC	0	Chu et al. 1990
	AC	0	Hassawi and Liang 1990
	AC	6.2	Hassawi et al. 1990
	AC	0-5.1	Konzak and Zhou 1991
	AC	2.3	Ball et al. 1993
	AC	<1	Stober and Hess 1997
	AC	4.1	Orshinsky and Sadasivaiah 1997
	MC	1	Gustafson et al. 1995
Hordeum vulgare	AC	<1	Powell et al. 1988
	AC	1.3	Finnie et al. 1989
	AC	0	Knudsen et al. 1989
	AC	<1	Muyuan et al. 1990
	AC	0	Jähne and Lörz 1991
	AC	<1	Szarejko and Kasha 1991
	AC	9	Kao et al. 1991
	AC	9.4	Li et al. 1993
	AC	18	Jähne et al. 1991
	MC	11	Ziauddin et al. 1990
	MC	34	Hoekstra et al. 1993
Zea mays	AC and MC	nc	Büter 1997
Oryza sativa	AC	1.5	Lentini et al. 1995
	AC	<1	Zhu et al. 1991
	MC	0	Xie et al. 1995
	MC	<1	Ogawa et al. 1995
	MC	1.7	Xie et al. 1996
Secale cereale	AC	0	Rakoczy-Trojanowska et al. 1997
Phleum pratense	AC	0-3.7	Abdullah et al. 1994
Dactylis glomerata	AC	<1	Christensen et al. 1997
Sinocalamus latiflora	AC	nc	Tsay et al. 1990
Avena sativa	AC	12	Sun et al. 1991
Pennisetum glaucum	AC	6	Choi et al. 1997 (review)
Sorghum	AC	nc	Liang et al. 1997
Triticosecale	AC	<1	Barcelo et al. 1994

consists in the embryogenic development of the microspore; 3/ the regeneration phase allowing the development of androgenetic embryos into haploid plantlets.

Performing androgenesis, anther or isolated microspore culture can be used, but anther culture is usually preferred because it is time saving working on dissected anther instead of using mechanical isolation of microspores. However, microspore culture enhances both the microspore response and the G/A ratio. For instance, in

Hordeum vulgare the regeneration rate can be exceptionally high using isolated microspore culture, reaching up to 2,700 green plantlets per 100 anthers (Hunter 1988). Besides, the G/A ratio has been recently increased from 1 in anther culture (Luckett and Smithard 1995) to 34 in microspore culture (Ziauddin et al. 1992; Cistué et al. 1995; Hoekstra et al. 1993).

The mechanism of androgenesis remains widely misunderstood, especially the occurrence of albinism in the Poaceae. Nevertheless, three parameters of the androgenetic process appear to be mainly involved in regards to the obtention of albino plantlets: 1/ the genotype of the mother plant (Finnie et al. 1989); 2/ the osmotic pressure during the pretreatment (Hoekstra et al. 1993); 3/ the carbohydrate source in the induction culture medium (Pickering and Devaux 1992).

Therefore, this chapter will successively consider the available literature focusing on these three parameters, before giving a tentative explanation of the origin of albinism.

20.2 Genotypic aspect of albinism

In Poaceae, there are large genotypic differences in both the microspore response and the relative proportion of green and albino regenerants following androgenesis (Finnie et al. 1989; Knudsen et al. 1989), even though the 2 characters do not fluctuate in parallel. The genotype effect is particularly important in *Triticum aestivum* (Lazar et al. 1984; Andersen et al. 1987) and *Hordeum vulgare* (Foroughi-Wehr et al. 1982; Knudsen et al. 1989). In both species, the G/A ratio is known to vary from 5% to 97% according to the genotype (Knudsen et al. 1989; Tuvesson et al. 1989; Zhou and Konzak 1989).

In *Triticum aestivum*, the G/A ratio has been significantly increased for years independently of the microspore response. Indeed, it currently reaches 16 in the genotype $F_1$88292 (Orshinsky et al. 1990) and Chu et al. (1990) even obtained 100% of green androgenetic plantlets with the genotype F_1Erik/HY320 (Table 2). At the opposite, some genotypes remain unable to produce any green plantlet such as the genotypes F188067, T80 and the cultivar Edwall (respectively Orshinski et al. 1990; Hassawi et al. 1990; Konzak and Zhou 1991). Also, some genotypes exhibit high androgenetic potential such as Ciano 067, Pavon 076 and Dirkwin, giving rise respectively to 70, 72 and 357 green haploid plants per 100 anthers (Ouyang et al. 1983; Zhou et al. 1991; Orshinsky and Sadasivaiah 1997), but unfortunately, the G/A ratio was not given by the authors, although it may be very useful in the knowledge of albinism. Tetraploid wheats are highly recalcitrant to androgenesis and most of the regenerated plants are further albinos (Zhu et al. 1979; Hadwieger and Heberle-Bors 1986; Foroughi-Wehr and Zeller 1990).

In *Hordeum vulgare*, the genotype is responsible for more than 60% of the variations obtained in anther cultures (Larsen et al. 1991) and a great part of regenerated plantlets is albinos. In this species, the cultivar Igri represents the model for the improvement of both anther and microspore cultures (Foroughi-

Wehr and Friedt 1984; Cistué et al. 1994), although it is no more used in breeding programs. In this cultivar, under anther culture (Table 2), the G/A ratio fluctuates according to the genotype from 1.7 to 9.4 (Li et al. 1993). The best results in the cultivar Igri were obtained by Hoekstra et al. (1993) with 5,000 green plantlets per 100 anthers and a G/A ratio of 34 by using microspore culture and modifying the osmotic pressure during the androgenetic process (see below).

A few studies reported that, in *Hordeum vulgare*, three kinds of genotypes can be distinguished related to the mean value of the G/A ratio (Knudsen et al. 1989; Jähne et al. 1991): genotypes leading to a G/A ratio in the range of 0.66 (30-40 % of green plants); lower than 0.01 (5-8% of green plants); or genotypes leading exclusively to albino androgenetic plantlets. These results show that the ability to produce high percentage of green androgenetic plantlets is dependent upon the genotypes and suggest that deeper genetic studies are undertaken in order to characterize this parameter.

Oryza sativa is the most albino producing species (Table 2). In anther culture, the influence of the genotype is particularly important: the genotype IR 43 does not regenerate any green plantlet, whereas the genotype INTI produces 100% green plantlets (Lentini et al. 1995). Microspore culture is also characterized by a genotype effect on green plantlet regeneration, i.e. Xiushui 117 and 02428 regenerate respectively 2.5 and 19.5 plantlets per 100 anthers with 80% and 100% of albino plantlets (Xie et al. 1995).

In *Zea mays*, anther and microspore cultures are not yet routine techniques, since only a few genotypes can produce microspore-derived embryos, and the efficiency is lower than in the species previously described (Genovesi and Collins 1982; Dieu and Beckert 1986). The first regenerated haploid plantlets under microspore culture were obtained in 1989 (Coumans et al. 1989; Pescitelli et al. 1989). Approximately 125 embryo like structures and 10 regenerated plantlets per 100 anthers are obtained under the best culture conditions with the genotype ETH-M 24 (Büter 1997) but no studies dealt with the genotypic effect on the G/A ratio (Table 2).

In *Secale cereale*, the regeneration of green androgenetic plantlets is difficult whatever the genotype (Wenzel et al. 1976; Hoffmann and Wenzel 1981; Sharma et al. 1982). Moreover, most of regenerated plantlets are albinos, i.e. 100% of albino plants with the genotype F1(5) (Rakoczy-Trojanowska et al. 1997). In the highly responding *Lolium perenne* clone 255, Andersen et al. (1997) could regenerate 182 green plantlets per 100 anthers using anther culture but the G/A ratio is only 1. In *Pennisetum glaucum*, the genotype plays also a key role in microspore response (Choi et al. 1989) and the G/A ratio can reach up to 7 in the Chungaecho hybrid (Choi et al. 1997). In other species of Poaceae such as *Phleum pratense*, *Dactylis glomerata*, *Sinocalamus latiflora*, *Avena sativa*, *Sorghum* and *Triticosecale*, correlations between the genotype and the G/A ratio have not been investigated since they were only recently used in androgenesis and the optimization of the microspore embryogenesis induction is still in progress.

Table 2. Influence of genotype on anther (AC) and microspore (MC) cultures in various cultivars and genotypes. Reg.: regenerated

Species	Culture	Reg. Plantlets /100 anthers	G/A ratio	References
Triticum aestivum				
F1 88292	AC	115	16	Orshinsky et al. 1990
F1 88067	AC	0	0	
F1 Thunderbird/ Norwick	AC	2	0	Chu et al. 1990
F1 Erik/HY320	AC	34	100	
Chris	AC	42	5.1	Konzak and Zhou 1991
Edwall	AC	12	0	
Pavon 76	AC	8	8	Hassawi and Liang 1990
T 81	AC	0	0	
Chris	AC	7.6	6.2	Hassawi et al. 1990
Centruk	AC	1.8	<1	
Hordeum vulgare	AC			
Princesse	AC	20	0	Jähne et al. 1991
Baronese	AC	20	0	
Tweed	AC	341	1.3	Finnie et al. 1989
Tyne	AC	56	<1	
Chernigovski	AC	86	2.7	Tiwari and Rahimbaev 1992
Dnieprovski	AC	19	8	
Igri	AC	420	9.4	Li et al. 1993
Atias	AC	120	1	
Igri	AC	1500	19	Cistué et al. 1994
Albacette	AC	100	1	
Bruce	AC	29	1.9	Cai et al. 1992
Klages	AC	15	<1	
H. spontaneum				
IB 122	AC	10	1.4	Piccirilli and Arcioni 1991
IB 74	AC	2	<1	
RS 170-47 (A)	AC	12	<1	Islam et al. 1992
1B-152B	AC	23	1.2	
Zea mays				
ETH-M24	AC	8.9	nc	Büter 1997
Oryza sativa				
Xiushi 117	MC	2.5	0.8	Xie et al. 1995
02428	MC	19.5	0	
INTI	AC	20	100	Lentini et al. 1995
Tox 1011	AC	-	<1	
Avena sativa				
	AC	13	12	Rines et al. 1997
Secale cereale				
DZ	AC	0	-	Rakoczy-Trojanowska et al. 1997
F1(5)	AC	3	0	
Phleum pratense				
29-92	AC	20	1.5	Abdullah et al. 1994
10-92	AC	0	0	

20.3 Influence of carbohydrates on the G/A ratio

Two types of carbohydrates are used during the androgenetic process in Poaceae: non metabolizable sugars such as mannitol adjust the osmotic pressure during anther pretreatment and metabolizable sugars such as sucrose or maltose are nutrient in the induction media for developing embryogenic microspores. The carbohydrate composition during both pretreatment and induction has been largely optimized in Poaceae but most of studies concerning the comparison between sugars have been carried out on anther culture (Table 3), although it should be also investigated using microspore culture (Hoekstra et al. 1993).

20.3.1 Osmotic pressure during the pretreatment

Pretreatment causes the prerequired stress for switching the pollen fate (Touraev et al. 1996; Hoekstra et al. 1996). Among various types of anther pretreatment, the adjustment of osmotic pressure seems to be the most appropriate in Poaceae (Hoekstra et al. 1993). It has a positive influence on the obtention of androgenetic green plantlets under both anther and microspore cultures (Hoekstra et al. 1992; Pickering and Devaux 1992; Cistué et al. 1994). Most of investigations on the impact of osmotic pressure were performed in *Hordeum vulgare* or in *Triticum aestivum*, although it could give good results when applied in anther/microspore culture of recalcitrant and high albino producing Poaceae species.

In *Hordeum vulgare*, a mannitol pretreatment (450 mOsm/kg) during 3 days is more efficient than a cold pretreatment at 4°C during 4 weeks (Roberts-Oehlschlager et al. 1990; Hoekstra et al. 1992; Luckett and Darvey 1992; Hou et al. 1993; Mordhorst and Lörz 1993; Cistué et al. 1994), enhancing plant production at least ten fold (Hoekstra et al. 1996). The increase of osmolality with mannitol enables microspores to enter embryogenesis with a higher frequency, but also positively influences the G/A ratio (Hoekstra et al. 1997). Adjusting the osmotic pressure at 450 mOsm/kg during anther pretreatment and 350 mOsm/Kg in the induction medium of isolated microsproes, the G/A ratio could reach 34 (Hoekstra et al. 1993).

In *Triticum*, microspores are also sensitive to the osmotic pressure. The addition of 200g/l Ficoll in the induction medium significantly increases the percentage of green plants from 38 to 62 % and leads to a G/A ratio of 2 (Zhou et al. 1991).

20.3.2 Sugars in the induction media

During Poaceae androgenesis, the nature of the sugars used in the induction media is more determinant than the growth regulators (Finnie et al. 1989; Roberts-

Oehlschlager et al. 1990; Olsen 1991; Cai et al. 1992; Karsai et al. 1994). Carbohydrates in the induction medium affect the success of androgenesis as well as the G/A ratio under both anther and microspore culture (Finnie et al. 1989; Roberts-Oehschlager et al. 1990).

In *Hordeum vulgare* (Finnie et al. 1989), *Triticum* (Chu et al. 1990; Orshinsky et al. 1990; Zhou *et al.* 1991) and *Oryza sativa* (Deimling et al. 1992; Deimling and Flehinghaus-Roux 1997), the use of maltose instead of sucrose as a carbohydrate source has led to a substantial increase in the induction of microspore embryogenesis and plant regeneration.

Hordeum vulgare is the most sensitive Poaceae to the influence of the carbohydrate source on androgenesis (Table 3). Sucrose, glucose and fructose, usually used as carbohydrate source in other in vitro culture systems are toxic for barley androgenesis (Scott and Lyne 1994b). They are rapidly assimilated by the microspores into starch reserves and further lead to an hypoxy of the medium (Scott et al. 1995). Pure maltose from 0.17 M to 0.26 M (Thompson et al. 1991; Luckett and Smithard 1992; Lührs and Nielsen 1992; Pickering and Devaux 1992; Tiwari and Rahimbaev 1992; Ziauddin et al. 1992; Salmenkallio-Marttila and Kauppinen 1995) or maltose 0.17 M in association with cellobiose and/or melibiose 0.17 M (Roberts-Oehschlager et al. 1990; Cai et al. 1992; Luckett and Darvey 1992; Pickering and Devaux 1992) gave the best results. Maltose is slowly assimilated by microspores (Scott et al. 1995) and induced a high frequency of microspores embryogenic development as well as a great proportion of green plantlets versus albino plantlets (Sorvari and Schieder 1987; Finnie et al. 1989; Piccirilli and Arcioni 1991; Cai et al. 1992). The combination of microspore culture with mannitol anther pretreatment and maltose in the induction medium is currently the most suitable protocol for the success of barley androgenesis (Hoekstra et al. 1993).

In *Triticum*, maltose also appears as the best sugar in the induction medium. Ouyang (1986) first reported that 0.26 M sucrose was the most appropriate sugar concentration for anther culture in terms of regeneration ability. But maltose progressively replaced sucrose and is now commonly used at 0.25 M for pollen embryogenesis (Last and Brettel 1990; Orshinsky et al. 1990).

The influence of sugar in the induction medium is highly genotype dependent (Table3), especially in *Triticum*. Using anther culture with the cultivar F188292, 18 green plantlets/100 anthers and a low G/A (not given) were obtained with 0.23 M sucrose, against 115 green plantlets/100 anthers and a G/A of 16 with 0.23 M maltose (Orshinsky et al. 1990). At the opposite, Chu et al. (1990) use monosaccharides instead of sucrose as the carbon source in the induction medium. With the genotype F1 Thunderbird/Norwin, replacing 0.2 M sucrose by 0.2 M glucose in the induction medium, the microspore response increases from 17 to 94 embryos/100 anthers, but the G/A ratio remains unchanged with a value of 3. These latter results show that, in this special case, the genotype effect is preponderant when compared to the nature of the sugar.

Table 3. Influence of sugars in anther (AC) and microspore (MC) cultures in various cultivars and genotypes. Conc.: concentration, Cult.: culture, nc: not calculated, RP/100A.: regenerated plantlets/100 anthers

Species	Cult.	Sugar	Conc (g/l)	RP/ 100 A	G/A ratio	References
Hordeum vulgare						
Dissa	AC	sucrose	80	13.6	1.7	Sorvari and Schieder
	AC	melibiose	80	27.5	26.5	1987
Tweed	AC	sucrose	60	8.6	0.2	Finnie et al. 1989
	AC	maltose	60	341	1.3	
Chernigovski	MC	sucrose	90	8	<1	Tiwari and Rahimbaev
	MC	maltose	90	86	2.7	1992
H930-36	AC	sucrose	60	27	<1	Cai et al. 1992
	AC	maltose	60	50	<1	
H. spontaneum						
Koch	AC	sucrose	80	0	nc	Piccirilli and
	AC	maltose	80	10	1.4	Arcioni 1991
Triticum aestivum						
F_1 88292	AC	sucrose	80	18	nc	Orshinsky et al. 1990
	AC	maltose	80	115	16	
F_1Thunderbird/	AC	sucrose	71	2	0	Chu et al. 1990
Norwin	AC	glucose	38	16	<1	
Pavon 76	AC	sucrose	90	106	1.1	Zhou et al. 1991
	AC	maltose	135	123	1.9	
Zea mays						
ETH-M24	AC	sucrose	90	8.9	nc	Büter 1997
	AC	maltose	90	0.9	nc	
Oryza sativa						
Taipei 309	AC	sucrose	60	0	-	Xie et al. 1995
	AC	maltose	60	32	<1	
Avena sativa						
	AC	sucrose	65	-	-	Rines et al. 1997
	AC	maltose	65	13	12	
Secale cereale						
DZ	AC	sucrose	30	0	-	Rakoczy-Trojanowska
	AC	maltose	120	9	<1	et al. 1997
Phleum pratense						
29-92	AC	sucrose	60	0	-	Abdullah et al.
	AC	maltose	90	20	1.5	1994

In *Oryza sativa*, maltose was largely used in the induction medium with an optimal concentration of 0.17 M. The microspore response was consequently enhanced up to 32 plantlets/100 anthers (Table 3), but the G/A ratio remains lower than 1 in all tested genotypes (Xie et al. 1995).

Zea mays is a particular case. Present works focus on increasing the plantlet regeneration yield and the G/A ratio still represents a secondary problem. In fact, the regeneration of albino plantlets occurs in defined genotypes but is not generally considered as the main obstacle. The N6 (Chu 1981) or YÜ-PEI (Ku *et al.* 1981) media are most commonly used (Brettel et al. 1981; Genovesi and Collins 1982; Pauk 1985; Dieu and Beckert 1986; Petolino and Jones 1986; Tsay et al. 1986; Pace et al. 1987; Genovesi 1990; Barloy and Beckert 1993; Pescitelli et al. 1994): sucrose concentration in a range of 0.26-0.35 M is the most appropriate for microspore response, but the value of the G/A ratio is not mentionned in both anther and microspore culture. The utilization of sucrose results in significantly higher overall plant production than cellobiose, maltose, fructose, glucose, trehalose, galactose. Only raffinose gave comparable embryo like structures production (Büter 1997).

In *Phleum pratense* and *Avena sativa* (Table 3), anther culture with 0.26 M and 0.19 M maltose can lead to only 15 and 13 plantlets/100 anthers (respectively Abdullah et al. 1994; Rines et al. 1997),, but the G/A reaches 15 with the genotype Phleeviola 15 (Abdullah et al. 1994) and 12 with a naked oat (Sun et al. 1991).

20.4 Towards a better understanding of albinism

A few information is available about the fundamental basis of albinism in androgenetic plants. Genetic studies testing various crosses of *Hordeum vulgare* and *Triticum aestivum* for the ability of microspores to undergo embryogenesis reveal that the genes controlling the microspore response as well as the G/A ratio are inherited chromosomally (Andersen et al. 1987; Sonnino et al. 1989; Tuvesson et al. 1989; Larsen et al. 1991).

Besides, plastids of albino androgenetic plantlets have been investigated in order to further characterize the albino phenotype. In *Hordeum vulgare*, plastids are present throughout the formation of microspore-derived embryo (Sunderland and Huang 1985). During the induction phase of the androgenetic process, plastids in the microspore-derived embryo are undifferentiated and just contain a few thylakoids and starch (Mlodzianowski and Idzikowska 1981), whatever the chlorophyllous or albino fate of the developing embryo. Later on, when the embryo regenerates an albino plantlet, plastids within are deprived of ribosomes (Sunderland and Huang 1985).

In this species, when androgenesis is performed via a callus phase, the plastid genome of albino androgenetic plantlets is affected and genes related to the albino phenotype of androgenetic plantlets have been identified (Dunford and Walden 1991): transcripts from the plastid genes *rbc L*, *psb D-psb C*, as well as the 16S and 23S rRNAs are present at reduced levels in albino plantlets when compared to those found in green plantlets. Alterations of plastid DNA are not entirely responsible of the albino phenotype: the nuclear genome is also involved in this process. Indeed, transcripts of the nuclear genes *rbc S* and *Cab*, which encoded chloroplast destined proteins are abnormally low in albino plants (Dunford and Walden 1991). Both nucleus and plastid genomes are then affected in callus-derived albino plantlets, leading to the lack of chlorophyll and Rubisco synthesis.

In *Triticum*, albino androgenetic plantlets obtained via a callus phase also exhibit modifications of the cytoplasmic genome such as deletions in plastid DNA;

in particular, an inverted repeated sequence arranged in a hairpin configuration was identified (Day and Ellis 1984; 1985; Ellis and Day 1986).

In both species, the alteration of plastid and nucleus genome was demonstrated in androgenetic albino plantlets obtained via a callus phase (Day and Ellis 1984; 1985; Ellis and Day 1986; Dunford and Walden 1991). But no information is available when microspore enters directly embryogenesis.

In *Oryza sativa*, albino androgenetic plantlets have common features with those in *Hordeum vulgare*. Albino microspore-derived plantlets exhibit immature plastids (Sun et al. 1974), with small amounts of 16S and 23S rRNAs, as well as reduced levels of Rubisco (Sun et al. 1979). The same genes may then be affected during microspore embryogenesis in *Hordeum vulgare* and *Oryza sativa*.

The nucleus or plastid DNA defects related to androgenetic albinism may be already determined in the microspore before sampling for culture. Indeed, the fate of pollen plastids in vivo is restricted in the Poaceae: they are usually destined to provide starch reserves during the pollen tube elongation and next disappear soon after fertilization (Mogensen 1996). These data suggest that a degradation of plastid DNA may occur during in vivo pollen development (Mogensen 1996). This idea was partly supported by recent studies. In *Hordeum vulgare*, DNA fragmentation naturally exists in the whole anther during microsporogenesis (Wang et al. 1999). A crucial point for the induction of microspore embryogenesis consists in stopping this phenomenom and restoring DNA integrity (Wang et al. 1999). This work does precise if DNA fragmentation affects nuclear or plastid but it is likely that, in Poaceae, alterations of plastid DNA in the microspore are not reversible during androgenesis, leading to the production of albino plantlets. Furthemore, the existence of a putative plastid signal has been suggested to control the expression of nuclear genes involved in the photosynthetic pathway and related functions in the chloroplast (Börner et al. 1986; Rajasekhar 1991). Altered plastids in microspore could therefore be responsible for nuclear genome deficiencies.

20.5 Conclusions

The occurrence of androgenetic albinism in Poaceae remains widely unclear and is still regarded as a problem in using anther/microspore culture in cereal breeding programs (Jähne and Lörz 1995). Considering the numerous cultivars and genotypes in each cultivated species of this family, no general rule can be drawn concerning microspore behaviour in vitro in regards to albinism.

The genotype appears to be the most important factor in androgenetic albinism, even though the improvement of the anther pretreatment and the culture conditions largely contributed to reduce the amount of albino regenerated plantlets in *Hordeum vulgare* and *Triticum aestivum*. DNA modifications such as deletions or rearrangements of both nucleus and plastid genomes are involved in the albino phenotype by reducing or removing the expression of genes coding for chlorophyll synthesis and photosynthetic related enzymes. Such alterations may due to the initial gametophytic program of the in vivo developing microspore in which the fate of plastids is limited.

Numerous papers deal with androgenesis in Poaceae but all the papers do not report the amount of albino regenerated plantlets. In order to establish a kind of data base of androgenetic albinism in Poaceae, the results section should systematically include the G/A ratio. This does not represent so much additional work and would provide useful information in comparing cultivars/genotypes or various culture conditions.

The osmolality during pretreatment and the sugar content in the medium have been largely optimized in regards to albino production (Orshinsky et al. 1990; Luckett and Smithard 1992; Barloy and Beckert 1993; Xie et al. 1995). Besides, it has been shown recently that sugars are involved in the regulation of gene expression (Hattori et al. 1990; Hattori et al. 1991; Yu et al. 1996). Consequently, the carbohydrates used during the pretreatment or the induction phase may influence positively the expression of some microspore genes, decreasing the amount of albino androgenetic plantlets (Sorvari and Schieder 1987; Finnie et al. 1989; Kuhlmann and Foroughi-Wehr 1989; Olsen 1991; Piccirilli and Arcioni 1991; Kao 1993; Scott and Lyne 1994b). Working in this optic may be of interest in understanding the impact of sugars on androgenetic albinism.

In *Zea mays*, a 32 kDa protein is specifically accumulated in the anther wall of responsive genotypes during the pretreatment (Vergne *et al.* 1990). The level of this protein is positively correlated to the frequency of androgenetic embryo and the authors finally suggest that this protein may represent a marker of androgenetic responsiveness. Similarly, genotypes producing exclusively albino plantlet should be investigated in order to look for markers of ability to provide albino androgenetic plantlets.

The data reported in this paper show that the rate of albinism cannot be dissociated from the frequency of microspore response. Therefore, in order to completely understand and further prevent albino plantlet formation during androgenesis, it is necessary to undertake fundamental studies dealing with the whole androgenetic process and not only focusing on albinism.

Acknowledgements

The authors thank Dr M Burrus for critically reading the manuscript.

References

Abdullah AA, Pedersen S, Andersen SB (1994) Triploid and hexaploid regenerants from hexaploid Thimothy (*Phleum pratense* L.) via anther culture. Plant Breed 112: 342-345

Andersen SB, Madsen S, Roulund N, Halberg N, Olesen A (1997) Haploidy in ryegrass. In Jain SM, Sopory SK, Veilleux RE (eds) In vitro haploid production in higher plants. 4. Kluwer Academic Publishers, The Netherlands, pp 133-147

Andersen SB, Due IK, Olesen A (1987) The response of anther culture in a genetically wide material of winter wheat (*Triticum aestivum* L.). Plant Breed 99: 181-186

Ball ST, Zhou HP, Konzak CF (1993) Influence of 2,4-D, IAA, and duration of callus induction in anther cultures of spring wheat. Plant Sci 90: 195-200

Barcelo P, Cabrera A, Hagel C, Lörz H (1994) Production of doubled-haploid plants from *tritordeum* anther culture. Theor Appl Genet 87: 741-745

Barloy D, Beckert M (1993) Improvement of regeneration ability of androgenetic embryos by early anther transfer in maize. Plant Cell Tiss Org Cult 33: 45-50

Börner T, Mendel RR, Schiemann J (1986) Nitrate reductase is not accumulated in chloroplast-ribosome deficient mutants of higher plants. Planta 169: 202-207

Brettel RI, Thomas E, Wernicke W (1981) Production of haploid maize plants by anther culture. Maydica 26: 101-111

Büter B (1997) In vitro haploid production in maize. In Jain SM, Sopory SK, Veilleux RE (eds) In vitro haploid production in higher plants. 4. Kluwer Academic Publishers, The Netherlands, pp 37-71

Cai Q, Szarejko I, Polok K, Maluszynski M (1992) The effect of sugars and growth regulators on embryoid formation and plant regeneration from barley anther culture. Plant Breed 109: 218-226

Choi BH, Park KY, Park RK (1989) Response to anther and tissue cultures of corn, pearl millet and buckwheat genotypes. Korean J Crop Sci 34: 142-146

Choi BH, Park KY, Park RK (1997) Haploidy in pearl millet (*Pennisetum glaucum* L.). In Jain SM, Sopory SK, Veilleux RE (eds) In vitro haploid production in higher plants. 4. Kluwer Academic Publishers, The Netherlands, pp 171-179

Christensen JR, Borrino E, Olesen A, Andersen SB (1997) Diploid, tetraploid, and octoploid plants from anther culture of tetraploid orchard grass, *Dactylis glomerata* L. Plant Breed 116: 267-270

Chu CC, Hill RD, Brule-Babel A (1990) High frequency of pollen embryoid formation and plant regeneration in *Triticum aestivum* L. on monosaccharide containing media. Plant Sci 66: 255-262

Chu CC (1981) The N6-medium and its applications to anther culture of cereal crops. Proc Symp Plant Tissue Culture, Beijing. Pitman, Boston, pp 43-50

Cistué L, Ziauddin A, Simion E, Kasha KJ (1995) Effects of culture conditions on isolated microspore response of barley cultivar Igri. Plant Cell Tiss Org Cult 42: 163-169

Cistué L, Ramos A, Castillo AM, Romagosa I (1994) Production of large number of doubled haploid plants from barley anthers pretreated with high concentrations of mannitol. Plant Cell Rep 13: 709-712

Clapham D (1971) In vitro development of callus from the pollen of *Lolium* and *Hordeum*. Z Pflanzenzüchtg 65: 285-292

Clapham D (1973) Haploid *Hordeum* plants from anthers in vitro. Z Pflanzenzüchtg 69: 142-155

Coumans MP, Sohota S, Swanson EB (1989) Plant development from isolated microspores of *Zea mays* L. Plant Cell Rep 7: 618-621

Day A, Ellis THN (1984) Chloroplast DNA deletions associated with wheat plants regenerated from pollen: possible basis for maternal inheritance of chloroplasts. Cell 39: 359-368

Day A, Ellis THN (1985) Deleted forms of plastid DNA in albino plants from cereal anther culture. Curr Genet 9: 671-676

Deimling S, Flehinghaus-Roux T (1997) Haploidy in rye. In Jain SM, Sopory SK, Veilleux RE (eds) In vitro haploid production in higher plants. 4. Kluwer Academic Publishers, The Netherlands, pp 181-204

Deimiling S, Flehinghaus T, Bohn M, Geiger HH (1992) Sugary aspects of androgenesis-alterations of carbohydrates during haploid induction in gramineous species. In Angers University (ed) Proceedings XIIIth Eucarpia Congress, pp 153-154
Dieu P, Beckert M (1986) Further studies of androgenetic embryo production and plant regeneration from in vitro cultured anthers of maize (*Zea mays* L.). Maydica 31: 245-260
Dunford RP, Walden RM (1991) Plastid genome structure and plastid-related transcript levels in albino barley plants derived from anther culture. Curr Genet 20: 339-347
Ellis THN, Day A (1986) A hairpin plastid genome in barley. Embo J 5: 2769-2774
Finnie SJ (1990) The development and utilisation of anther culture technology in barley breeding. Thesis, University of Edimburgh
Finnie SJ, Powell W, Dyer AF (1989) The effect of carbohydrate composition and concentration on anther culture response in barley (*Hordeum vulgare* L.). Plant Breed 103: 110-118
Foroughi-Wehr B, Friedt W (1984) Rapid production of recombinant barley yellow mosaic virus resistant *Hordeum vulgare* lines by anther culture. Theor Appl Genet 67: 377-382
Foroughi-Wehr B, Zeller FJ (1990) In vitro microspore reaction of different German wheat cultivars. Theor Appl Genet 79: 77-80
Foroughi-Wehr B, Friedt W, Wenzel G (1982) On the genetic improvement of androgenetic haploid formation in *Hordeum vulgare* L. Theor Appl Genet 62: 233-239
Fujii T (1970) Callus formation in wheat anther. Wheat Inf Serv 31: 1-2
Garrido D, Vicente O, Hebverle-Bors E, Rodriguez-Garcia I (1995). Cellular changes during the acquisition of embryogenic potential in isolated pollen grains of *Nicotiana tabacum*. Protoplasma 186: 220-230
Genovesi AD (1990) Maize (*Zea mays* L.): In vitro production of haploids. In Bajaj YPS (ed) Biotechnology in Agriculture and Forestry. 12. Haploids in Crop Improvement. Springer-Verlag, Berlin, pp 176-203
Genovesi AD, Collins GB (1982) In vitro production of haploid plants of corn via anther culture. Crop Sci 22: 1137-1144
Gustafson VD, Baenziger PS, Wright MS, Stroup WW, Yen Y (1995) Isolated wheat microspore culture. Plant Cell Tiss Org Cult 42: 207-213
Hadwieger MA, Heberle-Bors E (1986) Pollen plant production in *Triticum turgidum*. In Genetic Manipulation in Plant Breeding, Proc. Int. Symp. Eucarpia. Berlin, Germany, pp 303-305
Hassawi DS, Liang GH (1990) Effect of cultivar, incubation temperature, and stage of microspore development on anther culture in Wheat and Triticale. Plant Breed 105: 332-336
Hassawi DS, Qi J, Liang GH (1990) Effects of growth regulator and genotype on production of wheat and triticale polyploids from anther culture. Plant Breed 104: 40-45
Hattori T, Nakagawa S, Nakamura K (1990) High-level expression of tuberous root storage protein genes of sweet potato in stems of plantlets grown in vitro on sucrose medium. Plant Mol Biol 14: 595-604
Hattori T, Fukumoto H, Nakagawa S, Nakamura K (1991) Sucrose-induced expression of genes coding for the tuberous root storage protein, sporamine, of sweet potato in leaves and petioles. Plant Cell Physiol 32: 79-86
Hoekstra S, van Zijderveld MH, Louwerse JD, Heidekamp F, van der Mark F (1992) Anther and microspore culture of *Hordeum vulgare* L. cv. Igri. Plant Sci 8: 89-96

Hoekstra S, van Zijderveld MH, Heidekamp F, van der Mark F (1993) Microspore culture of *Hordeum vulgare* L.: the influence of density and osmolality. Plant Cell Rep 12: 661-665

Hoekstra S, van Bergen S, van Brouwershaven IR, Schilperoort RA, Heidekamp F (1996) The interaction of 2,4-D application and mannitol pretreatment in anther and microspore culture of *Hordeum vulgare* L. cv. Igri. J Plant Physiol 148: 696-700

Hoekstra S, van Bergen S, van Brouwershaven IR, Schilperoort RA, Wang M (1997) Androgenesis in *Hordeum vulgare* L.: effects of mannitol, calcium and abscissic acid on anther pretreatment. Plant Sci 126: 211-218

Hoffmann F, Wenzel G (1981) Selfcompatibility of microspore-derived double-haploid rye lines and single grain selection for alkylresorcinol content. Theor Appl Genet 60: 129-133

Hou L, Ullrich SE, Kleinhofs A, Stiff CM (1993) Improvement of anther culture methods for doubled haploid production in barley breeding. Plant Cell Rep 12: 334-338

Hunter CP (1988) Plant regeneration from microspores of barley (*Hordeum vulgare* L.). Thesis, University of London

Islam MR, Kintzios S, Fishbeck G (1992) Anther culture responsiveness of *Hordeum spontaneum* derived spring barley lines and a genetic analysis of plant regeneration. Plant Cell Tiss Org Cult 29: 235-239

Jähne A, Lörz H (1995) Cereal microspore culture. Plant Sci 109: 1-12

Jähne A, Lazzeri PA, Jäger-Gussen M, Lörz H (1991) Plant regeneration from embryogenic cell suspensions derived from anther cultures of barley (*Hordeum vulgare* L.). Theor Appl Genet 82: 74-80

Jardinaud MF, Souvré A, Alibert G, Beckert M (1995a) *uidA* gene transfer and expression in maize microspores using the biolistic method. Protoplasma 187: 138-143

Jardinaud MF, Souvré A, Alibert G, Beckert M (1995b) Optimisation of DNA transfer and transient β-glucuronidase expression in electroporated maize (*Zea mays* L.) microspores. Plant Cell Rep 15: 55-58

Kao KN (1993) Viability, cell division and microcallus formation of barley microspores in culture. Plant Cell Rep 12: 366-36

Kao KN, Saleem M, Abrams S, Pedras M, Horn D, Mallard C (1991) Culture conditions for induction of green plants from barley microspores by anther culture methods. Plant Cell Rep 9: 595-601

Karsai I, Bedö Z, Hayes PM (1994) Effect of induction medium pH and maltose concentration on in vitro androgenesis of hexaploid winter triticale and wheat. Plant Cell Tiss Org Cult 39: 49-53

Kasha KJ, Yao Q, Simion E, Hu T, Oro R (1995) Production and applications of haploids in crops. In Proc International Symposium, Use of Induced Mutation and Molecular Techniques of Crop Improvement, International Atomic Energy Agency (IAEA), Vienna, Austria, pp 23-37

Kasha KJ, Ziauddin A, Cho UH (1990) Haploids in cereal improvement: anther and microspore culture. In Gustafson JP (ed) Gene Manipulation in Plant Improvement. Plenum Press, New York, pp 213-235

Konzak CF, Zhou H (1991) Anther culture methods for doubled haploid production in wheat. Cereal Res Comm 19: 147-164

Knudsen S, Due IK, Andersen SB (1989) Components of response in barley anther culture. Plant Breed 103: 241-246

Ku MK, Cheng WC, Juo LC, Kuan YL, An HP, Huang CH (1981) Induction factors and morpho-cytological characteristics of pollen derived plants in maize (*Zea mays* L.). In Proc Symp Plant Tissue Culture. Science Press, Beijing, pp 35-42

Kuhlmann U, Foroughi-Wehr B (1989) Production of doubled haploid lines in frequencies sufficient for barley breeding programs. Plant Cell Rep 8: 78-81

Larsen ET, Tuvesson IKD, Andersen SB (1991) Nuclear genes affecting percentage of green plants in barley (*Hordeum vulgare* L.) anther culture. Theor Appl Genet 82: 417-420

Last D, Brettell RIS (1990) Embryo yield in wheat anther culture is influenced by choice of sugar in the culture medium. Plant Cell Rep 9: 14-16

Lazar MD, Schaeffer GW, Baenziger PS (1984) Cultivar and cultivar x environment effects on the development of callus and polyhaploid plants from anther cultures of wheat. Theor Appl Genet 67: 273-277

Lentini Z, Reyes P, Martinez CP, Roca WM (1995) Androgenesis of highly recalcitrant rice genotypes with maltose and silver nitrate. Plant Sci 110: 127-138

Li WZ, Jing JK, Yao GH, Hu H (1993) The effects of genotype and mannitol pretreatment on high frequency androgenesis in barley. Chin Sci Bul 38: 151-155

Liang GH, Gu X, Yue G, Shi ZS, Kofoid KD (1997) Haploidy in *Sorghum*. In Jain SM, Sopory SK, Veilleux RE (eds), In vitro haploid production in higher plants. 4. Kluwer Academic Publishers, The Netherlands, pp 149-161

Loeb TA, Reynolds TL (1994) Transient expression of the uidA gene in pollen embryoids of wheat following microprojectile bombardment. Plant Sci 104: 81-91

Luckett DJ, Darvey NL (1992) Utilisation of microspore culture in wheat and barley improvement. Aust J Bot 40: 807-828

Luckett DJ, Smithard RA (1992) Doubled haploid production by anther culture for australian barley breeding. Aust J Agric Res 43: 67-78

Luckett DJ, Smithard RA (1995) Barley anther culture using membrane rafts. Plant Cell Tiss Org Cult 42: 287-290

Lührs R, Nielsen K (1992) Microspore cultures as donor tissue for the initiation of embryogenetic cell suspensions in barley. Plant Cell Tiss Org Cult 31: 169-178

Mlodzianowski F, Idzikowska K (1981) Fine structure of plastids during androgenesis in *Hordeum vulgare* L. Acta Soc Bot Pol 49: 205-210

Mogensen HL (1996) The hows and whys of cytoplasmic inheritance in seed plants. Am J Bot 83: 383-404

Mordhorst AP, Lörz H. (1993) Embryogenesis and development of isolated barley (*Hordeum vulgare* L.) microspores are influenced by the amount and composition of nitrogen sources in culture media. J Plant Physiol 142: 485-492

Muyuan Z, Abing X, Miabo Y, Chunnong H, Zhilong Y, Lingi W, Jianjun Y (1990) Effects of amino acids on callus differentiation in barley anther culture. Plant Cell Tiss Org Cult 22: 201-204

Ogawa T, Fukuoka H, Ohkawa Y (1995) Plant regeneration through direct culture of isolated pollen grains in rice. Breed Sci 45: 301-307

Olsen FL (1991) Isolation and cultivation of embryogenic microspores from barley (*Hordeum vulgare* L.). Hereditas 115: 255-266

Orshinsky BR, McGregor LJ, Johnson GIE, Hucl P, Kartha KK (1990) Improved embryoid induction and green shoot regeneration from wheat anthers cultured in medium with maltose. Plant Cell Rep 9: 365-369

Orshinsky BR, Sadasivaiah RS (1997) Effect of plant growth conditions, plating density, and genotype on the anther culture response of soft white spring wheat hybrids. Plant Cell Rep 16: 758-762

Ouyang JW (1986) Induction of pollen plants in *Triticum aestivum*. In Hu H, Yan H (eds), Haploids of higher plants in vitro. Springer-Verlag, Berlin, pp 26-41

Ouyang JW, Zhou SM, Jia SE (1983) The response of anther culture temperature in *Triticum aestivum*. Theor Appl Genet 66: 101-109

Pace GM, Reed JN, Ho LC, Fahey JW (1987) Anther culture of maize and the visualization of embryogenic microspores by fluorescent microscopy. Theor Appl Genet 73: 863-869

Pauk J (1985) Production of haploid plants of maize (*Zea mays* L.) through androgenesis. Cereal Res Comm 13: 47-53

Pescitelli SM, Mitchell JC, Jones AM, Pareddy DR, Petolino JF (1989) High frequency androgenesis from isolated microspores of maize. Plant Cell Rep 7: 673-676

Pescitelli SM, Johnson CD, Petolino JF (1994) Isolated microspore culture of maize. In Bajaj YPS (ed) Biotechnology in Agriculture and Forestry. 25. Maize. Springer-Verlag, Berlin, pp 186-200

Petolino JF, Jones AM (1986) Anther culture of elite genotypes of maize. Crop Sci 26: 1072-1074

Piccirilli M, Arcioni S (1991) Haploid plants regenerated via anther culture in wild barley (*Hordeum spontaneum* C. Kock). Plant Cell Rep 10: 273-276

Pickering RA, Devaux P (1992) Haploid production: approaches and use in plant breeding. In Shewry PR (ed) Barley: Genetics, Biochemistry, Molecular Biology and Biotechnology. CAB International, Oxford, pp 519-547

Powell W, Borrino EH, Goodal V (1988) The effect of anther orientation on microspore-derived plant production in barley (*Hordeum vulgare* L.). Euphytica 38: 159-163

Rakoczy-Trojanowska M, Smiech M, Malepszy S (1997) The influence of genotype and medium on rye (*Secale cereale* L.) anther culture. Plant Cell Tiss Org Cult 48: 15-21

Rajasekhar VK (1991) Regulation of nuclear gene expression for plastidogenesis as affected by the developmental stage of plastids. Biochem Physiol Pflanzen 187: 257-271

Rines HW, Ricra-Lizarazu O, Nunez VM, Davis DW, Plilips RL (1997) Oat haploids from anther culture from wide hybridization. In Jain SM, Sopory SK, Veilleux RE (eds) In vitro haploid production in higher plants. 4. Kluwer Academic Publishers, The Netherlands, pp 205-221

Roberts-Oehlschlager SL, Dunwell JM, Faulks R (1990) Changes in the sugar content of barley anthers during culture on different carbohydrates. Plant Cell Tiss Org Cult 22: 77-85

Salmenkallio-Marttila M, Kauppinen V (1995) Efficient regeneration of fertile plants from protoplasts isolated from microspore cultures of barley (*Hordeum vulgare* L.). Plant Cell Rep 14: 253-256

Scott P, Lyne RL (1994a) Initiation of embryogenesis from cultured barley microspores: a further investigation into the toxic effects of sucrose and glucose. Plant Cell Tiss Org Cult 37: 61-65

Scott P, Lyne RL (1994b) The effect of different carbohydrate sources upon the initiation of embryogenesis from barley microspores. Plant Cell Tiss Org Cult 36: 129-133

Scott P, Lyne RL, Aprees T (1995) Metabolism of maltose and sucrose by microspores isolated from barley (*Hordeum vulgare* L.). Planta 197: 435-441

Sharma GC, Wang WC, Sapra VT (1982) Effect of genotype, media and temperature pretreatment on callus initiation in triticale, wheat, and rye anther cultures. Cereal Res Comm 10: 143-150

Sonnino A, Tanaka S, Iwanaga M, Schilde-Rentschler L (1989) Genetic control of embryo formation in anther culture of diploid potatoes. Plant Cell Rep 8: 105-107

Sorvari S, Schieder O (1987) Influence of sucrose and melibiose on barley anther cultures in starch media. Plant Breed 99: 164-171

Stober A, Hess D (1997) Spike pretreatments, anther culture conditions, and anther culture response of 17 German varieties of spring wheat (*Triticum aestivum* L.) Plant Breed 116: 443-447

Sun CS, Lu TG, Sondahl MR (1991) Anther culture of naked oat and the establishment of its haploid suspension cell. Acta Bot Sin 33: 417-420

Sun CS, Wang CC, Chu CC (1974) The ultrastructure of plastids in the albino pollen-plants of rice. Sci Sin 17: 793-802

Sun CS, Wu SC, Wang CC, Chu CC (1979) The deficiency of soluble proteins and plastid ribosomal RNA in the albino plantlets of rice. Theor Appl Genet 55: 193-197

Sunderland N, Huang B (1985) Barley anther culture: the switch of programme and albinism. Hereditas Suppl 3: 27-40

Szarejko I, Kasha KJ (1991) Induction of anther culture derived doubled haploids in barley. Cereal Res Comm 19: 219-237

Thompson DM, Chalmers K, Waugh R, Forster BP, Thomas WTB, Caligari PDS, Powell W (1991) The inheritance of genetic markers in microspore-derived plants of barley *Hordeum vulgare* L. Theor Appl Genet 81: 487-492

Tiwari S, Rahimbaev I. (1992) Comparison of glucose, sucrose and maltose for *Hordeum vulgare* L. isolated microspore culture using different methods. Ind J Exp Biol 30: 624-627

Touraev A, Pfosser M, Vicente O, Heberle-Bors E (1996) Stress as the major signal controlling the developmental fate of tobacco microspores; towards a unified model of induction of microspore/pollen embryogenesis. Planta 200: 144-152

Tsay HS, Miao SH, Widholm JM (1986) Factors affecting haploid plant regeneration from maize anther culture. J Plant Physiol 126: 33-40

Tsay HS, Yeh CC, Hsu JY (1990) Embryogenesis and plant regeneration from anther culture of bamboo (*Sinocalamus latiflora* (Munro) McClure). Plant Cell Rep 9: 349-351

Tuvesson IKD, Pedersen S, Andersen SB (1989) Nuclear genes affecting albinism in wheat (*Triticum aestivum* L.) anther culture. Theor Appl Genet 78: 879-883

Vergne P, Riccardi F, Beckert M, Dumas C (1990) Detection of androgenesis-related proteins in maize. In Nijkamp HJJ, van der Plas LHW, van Aartrijk J (eds) Progress in Plant Cellular and Molecular Biology. Kluwer Academic Publishers, Dordrecht, pp 416-421

Wang M, van Bergen S, Lamers GEM, Oppedjik BJ, Schilperroort RA (1999) Programmed cell death during androgenesis in *Hordeum vulgare* L. Ibid, pp 201-209

Wenzel G, Hoffmann F, Thomas E (1976) Heterozygous microspore-derived plants in rye. Theor Appl Genet 48: 204-208

Xie JH, Gao MW, Cai QH, Cheng X, Shen Y, Liang ZQ (1995) Improved isolated microspore culture efficiency in medium with maltose and optimized growth regulator combination in japonica rice (*Oryza sativa*). Plant Cell Tiss Org Cult 42: 245-250

Xie JH, Gao MW, Cai QH, Liang ZQ, Xue QZ (1996) Effects of donor plant growth status and preculture temperature on isolated microspore ability in japonica rice (*Oryza sativa* L.). Cereal Res Comm 24: 133-138

Yao QA, Simion E, William M, Krochko J, Kasha KJ (1997) Biolistic transformation of haploid isolated microspores of barley (*Hordeum vulgare* L.) Genome 40: 570-581

Yu SM, Lee YC, Fang SC, Chan MT, Hwa SF, Liu LF (1996) Sugars act as signal molecules and osmotica to regulate the expression of α–amylase genes and metabolic activities in generating cereal grains. Plant Mol Biol 30: 1277-1289

Zhou H, Konzak CF (1989) Improvement of anther culture methods for haploid production in wheat. Crop Sci 29: 817-821

Zhou H, Zheng Y, Konzak CF (1991) Osmotic potential of media affecting green plant percentage in wheat anther culture. Plant Cell Rep 10: 63-66

Zhu BY, Ding XH, Yi JH (1991) Anther culture and breeding in indica rice: In Yan CJ (ed) Tissue Culture of Field Crops. Shangai Press, Shangai, pp 153-169

Zhu ZQ, Warg JJ, Sun JS (1979) The introduction of the albinos pollen plants and preliminary observation of their ploidy in *Triticum durum*. Acta Bot Sin 21: 295-298

Ziauddin A, Marsolais A, Simion E, Kasha KJ (1992) Improved plant regeneration from wheat anther and barley microspore culture using phenylacetic acid (PAA). Plant Cell Rep 11: 489-498

Ziauddin A, Simion E, Kasha KJ (1990) Improved plant regeneration from shed microspore culture in barley (*Hordeum vulgare* L.) cv. Igri. Plant Cell Rep 9: 69-72

21 Green and albino plant regeneration in anther culture of alloplasmic wheat lines

Orlov* PA and Pyko VI

21.1 Introduction

The phenomenon of pollen embryogenesis in anther culture is widely used in fundamental and applied investigations. In some plant species its induction in tissue culture may be limited for certain difficulties. The most important of them are: the impossibility to induce embryogenesis from the most of the explant cells, low embryogenic potential of many explants, undesirable somaclonal variation, dying of embryoids, albino plant appearance and others.

Such important technologies as plant transformation, selection in vitro, obtaining secondary metabolites include as one of their step the cultivation of plant cells and tissues in vitro conditions. All the above-mentioned problems may remarkably decrease the efficiency of their practical application. The detail investigation of the genetic mechanisms of the in vitro induced embryogenesis is necessary to overcome these difficulties.

The genetic control of the pollen embryogenesis parameters in cereals is not exactly determined. The most established fact is that callus induction and plant regeneration show independent inheritance (Henry and De Buyser 1985; Agache et al. 1988; 1989). Comparative estimation of the dihaploids, produced from the first generation hybrids and their parental forms revealed the presence of the additive gene action on the character embryoid yield and epistatic control of the regeneration ability (Agache et al. 1988). 1-2 dominant genes is suggested to take part in controlling regeneration ability (Tuvesson et al. 1989). Essential increase in the responding anther number was noted in the presence of 1B/1R recombination in the chromosome set (Henry and De Buyser 1985). However, the number of nuclear genes and their localisation is still unknown.

Cytoplasmic genes modify expression of nuclear genes determining pollen embryogenesis parameters in certain cases. As a result the reciprocal differences in these characters in interspecific hybrids of triticale, barley and wheat were found in some investigations (Charmet and Bernard 1984; Powell 1988; Ekiz and

---

* for correspondence: Institute of Genetics and Cytology NASB; Akademicheskaya Str 27; Minsk 220072; Belarus

Konzak 1994). In alloplasmic lines which have a substituted cytoplasm these effects may be both positive and negative (Ekiz and Konzak 1991; Sagi and Barnabas 1989; Becraft and Taylor 1989).

The importance of cytoplasm is indirectly determined by the fact that the anther development in angiosperm plants is under joint control of nuclear and cytoplasmic genetic systems. Cytoplasm substitution or mutational alteration may lead to change in development of sporophytic and gametophytic tissues (Orlov 1992). The extreme case of this phenomenon is the cytoplasmic male sterility (Palilova 1986).

Switching over the microspore developmental program from gametophytic to sporophytic way takes place during the anther cultivation. Radical reorganisation of microspore cytoskeleton and cytoplasm organelles occurs (Reynolds 1997). The genetic role of cytoplasm in these processes and further embryogenesis has not been studied enough. At the same time it is very attractive to use the cytoplasmic effects for increasing the embryogenic potential of the microspore derived cultures in wheat. The examination of this suggestion was the main aim of our work.

21.2 Material and methods

21.2.1 Experimental model

To extract the cytoplasmic and nuclear cytoplasmic effects it is necessary to have the appropriate model. The reciprocal hybrids of first generation used in some cases provide only preliminary information on possible cytoplasmic involvement in genetic control of phenotypic parameters because the maternal effect may exist (Palilova 1986). Series of alloplasmic lines are quite informative model. Such lines have the same nucleus obtained from one species and cytoplasms from another one. Any differences between these genotypes are determined by cytoplasm or nuclear-cytoplasm effects (Tsunewaki 1980).

We have developed a new model that includes series of alloplasmic wheats of 20 lines obtained on the basis of five species - *T. aestivum* L., *T. turgidum* L., *T. durum* Desf., *T. dicoccum* Schrank. and *T. aethiopicum* Jakubz., combining all possible combinations of genome and plasmon. These genotypes and their parental forms have been used as an experimental material in our recent work.

21.2.2 Methods

Plant was grown under field condition. Light and electron microscopic investigation was carried out according the previously described protocol (Orlov 1992).

Temperature pre-treated and not pre-treated plant material was used in our experiments. For plant pre-treatment cut spikes were kept at 5-7°C for 5-7 days. Anthers were aseptically isolated at the stage of late mononuclear microspores and plated on nutrient media PE-2 (Chuang et al. 1978). The plates were sealed with parafilm and kept in the dark for 2 weeks at 31°C and then at 26°C. The embryoids were transferred to modified medium and in a month to regeneration medium (Orlov et al. 1995).

DNA isolation, restriction analysis and endogenous hormone analysis were carried out, as was earlier described (Orlov et al. 1998).

A statistical evaluation of the results was performed according to Rokitskii (1973).

21.3 Results and discussion

Most of the lines showed negative reaction to pollen embryogenesis induction in experiments where isolated anthers were grown at 26°C without temperature pre-treatment. Plant regenerants were obtained only in (*durum*) x *T. aestivum* (0.17%) and (*turgidum*) x *T. aestivum* (0.12%) lines (data not presented).

In the experiments carried out with temperature pre-treated material the regenerants were obtained in four of five species (with the exception of *T. aethiopicum* species). All plants were green and only in *T. aestivum* insignificant appearance of albino plants (0.12%) was observed.

Cytoplasm substitution increased the green regenerants yield in most combinations of genome and plasmon: for example in lines (*durum*) x *T. aestivum* (from 1.16 to 3.46%), (*durum*) x *T. dicoccum* (from 0.07 to 1.07%). In plants of *T. turgidum* species substitution of own cytoplasm by alien one decreased the number of regenerants in four of five lines. In some alloplasmic lines the increase in the etiolated regenerants as compared to the initial euplasmic forms was observed. Bifactorial dispersion analysis has shown a reliable involvement of the cytoplasm and nuclear-cytoplasm interaction in determination of green regenerants number (Orlov et al. 1995).

The most important was the fact that substitution of own cytoplasm by alien one in many cases increased all parameters defining the plant ability for morphogenetic process induction in anther culture as compared to the initial parental forms. This was noted for example in (*durum*) x *T. aestivum*, (*dicoccum*) x *T. aestivum*, (*durum*) x *T. dicoccum* and other lines. It is especially important that cytoplasmic effects reproduced during several years and consequently it may be used for the purposeful alteration of the pollen embryogenesis characters without nuclear genes changing.

The analysis of the results obtained during experiments indicated that in certain alloplasmic lines for example (*durum*) x *T. aestivum*, (*durum*) x *T. dicoccum* and some others the parameters of the pollen embryogenesis has been strongly differed as compared with parental forms and corresponding reciprocal alloplasmic combination. This fact indicated to our mind on specific nuclear-cytoplasmic interaction during the genetic control of these parameters in some combinations of genome and plasmon.

Two most unusual alloplasmic lines combining genome and plasmon of *T. aestivum* and *T. durum* species and also their parental forms were investigated more in detail.

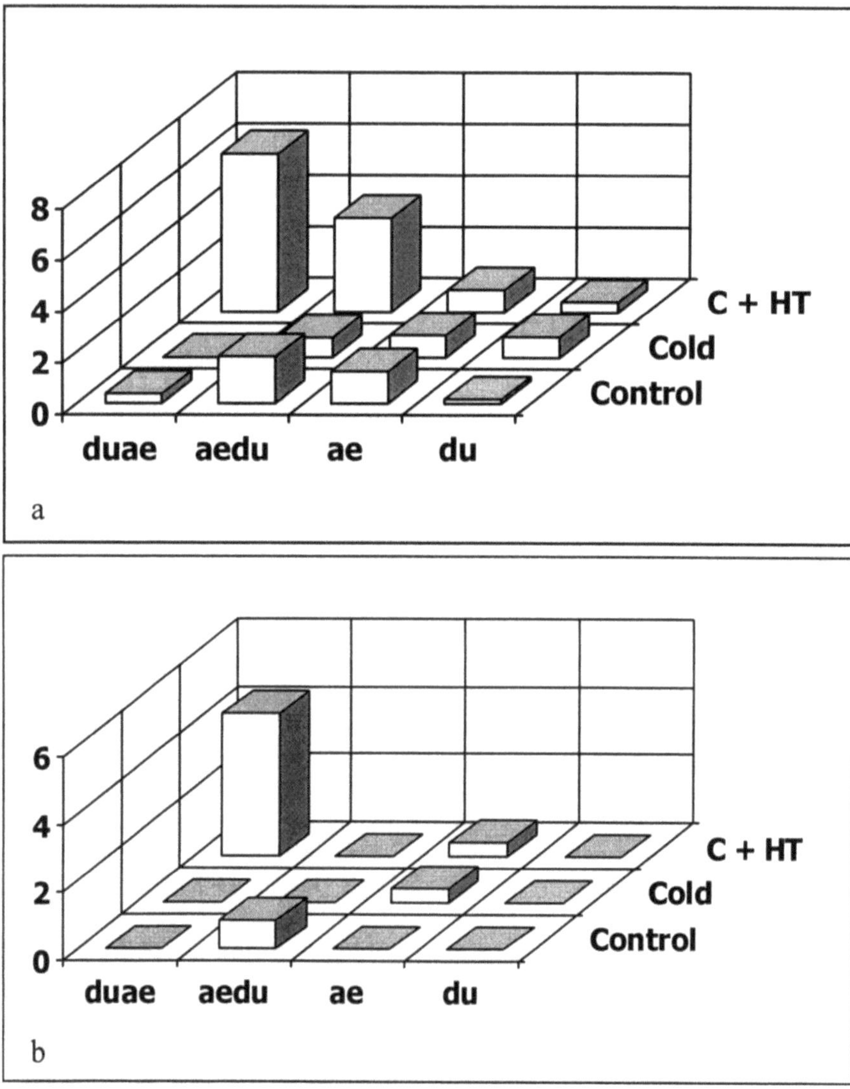

Fig. 1. The number of responding anthers (a) and green regenerants (b) in eu- and alloplasmic wheat lines under different regimes of temperature pre-treatment Cold: pre-treatment at 5-7°C for 5-7 days; C + HT: pre-treatment at 5-7°C and than 10 days at 31°C; Duae: (*durum*) x *T. aestivum*; aedu - (*aestivum*) x *T. durum*; ae - *T. aestivum*; du- *T. durum*

The results showed that this specific effect expressed in different way depending on type of temperature pre-treatment. Negative reaction to pollen embryogenesis induction in experiments where isolated anthers were grown at 26°C without temperature pre-treatment. Plant regenerants were obtained only in (*durum*) x *T. aestivum* (0.17%) line (Fig. 1). After combined pre-treatment with cold and high temperature the effect expressed in highest degree.

Cytoplasmic genes takes place in genetic control of plant response to action of biotic and abiotic factors (Palilova 1986). Such factors as temperature stress, light regime, starvation and others are widely used for plant explant pre-treatment (Moieni and Sarrafi 1996; Kudirka et al. 1986). Our data testify the important role of plasmon in reaction plant to temperature pre-treatment.

We suggested that changes in cell organelles might be responsible for these differences. Different methods such as analysis of the restriction fragment polymorphism of mitochondrial DNA, quantitative electron microscopy, immunoferment analysis of the endogenous hormones were used to detect possible mechanism of this phenomenon.

Six endonucleases were used for organelle DNA digestion. The data obtained during experiments did not reveal any difference in mitochondrial DNA structure at the level of restriction fragment polymorphism (Orlov et al. 1998). This may testify the fact that the mitochondria does not involved in specific nuclear-cytoplasm interaction in determination of the abnormal expression of pollen embryogenesis parameters. Another explanation – not enough sensitivity of the method used in our experiments to detect small differences in mitochondrial DNA structure.

Quantitative electron microscopy of cell organelle structure is a quite sensitive method that allows detecting the cytoplasmic differences. It was successfully applied to characterise the cytoplasmic diversity among the *Tritium* and *Aegilops* species (Davidenko 1984), differences in cell organelles contrast on plant resistance to virus infection (Palilova 1986), potential productivity of heterotic hybrids (Khotylyova et al. 1986).

In our experiments quantitative electron microscopy was used to compare the cell organelles structure in alloplasmic lines differing in pollen embryogenesis induction. Thirteen parameters such as number of grana per chloroplast section, number of lamella, number of starch inclusions, perimeter of chloroplast, ratio of total lamella length to chloroplast perimeter, ratio of internal membranes length to chloroplast perimeter and others have been studied.

It was found that certain deviations exist in expression of structural parameters as compared with expected ones. For instance, the number of grana in line *T. durum* with *aestivum* cytoplasm exceeded both parental forms (21.38 against 16.73 and 18.50). The same parameter in alloplasmic line of *T. aestivum* with durum cytoplasm was similar to donors of nucleus and plasmon (17,18). The ratio of internal membranes length to chloroplast perimeter in line of *T. durum* with *aestivum* cytoplasm was higher than in initial forms (271.59), but in line of *T. aestivum* with durum cytoplasm had the intermediate value (199.69). Some another parameters also expressed different inheritance of structural chloroplast parameters (examples presented in Table 1).

These data may indirectly indicate the involvement of the chloroplast genome in determining specific reaction of alloplasmic lines to induction of pollen embryogenesis. The further experimental testing is necessary to confirm this suggestion.

The mechanisms of positive influence of temperature stress may be connected with the changes in endogenous hormones during the pre-treatment. To detect their content in microspores was not possible.

Table 1. Quantitative characteristics of chloroplast structural parameters in eu- and alloplasmic

¹	Structural parameters	*(aestivum)* x *T. aestivum*	*(aestivum)* x *T. durum*	*(durum)* x *T. aestivum*	*(durum)* x *T. durum*
1	Number of grana	16.73	21.38ab	17.18	18.50
2	Number of starch inclusions	1.03	0.25ab	0.59ab	1.05
3	Perimeter of chloroplast	31.07	35.71b	36.18a	35.10
4	Total grana width	11.21	14.30b	15.21a	14.86
5	Ratio of total lamella length to chloroplast perimeter	0.93	1.09a	1.04b	0.85
6	Ratio of internal membranes length to chloroplast perimeter	144.38	271.59b	199.69a	231.76

Cytoplasm donors are given in brackets
Differences of frequencies among appropriate indices are significant at $P<0.05$ between:
a - alloplasmic line and genome donor; b - alloplasmic line and cytoplasm donor

The determination of the endogenous hormone content in leaves of eu- and alloplasmic lines after cold pre-treatment and in control was carried out. Untypical nuclear-cytoplasmic effect on these parameters in (*durum*) x *T. aestivum* line has been also observed. This line has the highest level of pollen embryogenesis induction (Orlov et al. 1998). These data may testify to our mind the involvement of cytoplasmic genetic system in hormonal regulation of morphogenetic processes induced in anther culture.

Generally the four-year investigation of the series of alloplasmic wheat lines and parental forms give the possibility to make the following main conclusions.

The real cytoplasmic and nuclear cytoplasmic effects during early stages of pollen embryogenesis induction and plant regeneration in wheat anther culture exist.

Cytoplasmic effects in anther culture of alloplasmic wheat lines depend on type of temperature pre-treatment.

In certain combinations of genome and plasmon the stable cytoplasmic effects re-produced during 4 years take place and consequently, they may be used as additional reserve to improve the in vitro manipulation in wheat plants.

References

Agache S, De Buyser J, Henry Y, Snape JW (1988) Studies of the genetic relationship between anther culture and somatic tissue culture abilities in wheat. Plant Breed 100: 6-33

Agache A, Bachelier B, Buyser J, Henry Y, Snape S (1989) Genetic analysis of anther culture response in wheat using aneuploid, chromosome substitution and translocation lines. Theor Appl Genet 77: 7-11

Becraft PW, Taylor GA (1989) Effects of nucleus, cytoplasm and male sterile nucleus-cytoplasm combinations on callus initiation in anther culture of wheat. Euphytica 44: 235-240

Charmet G, Berbard S (1984) Diallel analysis of androgenetic plant production in hexaploid Triticale (*x triticisecale,* Wittmack). Theor Appl Genet 69: 55-61

Chuang C, Ouyang JW, Ching CK (1978) A set of potato medium for wheat anther culture. In: Proc. Symp. Plant Tissue Culture, Peking, pp 51-56

Davidenko O (1984) Nonchromosomal mutations. Nauka i Technika Press, Minsk, 156 p

Ekiz H, Konzak CF (1991) Nuclear and cytoplasmic control of anther culture response in wheat: 1. Analysis of alloplasmic lines. Crop Sci 31: 1421-1427.

Ekiz H, Konzak CF (1994) Anther culture response of some spring wheat (*Triticum aestivum* L.) cultivars, lines and F1 crosses. Cereal Res Comm 22: 165-171

Heberle-Bors E, Odenbach W (1985) In vitro pollen embryogenesis and cytoplasmic male sterility in *Triticum aestivum* L. Z Pflanzenzucht 95: 14-22

Henry Y, De Buyser J (1985) Effect of the 1B/1R translocation on anther culture ability in wheat (*Triticum aestivum* L.). Plant Cell Rep 4: 307-310

Khotylyova LV, Ilchenko VP, Atrashonok NV, Trukhanovetz NL (1986) Chloroplast structure and duration of thir functioning during selection on productivity in triticale. In: Zhuchenko A (Ed) Genetic methods of selection process acceleration. Shtiintza, Kishinev, pp. 127-140

Kudirka DT, Schaeffer GW, Baenziger PS (1986) Wheat: genetic variability through anther culture. In: Bajaj YBS (Ed) Biotechnology in agriculture and forestry. Springer-Verlag, Berlin, Heidelberg. V 2. Crops 1, pp 52-75

Moieni A, Sarrafi A (1996) The effects of gibberellic acid, phenyllethylamine, 2,4-D and genotype on androgenesis in hexaploid wheat (*Triticum aestivum*). Cereal Res Comm 24: 139-145

Orlov PA (1992) Structural and functional aspects of anther development in alloplasmic wheats. In: Proc XI Int Symp ìEmbryology and Seed Reproduction, Nauka Press, St. Petersburg, pp 415-416

Orlov PA, Mavrishcheva EB, Palilova AN (1995) Genome and plasmon interaction in induction of morphogenetic response of wheat plants in anther culture. Russ J Genet 31: 333-336

Orlov PA, Pyko VI, Ryzhkov VV, Mikhalevich EA (1998) Application of cytoplasmic and nuclear-cytoplasmic effects in plant bitechnology In: Kartel NA, Makeeva EN (Eds) Molecular genetics and biotechnology. Acad Sci Press, Minsk, pp 238-240

Orlova IN (1994) Nuclear-cytoplasmic interaction and partial instability of alien grass hybrids. Genetica (Russia) 30: 1423-1431

Palilova AN (1986) Genetic Systems in Plants and Their Interaction. Nauka i Technika, Minsk, 160 p

Powell W (1988) Diallel analysis of barley anther culture response. Genome 30: 152-157

Pickard E and J De Buyser (1973) Obtention de plantules haploides de *Triticum aestivum* L. a partir de culture d'antheres in vitro. C R Acad Sci Paris, 277D: 1463-1466

Reynolds TL (1997) Pollen embryogenesis. Plant Mol Biol 33: 1-10
Rokitskii PF (1973) Biological statistics. Vysheisha Shkola, Minsk, 320 p
Sagi L, Barnabas B (1989) Evidence for cytoplasmic control of in vitro microspore embryogenesis in the anther culture of wheat (*Triticum aestivum* L.). Theor Appl Genet 78: 867-872
Tsunewaki K (1980) Genetic diversity on the cytoplasms in *Triticum* and *Aegilops*. Japan Soc Prom Sci, Tokyo
Tuvesson IKD, Pedersen S, Andersen D (1989) Nuclear genes affecting albinism in wheat (*Triticum aestivum* L.). Theor Appl Genet 78: 879-883

22 Assessment of green regenerants from rye and triticale anther cultures

Immonen* S, Tauriainen A and Manninen O

22.1 Introduction

Doubled haploids (DH) are a potential tool for plant breeding, gene technology and basic research. With self pollinated species, the possibility to fix a homozygote in an early generation facilitates variety breeding directly (Snape 1986). The prerequisites for using DH-technology are that a) the production method can be applied to a large number of genotypes, b) large numbers of green DH regenerants can be produced at reasonable cost and c) the green regenerants are of high quality, displaying the genetic variability of the cross. The methods for producing DH include anther and isolated microspore culture, and ovary culture after pollination with alien pollen. With cereals, success in anther culture still depends largely on the genotype despite methodological developments to circumvent this.

22.1.1 DH in rye (*Secale cereale* L.) breeding

With cross pollinated species, DH can be used in breeding synthetic and hybrid varieties and for population improvement in recurrent selection schemes (Snape 1986). The latter requires high regeneration rate and low production cost. Even with a limited success rate, DH could be tested for improving particular traits selectable at the plantlet stage before cross pollination. Rye has been considered a recalcitrant cereal in anther culture and production of green regenerants has been successful mainly with material containing wild *S. vavilovii* L. in the pedigree (Flehinghaus-Roux 1995, Deimling and Flehinghaus-Roux 1996). We identified Finnish spring rye material which gave sufficiently high androgenetic response to

* for correspondence: Agricultural Research Centre; Plant Production Research; Crops and Soil; 31600 Jokioinen; Finland

allow further development of the method (Immonen and Anttila 1996). Results from the methodological studies have been applied to a large number of breeder's lines, and to some elite winter rye varieties which have shown relatively good anther culture ability.

22.1.2
DH in triticale (X *Triticosecale* Wittmack) breeding

Triticale is a basically self pollinated cereal, the cultivated forms of which are mainly hexaploid, containing the A and B genomes of wheat (bread wheat, *Triticum aestivum* L., durum wheat, *T. durum* Desf.) and the R genome of rye. The gene pools available for triticale breeding include those of durum and bread wheat and rye. Anther culture methods for DH production of triticale have been developed (review by Ryöppy 1996), although low green plant yield and albinism still remain bottlenecks. DH can be used to improve selection efficiency in a normal crossing programme (Charmet and Branlard 1985). They could also be used to facilitate introgression of distant genetic material (Wang et al. 1993). Anther culture is used to some extent in practical triticale breeding programmes, although the relatively high cost remains a limiting factor (Bernard et al. 1996, Fossati et al. 1996).

22.1.3
Quality of androgenetic green plants

A prerequisite for using DH in breeding and research is production of completely homozygous green fertile regenerants. Doubling of the genome may occur spontaneously, as is common with rice (*Oryza sativa* L.) (Nishi and Mitsuoka 1969), via endoreduplication or endomitosis, or through chemical doubling (DeBuyser and Henry 1986). High spontaneous doubling would be advantageous for saving time and for avoiding use of toxic chemicals such as colchicine. Spontaneous doubling may vary according to the genotype as Finnie et al. (1991) reported for barley and Löschenberger et al. (1995) for wheat. Heterozygous regenerants could derive from unreduced microspores or from cell fusion. Gametoclonal variation may give rise to plants with chromosomal aberrations or minor mutations. Most of these changes are deleterious. Multiplicity of chromosome sets in the genomes of polyploid species may buffer against variability, which at the haploid or diploid level might be deleterious (Karp 1995). More gametoclonal variation could therefore be expected in the regenerants from hexaploid triticale than in the regenerants from diploid rye. The aims of this work were a) to study ploidy level of green rye and triticale regenerants derived from anther culture and b) to study the homogeneity of the rye regenerants using molecular markers. For ploidy analysis we used flow cytometry, which is an efficient method to study large numbers of samples.

22.2
Materials and methods

22.2.1
Plant material and anther culture

The 135 rye regenerants analysed included plants from four rye cultivars (Table 1). The regenerants of Amilo included also very small plantlets of poor vigor and albino plants. The triticale regenerants, 378 in total, were obtained from a study initiated to assess anther culture ability and impact of stress among 10 triticale varieties and cultivars representing the different sources of breeding material (Table 1). All seed was kindly provided by Boreal Plant Breeding. All mother plants were vernalised and grown in the greenhouse. Eight hundred anthers per line x treatment were plated on W14-medium (Ouyang et al. 1988) containing MS-vitamins and iron (Murashige and Skoog 1962), 9 μM 2,4-D, 2.3 μM KIN, 9% maltose and 3% Phytagel. The basic stress treatment applied for all ryes and for all triticales was 2 weeks cold pretreatment of tillers at +4°C. Additional stress was applied to some triticales and some were cultured on liquid medium supplemented with 100 mg/l of Ficoll® 400 (Table 1). For six triticale lines, tillers were also collected from field-grown material. After 6-10 weeks embryoids/calli were transferred into regeneration medium ($\frac{1}{2}$MS-salts, MS-vitamins and iron, 2.85 μM IAA, 4.65 μM KIN, 3% succrose, 3% Phytagel). After 4-5 weeks, green plants were potted in soil-peat mixture and ploidy level was determined from leaf protoplasts using flow cytometry. Chromosome count from root-tip samples was also done for 48 triticale plants.

22.2.2
Flow cytometry

Thc principle in determining DNA-ploidy with flow cytometry is to measure the fluorescence associated with each nucleus following its stoichiometric staining with a DNA-specific fluorochrome (Brown et al. 1991). The leaf samples for ploidy analysis (max. 50 mg of leaf per sample) were taken when the plantlets had reached vigorous growth in soil. The sample consisted of leaf segments from at least two tillers. For the analysis of the small rye plantlets and albino plants all leaf mass available (1-20 mg) was used. For isolation of protoplasts the method described by Hülgenhof et al. (1988) was followed. We used "Onozuka" Cellulase and agitated the mixtures gently over night. After rinsing and centrifugation the pellet with the protoplasts was suspended in 1 ml of solution A containing nuclear buffer I (Galbraith et al. 1983), 1 % Triton X-100, 5 μm/ml of β-mercaptoethanol, as described by Sangwan and Sangwan (1997), and 20 μm/ml of propidium iodide for staining. The samples were allowed to stain for 15 minutes after which they were centrifuged for 10 seconds at 14000 rpm. The supernatant was discarded and the pellet suspended in 200 μl of solution B, which consisted of solution A and 2.4 μl/ml RNAase (Arumuganathan and Earle 1991). The tubes were incubated at 37°C for 15 minutes and then returned to ice until analysis. The samples were analysed using a Becton Dickinson FACSort flow cytometer with an argon-ion-laser at excitation wave length of 488 nm. The results were analysed with the CELLQuest program. Barley (*Hordeum vulgare* L.) variety 'Sultan' (2C = 11.12 pg according to Bennett et al. 1982) was used as an external standard included in each batch analysed. The aim of this work was to study variation in DNA-ploidy to compare varieties and plant groups within the species.

Table 1. Ten winter triticales and four ryes and their origins. Treatments applied in triticale anther culture tests from which green regenerants were assessed are indicated with x.

		Origin of mother plants, media, stress treatment					
		Field	Greenhouse				
		Solid medium					Ficoll
	Origin	2 wk[a]	2 wk[a]	3 wk[a]	mannitol[b]	heat[b]	2 wk[a]
Rye							
Amilo	Poland	-	-	x	-	-	-
Jo02	Finland	-	-	x	-	-	-
Sibirskaja	Russia	-	-	x	-	-	-
Metsnaja Nizkobelnaja	Russia	-	-	x	-	-	-
Triticale							
Asmus	Germ.	x	x	x	x	-	-
LP4496.5.9	Germ.	x	-	x	-	x	-
Bor 96151	Finland	-	x	x	-	-	x
Bor 96152	Finland	-	x	x	x	-	-
Paljus	Belarus	x	x	x	-	x	-
Modul	Belarus	x	x	x	-	-	x
Ulrika	Sweden	x	-	x	x	-	-
Pika	Canada	-	x	x	-	-	-
Trillium	Canada	x	x	-	-	x	-
Wintri	Canada	-	x	x	-	-	x

[a] Length of cold pretreatment
[b] In addition to 2 weeks cold pretreatment

22.2.3 Analysis with molecular markers

Clones from A. Graner, Gatersleben, Germany and from the North American Barley Genome Mapping Project were used as probes in RFLP. Southern blotting and hybridization were performed according to the manufacturer's instructions for the Hybond N+ -membranes (Amersham LIFE SCIENCES, Version 2.0).

22.3
Results and Discussions

22.3.1
Spontaneous doubling among rye regenerants

The rye plants were analysed in five groups including three varieties and the groups of small plantlets and albino plants. The rye cultivar Jo02 yielded only five plants for the analysis (one C, one 2C and three 3C plants, Fig. 1), and they are not considered in the discussion of doubling rates. Spontaneous doubling for all the groups was similar, ranging from 65 % to 80 % (Table 2). These results are in accordance with earlier observations by Friedt et al. (1983), Wenzel and Foroughi-Wehr (1984) and Deimling and Flehinghaus-Roux (1996) on spontaneous doubling of approximately 70 % for rye. The highest spontaneous doubling observed in our studies was with the small plantlets, among which also one 3C and four 4C plants were found. Most variability in the DNA estimates for 2C plants was observed among the small plantlets as compared with the other rye groups. The mean 2C value for the five rye varieties which were analysed for comparison was 16.58 (s.e. ± 0.38) and for the 2C regenerants from the variety Amilo 16.24 (s.e. ± 0.59). For the small plantlets, the mean 2C value was 16.54 (s.e. ± 2.03). Chromosomal aberrations, indicated by the large variability in the DNA estimates and changes in ploidy may be common among the less viable regenerants and a reason for their reduced vigour. Their lack of vigour seemed not to be caused by haploidy. Attempts to nurse the small plantlets in another medium prior to potting into soil may be worthwhile. The group of albino plants included also two 3C plants. In this group the proportion of haploid plants was thus smallest of all, only 10 %. The albino plants had, on average, lower DNA content than the other groups (2C plants: 15.38, s.e. ± 0.36). One mixoploid individual was detected among the rye plants (Fig. 1).

22.3.2
Spontaneous doubling among triticale regenerants

A total of 378 green triticale plants was analysed using flow cytometry. Each cultivar was included in 2-3 different tests, excluding the field test. No treatment x cultivar interaction was evident for the doubling rate, with the exception of Paljus, with which the highest doubling rate was significantly higher than the mean for the other tests (65.6 % of 32 plants vs. 10.9 % of 46 plants, respectively). The data for spontaneous doubling from the different anther culture tests was pooled with the exception of the test investigating field grown mother plants, as doubling rates in that test were considerably lower than in the other tests (Table 2). The tested lines fell into three categories according to their tendency for spontaneous doubling: >60 % (Modul, LP4496.5.92, Paljus), 30-60 % (Wintri, Trillium, Bor 96152) and <30 % (Ulrika, Asmus, Bor151). Considering some cultivars, the regenerant numbers were quite low, and this was reflected in the wide confidence limits for those observed rates (Table 2). With this amphiploid cereal variability can be expected as one of its parents, rye, has shown high spontaneous doubling and for bread wheat, included in the rest of the genome, rather variable doubling rates have been reported (Ouyang et al. 1993: 0-35 %; Löschenberger et al. 1995: 26.1-54.8 %). Tetraploid durum wheat, also contributing to the triticale genome, has not been much studied for haploid production (Otani and Shimada 1995).

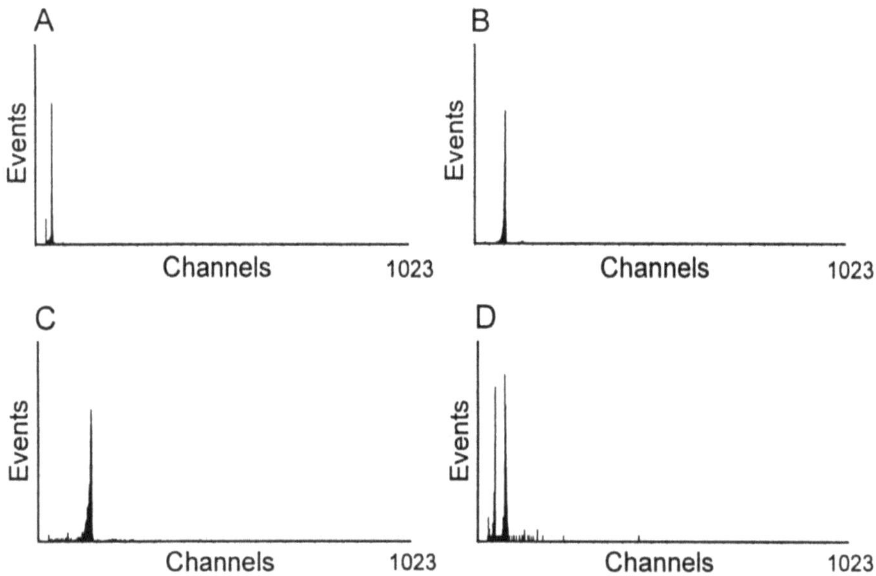

Fig. 1 A-D Flow cytometer peaks for rye. **A** Peak for a C plant; total events 2265. **B** Peak for a 2C plant; total events 8925. **C** Peak for a 3C plant; total events 2085. **D** C and 2C peaks for a mixoploid plant; total events 285

With triticale, the DNA estimates varied in both C and C2 groups. The results for the root-tip chromosome counts were compared with the corresponding DNA estimates. The DNA estimates for the C plants containing 21 chromosomes (24 of 48) had a mean of 19.8 pg, (s.e. ± 0.82) and the estimates for the 2C plants with 42 chromosomes (11 of 48) had a mean of 37.4 pg (s.e. ± 0.86). The remaining plants had chromosome numbers deviating from the normal n and 2n values, which was, in most cases, associated with an exceptionally low or high DNA estimate within the C or 2C category. Large deviations observed in these triticale data may thus reflect chromosomal aberrations, which merit further studies.

22.3.3
Effect of anther culture treatment on spontaneous doubling

The triticale cultivars were subjected to different treatments in anther culture (Table 1). Field growth conditions of the mother plants were associated with lower spontaneous doubling in comparison with the other anther culture treatments. The spring weather preceding anthesis was exceptionally cool. Therefore, this result was in contrast with our observation that prolonged cold pretreatment of the detached tillers increased spontaneous doubling. When treatments were compared in pairs, 3 week cold pretreatment was found to significantly increase spontaneous doubling in comparison with 2 weeks in cold (χ^2-test, $p<0.05$), heat treatment of anthers ($p<0.0001$) and field conditions ($p<0.0001$). Cold treatment has been found to increase spontaneous doubling also with wheat (DeBuyser and Henry 1986). In our study, mannitol starvation lead to an increase in spontaneous doubling in comparison with 2 or 3 weeks of cold, but this increase was not statistically significant. The effect of mannitol, which was also favourable in anther culture of triticale, should be further investigated in relation to spontaneous doubling. It seems possible to influence spontaneous doubling in a favourable manner through culture methods and stress treatment.

Table 2. Spontaneous doubling among anther culture derived green regenerants of ten triticale and four rye lines grown in the greenhouse

	green plants	% in 2C class[a]		confidence limits[b]
Rye				
Amilo	34	64.5	a	48.6-80.8
- small plants	21	61.9	a	41.1-82.7
- albino plants	20	80.0	a	62.5-97.5
Sibirskaja	14	75.0	a	45-100
Metsnaja				
Nizkobelnaja	8	71.4	a	47.8-95.1
Triticale				
Modul	15	80.0	a	59.8-100
LP4496.5.92	13	76.9	ab	54-99.8
Paljus	32	65.6[c]	abc	49.2-82.1
Wintri	39	48.8	bcd	33.5-64.1
Trillium	26	46.2	bcd	27-65.3
Bor 96152	16	37.5	cde	13.8-61.2
Ulrika	12	25.0	def	0.5-49.5
Asmus	64	17.2	ef	7.9-26.4
Bor 96151	85	9.5	f	3.2-15.8
Pika	3	-		-

[a] Numbers in the columns with the same letter are not significantly different, χ^2, $p>0.05$; separate for triticale and rye. For triticale results are pooled and field data are excluded

[b] Snedecor and Cochran 1967

[c] Only plants regenerated from heat treatment test

22.3.4
Homozygosity of doubled haploid rye regenerants

Regenerants of four rye mother plants, one Jo02 and 3 Amilo plants, were studied using RFLPs. Hybridization was successful using six probes. According to the RFLP patterns all 2C regenerants appeared different from the parent and homozygous for the investigated loci and thus they did not derive from unreduced microspores or fuced cells, which would cause heterozygosity in androgenetic regenerants. These results are considered preliminary and further investigation with rye and triticale will be done with F_1 crossing progeny.

Acknowledgements

The authors wish to thank Ms Leena Ramstedt and Ms Leena Lohermaa for excellent technical assistance.

References

Arumuganathan K, Earle ED (1991) Estimation of nuclear DNA content of plants by flow cytometry. Plant Mol Biol Rep 9: 229-233

Bennett MD, Smith JB, Heslop-Harrison JS (1982) Nuclear DNA amounts in angiosperms. Proc Royal Soc London B216: 179-199

Bernard M, Bernard S, Bonhomme H, Faurie C, Gay G, Jestin L (1996) Triticale research and breeding programmes in France: recent developments. In Guedes-Pinto H, Darvey N, Carnide VP (eds) Triticale: Today and tomorrow. Kluwer Academic Publishers, Dordrecht, vol 5, pp 643-647

Brown SC, Bergounioux C, Tallet S, Marie D (1991) Flow cytometry of nuclei for ploidy and cell cycle analysis. In Negrutiu I, Gharti-Chhetri GB (eds) A laboratory guide for cellular and molecular plant biology. Birkhäuser Verlag, Basel, pp 326-345

Charmet G, Branlard G (1985) A comparison of androgenetic doubled-haploid and single seed descent lines in Triticale. Theor Appl Genet 71: 193-200

DeBuyser J, Henry Y (1986) Wheat: Production of haploids, performance of doubled haploids and yield trials. In Bajaj YPS (ed) Biotechnology in agriculture and forestry, vol 2, Crops I, Springer-Verlag, Berlin Heidelberg, pp 73-88

Deimling S, Flehinhaus-Roux T (1996) Haploidy in rye. In: Jain SM, Sopory SK, Veilleux RE (eds) In vitro haploid production in higher plants. Kluwer Academic Publishers, Dordrecht, vol 4, pp 181-204

Finnie SJ, Foster BP, Chalmers KJ (1991) Genetic stability of microspore-derived doubled haploids of barley: a cytological, biochemical and, molecular study. Genome 34: 923-928

Flehinghaus-Roux T, Deimling S, Geiger HH (1995) Anther-culture ability in *Secale cereale* L. Plant Breed 114: 259-261

Fossati A, Fossati D, Kleijer G (1996) Triticale breeding in Switzerland. In Guedes-Pinto H, Darvey N, Carnide VP (eds) Triticale: Today and tomorrow. Kluwer Academic Publishers, Dordrecht, vol 5, pp 649-653

Friedt W, Lind V, Walther H, Foroughi-Wehr B, Züchner S, Wenzel G (1983) The value of inbred lines derived from *Secale cereale* x *S. vavilovii* via classical inbreeding and androgenetic haploids. Z Pflanzenzüchtg 91: 89-103

Galbraith DW, Harkins KR, Maddox JM, Ayres NM, Sharma DP, Firoozabady E (1983) Rapid flow cytometric analysis of the cell cycle in intact plant tissues. Science 220: 1049-1051

Immonen S, Anttila H (1996) Success in anther culture of rye. In Proceedings, EUCARPIA: International Symposium on Rye Breeding & Genetics. Vortr Pflanzenzüchtg 35: 237-244

Hülgenhof E, Weidhase RA, Schlegel R, Tewes A (1988) Flow cytometric determination of DNA content in isolated nuclei of cereals. Genome 30: 565-569

Karp A (1995) Somaclonal variation as a tool for crop improvement. Euphytica 85: 295-302

Löschenberger F, Pfosser M, Heberle-Bors E (1995) Genetic variability for the frequency of doubled haploid green plants is correlated with the ratio of green to albino plants in wheat (*Triticum aestivum* L.) microspore-derived plants. J Genet Breed 49: 37-44

Murashige T, Skoog F (1962) A revised medium for rapid growth and bioassays with tobacco tissue cultures. Physiol Plant 15: 473-497

Nishi T, Mitsuoka S (1969) Occurrence of various ploidy level plants from anther and ovary cultures of rice plant. Japan J Genet 44: 341-346.

Otani M, Shimada T (1995) Effect of synthetic medium on microspore-derived embryoid formation of tetraploid wheat species. Cer Res Commun 23: 345-350

Ouyang J-W, Jia S-G, Zhang, C, Chen X-D, Feng G-H (1988) A new synthetic medium (W14 medium) for wheat anther culture. Ann Rep Inst Genet Aca Sci, Beijing, pp 91-92

Ouyang J-W, Liang H, Jia S-E, Zhang C, Zhao T-H, He L-Z, Jia X (1994) Studies on the chromosome doubling of wheat pollen plants. Plant Sci 98: 209-214

Ryöppy PH (1996) Haploidy in triticale. In: Jain SM, Sopory SK, Veilleux RE (eds) In vitro haploid production in higher plants. Kluwer Academic Publishers, Dordrecht, vol 4, pp 117-131

Sangwan RS, Sangwan BS (1997) Flow cytometry for ploidy analysis in higher plants. In Abstracts COST-824 Gametic Embryogenesis. Working Group 2: Anther, microspore and ovule culture in monocots. Sjusjoen, Norway, p 34

Snedecor GW, Cochran WG (1967) Statistical methods. 6th ed. Ames, The Iowa State Univ Press, p 211

Snape JW (1986) Doubled haploid breeding, theoretical basis and practical applications. In Mujeeb-Kazi A, Sitch LA (eds) Review of advances in plant biotechnology 1985-88. 2nd International Symposium on Genetic Manipulation in Crops. Mexico, DF, Mexico and Manila, Philippines: CIMMYT and IRRI, pp 19-30

Wang G, Ji J, Wang Y-B, Hu H, King IP, Snape JW (1993) The genetic characterisation of novel multi-addition doubled haploid lines derived from triticale x wheat hybrids. Theor Appl Genet 87: 531-536

Wenzel G, Foroughi-Wehr B (1984) Anther culture of cereals and grasses. In Vasil IK (ed) Cell culture and somatic cell genetics of plants, vol 1, Academic Press Inc, New York, pp 311-327

23 Development and use of microspore culture applied to vegetable *Brassica oleracea* breeding

Chatelet*[1,2] P, Gindreau[1] K and Hervé[2] Y

23.1 Introduction

Most of the commercialized varieties of cauliflower and broccoli (two cultigroups belonging to the same species *Brassica oleracea*) are based on the F1 hybrid system which meets most accurately the growers needs for homogeneous marketable products, a reduced cropping period and a wide range of precocities. Among other advantages can be stressed the stability of these F1 varieties, better protection of the breeder's rights and of the market of seed-producing companies.

To insure obtaining suitably homogenous hybrid varieties, development of parental lines exhibiting a sufficient degree of homozygosity is an important requisite. These lines are most frequently obtained through selfing and step-wise selection of putative parents through at least six to eight generations; still, use of haplodiploidization methods can result in a quicker production of homozygous lines. This could be of importance in the case of winter cauliflower or when initiating a breeding programmeIn broccoli, doubled haploids (DH) can readily be obtained through anther or microspore culture and DH parental lines have already been developed (Farnham 1998). In cauliflower, doubled haploids can be also be produced through anther culture (Yang et al. 1992) and cauliflower DH have also been introduced into breeding schemes. However, and though improvements of the existing protocoles have been recently published (Stipic and Campion 1997), yields can still be low enough to hamper the routine use of the technique as a breeding tool. In contrast to oilseed *Brassica* - e.g. *B. napus* (Lichter 1982), *B. campestris* (Baillie et al. 1992), microspore embryogenesis in *Brassica oleracea* has met variable success (Lichter 1989; Cao et al. 1990); it has however more recently been shown to be sufficiently successful in most *B. oleracea* crop types to allow use of microspore culture for breeding purposes (Duijs et al. 1992).

Adaptation of the existing protocoles was undertaken to improve doubled haploids production and test the use of DH in INRA-CERAFEL vegetable

*[1] for correspondence: Bretagne Biotechnologie Végétale (BBV); St Pol de Léon; France
[2] INRA; Laboratoire d'Amélioration des Plantes Maraîchères; Plougoulm; France

Brassica breeding schemes. Will be presented here results concerning various factors affecting embryogenic response: genotypic influence and interaction with medium/heat stress treatments on microspore culture, together with the influence of initial developmental stage of the microspore population. Action of an activated charcoal suspension was also studied. A few characteristics of microspore-derived plants are also presented.

23.1.1
Influence of microspore population stage

Developmental stage is an important factor in the induction of microspore embryogenesis or callus formation since in many species, haploid plant obtention is strongly dependent on careful staging and selection of microspore population. In *Brassica oleracea* (including cauliflower and broccoli), anther culture is conducted preferably with anthers containing microspores at the late uninucleate stage (Yang et al. 1992). However, while in *Brassica napus* microspore culture results are optimal when culturing unicellular to early bicellular microspores (Pechan and Keller 1988), various authors (Cao et al. 1990; Duijs et al. 1992; Takahata et al. 1993) have stated that higher levels of microspore embryogenesis in *B. oleracea* could be obtained through the use of microspore populations containing high numbers of bicellular pollen. In our conditions, previous observations had led in contrast to the conclusion that the late uninucleate stage was to be used to obtain sufficient reactivity. Moreover bicellular pollen has been found inhibitory in *B. napus* microspore culture (Kott et al. 1988). In order to ascertain either hypothesis and to try and determine possible explanations for the discrepancy of these different results, experiments were performed testing the embryogenetic potential of uninucleate *versus* later-stage microspore populations.

23.1.2
Influence of activated charcoal

In the various works already published, cauliflower microspore response and embryo yields were always low (Duijs et al. 1992; Vyvadilova et al. 1993). In order to attempt increasing in our conditions the number of responsive genotypes of broccoli and cauliflower and/or the number of developping embryos, experiments have been performed including activated charcoal in the culture medium.

23.2 Material and methods

23.2.1 Material

In these studies, plants from breeding stock or cultivars of broccoli and cauliflower were grown in the greenhouse (10°C minimum as growing temperature, vernalization as required at a 9°C/5°C day/night temperature with a 8-hour light period). Microspore cultures were performed as in Burnett et al. (1992) with minor modifications. Buds were selected after careful staging using Vergne et al. (1987) DAPI staining procedure resulting in homogeneous populations of microspores at the late uninucleate stage (usually 10-20% bicellular pollen). Differences between treatments were assessed by counting globular and cotyledonary embryos one month after culture initiation. Treatments were applied to a minimum of ten plates per condition; when more than this number was inoculated, the total number of embryos was expressed as the value corresponding to the mean for ten Petri dishes.

Experiments were repeated at least once unless otherwise indicated and statistical analysis realized through a χ^2 test at a 95% confidence level.

23.2.2 Microspore culture

Subculture of microspore-derived embryos was first performed in 90 mm Petri dishes, the embryos being placed on filter-paper on top modified B5 medium containing 0.46 µM kinetin. Embryos were further subcultured on regulator-free modified B5 medium and acclimatized in the greenhouse. After expansion of the first one or two leaves developped ex vitro, plants were tested for their ploidy level through flow cytometry using standard protocoles (Fox and Galbraith 1990) against a diploid *B. oleracea* external reference. Microspore-derived plants characterized as diploids were grown and self-pollinated in the greenhouse. Pollination was performed at bud stage to overcome a possible pollen rejection as self-incompatibility is widely present in these *Brassica* cultigroups.

23.2.3 Genotypic interaction with inductive conditions

Cross experiments were realized combining the use of two media with two inductive thermic stresses. Media used were NLND (Takahata and Keller 1991) and BOL (NLN medium with half-concentrated minerals) when the temperatures treatments were 30°C (T0) and 32.5°C (T2) both applied for 24h and further culture at 25°C.

23.2.4
Influence of activated charcoal

Activated charcoal (AC) was added to the culture plates at the time of inoculation after Gland et al. (1988) procedure: AC (Merck n°2186) , previously co-autoclaved with agarose (Sigma, type IX) - 1.0 g charcoal to 0.5 g agarose in 100 ml water- was included to the microspore suspension on the basis of 0.1 ml per 5 ml dish.

23.2.5
Influence of microspore population stage

To test the influence of microspore population stage on embryogenic response, late uninucleate microspore populations (see above) were compared with larger-bud-derived samples containing higher but variable numbers of bicellular pollen. At isolation time, samples of the purified microspore populations were preserved in glutaraldehyde (2.5% in a 0.2 M phosphate buffer, pH 7.2) for further reference and actual composition of the initial microspore population was checked at the end of the experiment. For instance, 2 temperature treatments were applied to the two differently staged populations.

23.3
Results and discussion

23.3.1
Embryogenic response

As usual in cauliflower and broccoli microspore culture, embryo yields varied greatly with the various inoculations even within a given genotype.

23.3.2
Genotypic interaction with culture conditions

Using the media/heat stress levels combinations described above, genotypic reactions could be classified in four different types *i.e.*

- genotypes with a response specific to one thermic treatment but indifferent to medium composition
- genotypes responsive to one type of medium whatever the temperature stress level
- genotypes highly responsive only to a specific combination of medium and thermic treatment

- genotypes without any apparent specificity in the embryogenic response

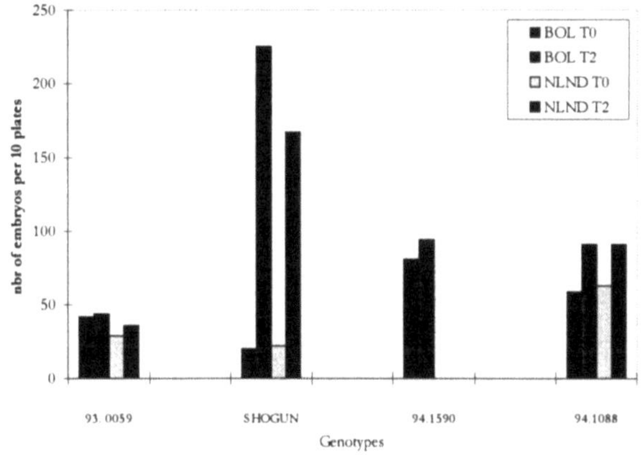

Fig. 1

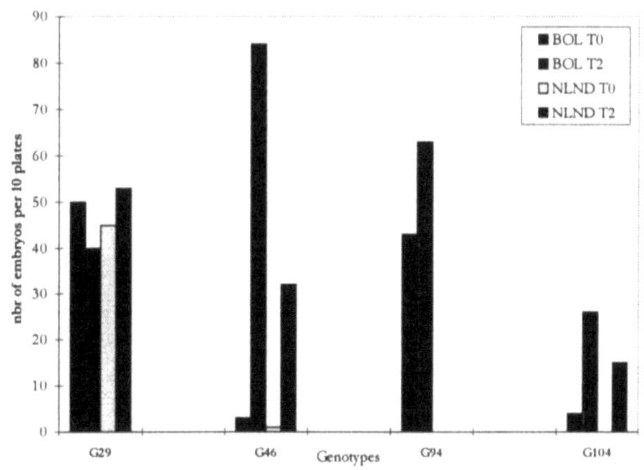

Fig. 2

Figs. 1 and 2. Classification of embryogenic response in broccoli (Fig. 1) and cauliflower (Fig. 2) genotypes to a combination of media (BOL/NLND) with two levels of heat stress (T0/T2). From left to right: - genotypes without any apparent specificity in the embryogenic response / genotypes highly responsive only to a specific combination of medium and thermic treatment / genotypes responsive to one type of medium (in these case BOL medium) whatever the temperature stress level/genotypes with a response specific to one thermic treatment but indifferent to medium composition

This distinction could be observed both in broccoli and in cauliflower genotypes (examples given in Figs. 1-2); in some cases, characterization could be observed to be strict (i.e. no embryo produced in non-optimal treatments) while in others, treatment differences (although statistically significant) only modulated embryogenic response. Such strict responses were surprising considering the seemingly modest differences between treatments (2.5°C variation for the heat stress and in the case of media, dilution of microelements only); however these cases have been observed in such a number of genotypes to be considered a general feature. All the different types of response could be observed (i.e. specificic response to BOL or NLN media, specific response to a 30 or 32.5 °C stress treatment). Considering this, it is quite likely that, had other combinations of treatments been used, more categories of response would have been identified.

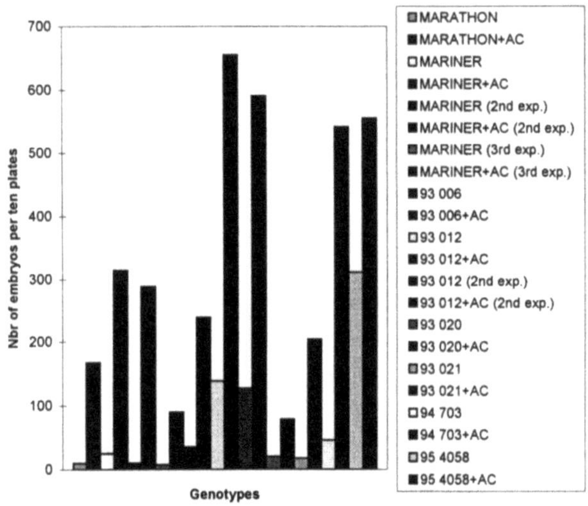

Fig. 3

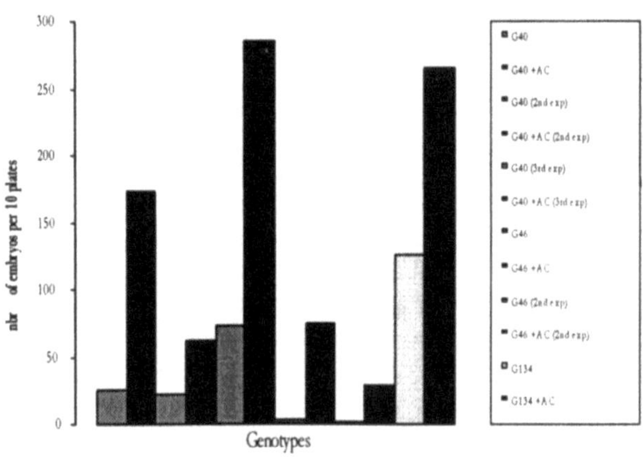

Fig. 4

Figs. 3 and 4. Effect of activated charcoal suspension on microspore-derived embryo obtention. Fig. 3. Case of broccoli genotypes- Fig. 4. Case of cauliflower genotypes. Results of activated charcoal (AC)-containing treatments indicated with striped bars

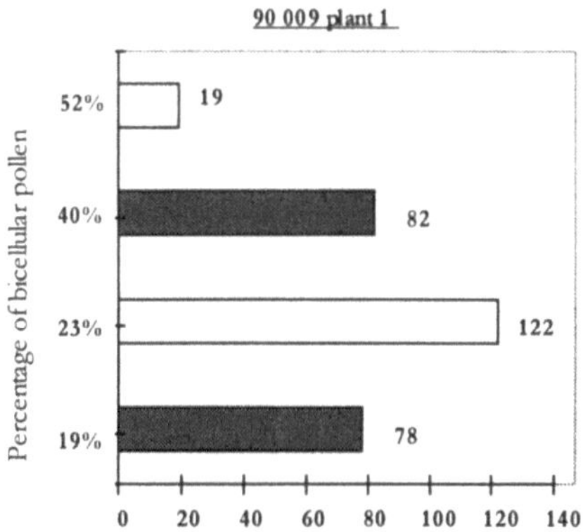

Fig. 5

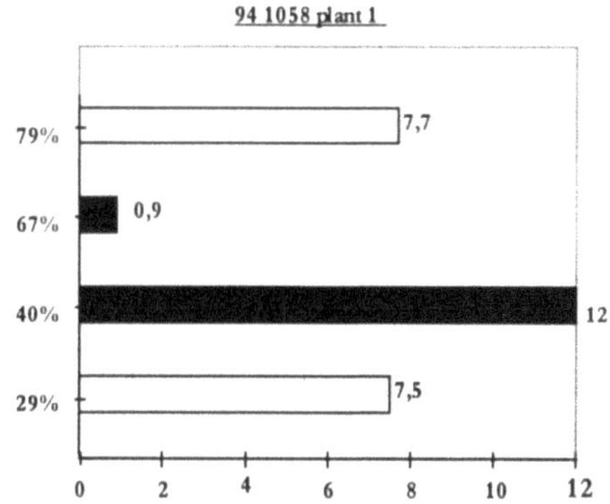

Fig. 6

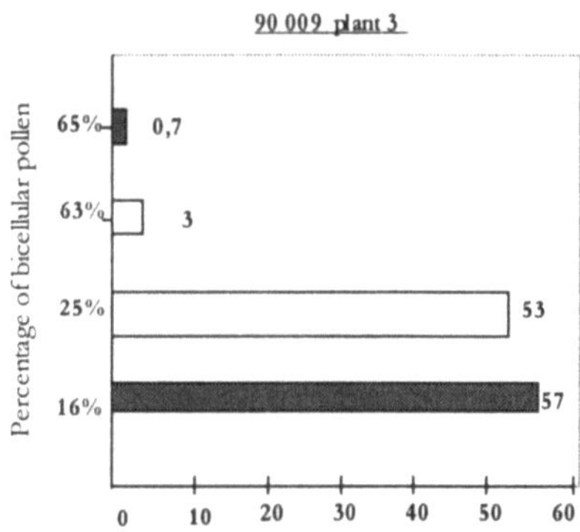

Fig. 7

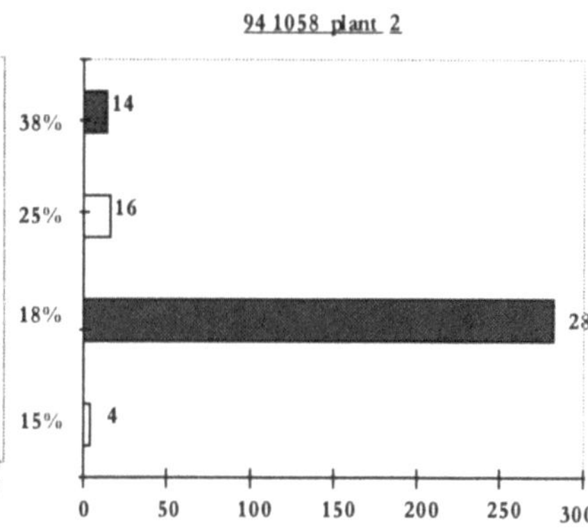

Fig. 8

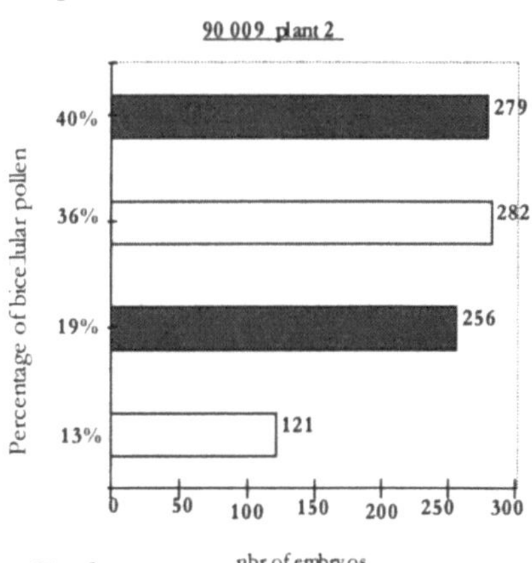

Fig. 9

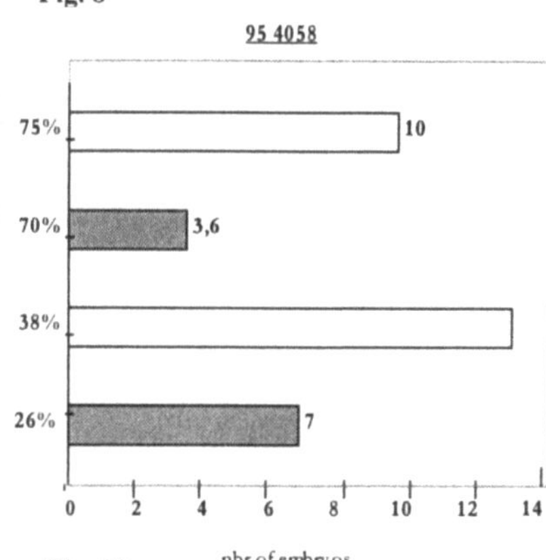

Fig. 10

Figs. 5 to 10. Effect of microspore developmental stage on embryogenesis rates in broccoli. Results from two repetitions (same plant): results from the same experiment are represented by empty or filled bars respectively

23.3.3
Influence of activated charcoal

Again, differences in embryo induction were observed depending of the genotype and between successive experiments within the same genotype. However, in all the experiments performed save two, addition of the charcoal suspension increased yields of microspore-derived embryos (MDE) (Figs. 3-4). Stimulation was clear for all genotypes tested (save one) and repetitions when performed confirmed AC positive action (though actual yields did fluctuate). In these cases, a highly significant augmentation of MDE numbers was observed. This is in contrast to Gland et al. (1988) findings in rapeseed microspore culture where AC did not increase the number of embryos formed but rather improved their subsequent development. In our case, a qualitative improvement was also notable as embryo morphology was closer to normal in the presence of AC. No clear effect of AC on embryo conversion to plant could be determined although embryos formed in AC-containing media showed in some cases a better development, perhaps only related to a better initial physiological state.

No definite explanation of charcoal effect is generally offered by authors having observed positive action of this compound. Activated charcoal could be expected to act through adsorption of ethylene which is known to have adverse effect on pollen embryo induction (Dunwell 1979; Yang et al. 1992). Mensuali-Sodi et al. (1993) confirmed AC retention of biotic and abiotic ethylene in their cultures although stimulatory efffect was in fine rather attributed to adsorption of phenolics. In anther cultures of *Datura* and *Anemone*, Johanson (1983) also demonstrated that adsorption of phenolics could explain AC stimulation of embryo production. But as in the case of *Brassica*, microspore cultures are essentially devoid of somatic cell fragments, very low levels only of phenolic compounds could be expected in the microspore suspension.

Medium pH modification could be suspected but AC suspension did not modify measurably the set value after a 4-hour continuous agitation. Possible interaction of AC with a medium modification by the microspores after some time in culture has however not been investigated. AC has also been shown to adsorb some iron chelates included in the medium (Heberle-Bors 1980) but not Fe-EDTA (used in the NLN medium).

The most probable answer might be by Gland et al. (1988) who suggest that AC acts by removing toxic substances released by non-reactive microspores thus allowing a larger number of embryogenic cells to develop. This explanation had also been offered by Kott et al. (1987) though AC did not exert any positive effect in their report (probably due to a different protocole as AC was not combined with agarose). Experiments are in progress to check whether AC would also have removed some inhibitory factor initially present in the medium.

No detrimental effect of charcoal addition has been observed so far; in the only instance where microspores cultured in charcoal-containing medium responded poorly in comparison to the control, repetition of the experiment showed that this adverse effect was not confirmed.

As already shown in *Brassica napus* (although in a somewhat different way), activated charcoal played a stimulatory role on microspore embryogenic response in the genotypes of broccoli and cauliflower tested. In contrast to other reactivity factors in *B. oleracea* microspore culture, stimulation by AC has not proved so far genotype-dependant with one possible exception. This protocole modification may very well allow obtention of doubled haploid plants from previously poorly or non-responsive genotypes.

23.3.4
Influence of microspore population stage at the time of inoculation

In many cases, though careful bud size selection was performed, resulting microspore population was found to contain a higher number of bicellular pollen than the forecast 20% upper limit. This might be related to the less dense younger-stage microspores being preferentially eliminated in the centrifugation steps of the isolation process. This could also account for the variation of composition of the microspore populations between successive experiments. In broccoli (Figs. 5-9), some genotypes responded better when using early stage microspores -e.g. 94 1058 plants 1 (Fig. 6) and 2 (Fig. 8), 90 009 plants 1 and 3 (Figs. 5-7)- whereas others reacted preferably when over 30% bicellular pollen was present (Fig. 9). Differences of reaction within the same genotype 90 009 can be accounted for by the genetic variability of this segregating line. A third type of response was indifference (within the tested range) to the ratio uni/bicellular pollen in the embryogenic yield (e.g. genotype 95 4058 in Fig. 10) although in this case embryo yields could have been too low to allow clear distinction. In cauliflower, and although actual proportion of bicellular pollen was not determined, the two experiments realised showed also a better response for younger stage microspores (data not shown).

Surprisingly however, optimal thermal treatment could be different according to initial microspore population composition (Figs. 11-12); for instance in Fig. 12, optimal temperature treatment was T0 for early-stage microspores and T2 for later staged population; in both cases observed, microspores at the later stage best reacted at the highest temperature treatment. This is consistent with Binarova et al. (1997) findings where *B. napus* microspores at an advanced, usually non responsive stage, could be drawn into embryogenic development through a high (40°C) temperature stress. However, in our case, difference of temperature levels were small in comparison, probably indicating a high sensitivity of *B. oleracea* microspores to small shifts of temperature.

The results provided here show that developmental stage is yet another factor for microspore embryogenic response which is under genotypic influence.

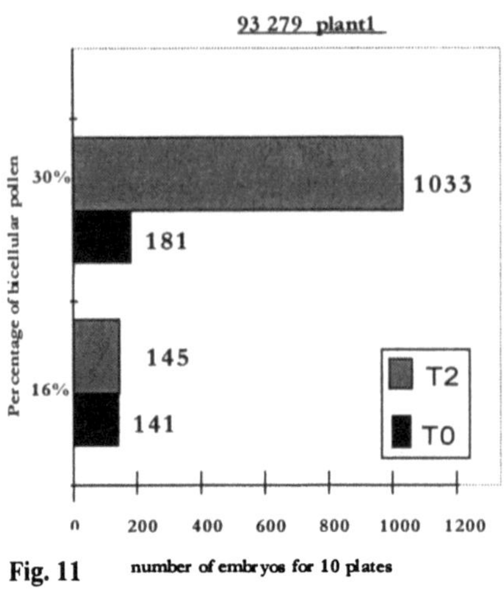

Fig. 11

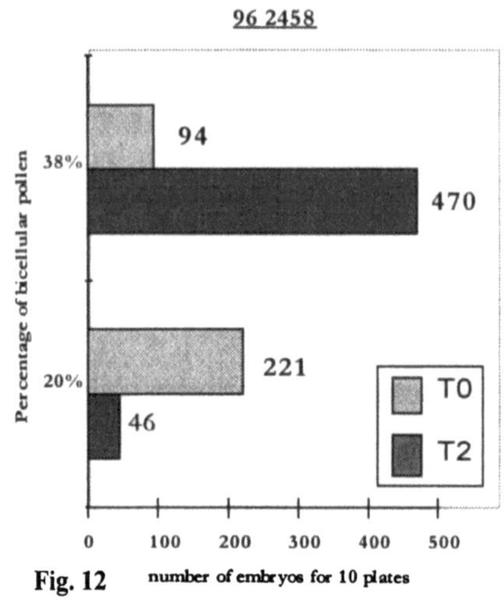

Fig. 12

Table 1

	Total	1C	2C	3C	4C	6C
Cauliflowers	1566	16%	56%	8%	19%	1%
Broccoli	7579	19%	55%	4%	21%	1%

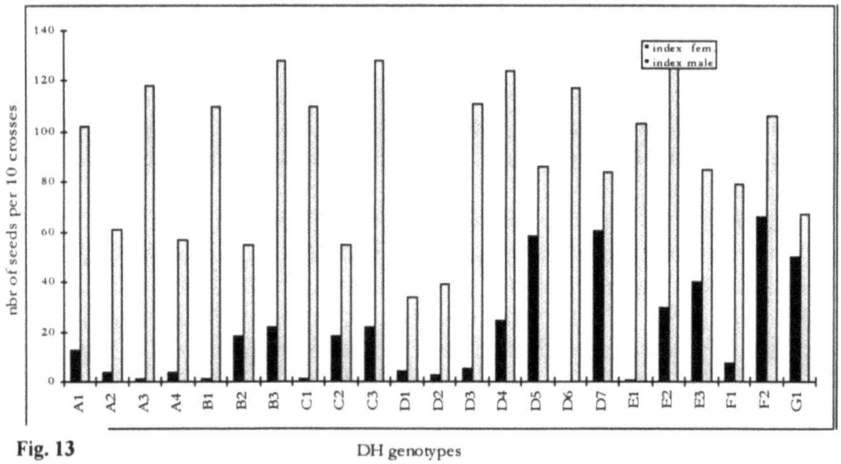

Fig. 13

Figs. 11-12. Interaction of microspore developmental stage with inductive temperature stress treatments (T0 and T2 as in text. **Table 1.** Distribution of ploidy levels of microspore-derived embryos from broccoli and cauliflower genotypes. **Fig. 13.** Evaluation of male and female fertility in doubled haploids of broccoli. A1, A2, B1, B2: DH individuals derived from mother plants A, B etc. Index established as number of seeds per 10 crosses using DH as male or female parent. Unpublished data by JP Lesaint and B Le Jeune

23.3.5
Embryo-to plant conversion and ploidy of microspore-derived plants

Plant regeneration from microspore-derived (MD) embryos was also observed to differ greatly in relation to the mother plant genotype (from less than 10% to over 70%). Plantlet regeneration proceeded either through development of the embryo apical meristem or through adventitious shoot growth. No histological studies were performed to determine whether the latter arose from preexisting meristems and/or meristem neoformation.

Flow cytometry allowed quick and reliable determination of the ploidy level of MD plants, resulting in a classification as shown in Table 1. Pooled results of all genotypes tested are presented since distribution of ploidy levels within microspore-derived progenies varied again depending on the mother plant genotype. It was usually, but not always, constant in the microspore-derived progeny for a given genotype. Although higher numbers of microspore-derived embryos have been obtained in broccoli, comparison with cauliflower showed that the distribution of diploids and tetraploids was not significally different for broccoli or cauliflower MD plants. The number of spontaneous doubled haploids is higher than the proportion generally observed in *B. napus* MD plants but similar to the DH percentage observed in anther-derived broccoli plantlets (Farnham 1998). A χ^2 homogeneity test (at a 95% confidence level on the untransformed data) however shows that the general distribution is not identical, due to a difference in the repartition of haploid and triploid plants (although the total percentage of 1C plus 3C plants is similar in both cultigroups). The flow cytometry protocole did not allow identification of aneuploid plants: such plants would be expected to occur in low numbers only (Chauvin et al. 1993; Farnham 1998); since no heterogeneity could be observed in self-pollinated DH progenies, it can be assumed that aneuploids formed (if any) had major fertility deficiencies and did not set seed.

23.3.6
Fertility of DH plants

Of the doubled haploids obtained and self-pollinated, the majority set seed, since 70% (broccoli) to 80% (cauliflower) yielded at least one seed, allowing further testing of the DH genotypes. Male and female fertility still fluctuated greatly depending on the original genotype and of each DH specimen (Fig. 13). As a rule, male fertility was less affected while female fertility showed greater variation : some fertility index were quite low (some plants being female sterile) while other DH individuals showed a fertility high enough to allow their use as female lines.

23.4 Conclusion

In most results presented here, genotype plays a key role in both embryogenic response and characteristics of MDE in broccoli and cauliflower microspore culture. Although this is not an uncommon feature of *in vitro* plant culture, genotypic specificity in these two groups seems extremely high since, save for activated charcoal effect, all factors tested in our conditions showed clear genotype-dependent action. This is of importance in breeding programmes where lines genetically quite different are maintained and combined, thus increasing the genetic variability of response to androgenetic techniques. This should emphasize the need of designing a protocole inducing average levels of response in most cultivars, bearing in mind that some genotypes might demand specific inductive conditions or combinations thereof to insure DH obtention. Although rather large numbers of DH lines have been obtained in both cultigroups, actual contribution of such genetic material to the breeding programmes has not been fully determined yet. Genetic and agronomic value of this perfectly homozygous breeding material is currently under study to assess further its characteristics.

Aknowledgements

The authors gratefully acknowledge the financial support of the GIP "Bretagne Biotechnologie Végétale" through its Director, Dr S Mabeau as well as of Comité Economique régional fruits et légumes of Brittany (CERAFEL); this work was partly funded by European Union (FEDER), by the Conseil Régional de Bretagne and Conseil Général du Finistère. Financial support of ONIFLHOR and French Ministry of Agriculture (MAPA) is also gratefully aknowledged. Initial plant material has been kindly provided by Dr V. Ruffio, J.P. Le Saint, C. Laviec (CERAFEL) as well as B Le Jeune (INRA, Plougoulm) and H. Bellis (INRA Rennes-Le Rheu). We are very grateful to Dr Renard and especially MO Lucas (INRA, Le Rheu) for introduction to microspore culture techniques and sharing of protocoles. Thanks are further extended to J Brochec and D Cloarec for excellent lab work and plant tending and to J Monfort for ploidy analyses.

References

Baillie AMR, Epp DJ, Hutcheson D, Keller WA (1992) In vitro culture of isolated microspores and regeneration of plants in *Brassica campestris*. Plant Cell Rep 11: 234-237

Binarova P, Hause G, Cenklovà V, Cordewener JHG, Van Lookeren Campagne MM (1997) A severe heat shock is required to induce embryogenesis in late bicellular pollen of *Brassica napus* L. Sex Plant Reprod 10: 200-208

Burnett L, Yarrow S, Huang B (1992) Embryogenesis and plant regeneration from isolated microspores of *Brassica rapa* ssp *oleifera*. Plant Cell Rep 11: 215-218

Cao MQ, Charlot F, Doré C (1990) Embryogenèse et régénération de plantes de chou à choucroute (*Brassica oleracea* L. *ssp capitata*) par culture *in vitro* de microspores isolées. CR Acad Sci Paris, 310, Sér III : 203-209

Chauvin JE, Yang Q, Le Jeune B, Herve Y (1993) Obtention d'embryons par culture d'anthères chez le chou-fleur et le broccoli et évaluation des potentialités du matériel obtenu pour la création variétale. Agronomie 13: 579-590

Chuong PV, Deslaurier C, Kott LS, Beversdorf WD (1987) Effects of donor genotype and bud sampling on microspore culture of *Brassica napus*. Can J Bot 66: 1653-1657

Duijs JG, Voorips RE, Visser DL, Custers JBM (1992) Microspore culture is successful in most crop types of *Brassica oleracea* L. Euphytica 60: 45-55

Dunwell JM (1979) Anther culture in *Nicotiana tabacum*: the role of the culture vessel atmosphere in pollen embryo induction and growth. J Exp Bot 66: 419-428

Farnham MW (1998) Doubled-haploid broccoli production using anther culture: effect of anther source and seed set characteristics of derived lines. J Amer Soc Hort Sci 123 (1): 73-77

Fox MH, Galbraith DW (1990) Application of flow cytometry and sorting to higher plant systems. In Flow Cytometry and Sorting,Wiley-Liss Inc, pp 633-650

Gland A, Lichter R, Schweiger HG (1988) Genetic and exogenous factors affecting embryogenesis in isolated microspores cultures of *Brassica napus* L. J Plant Physiol 132: 613-617

Heberle-Bors E (1980) Interaction of activated charcoal and iron chelates in anther cultures of *Nicotiana* and *Atropa belladonna*. Z Pflanzenphysiol 99: 339-347

Johansson L (1983) Effects of activated charcoal in anther cultures. Physiol Plant 59: 397-403

Kott LS, Polsoni L, Ellis B, Beversdorf WD (1988) Autotoxicity in isolated microspore culture of *Brassica napus*. Can J Bot 66: 1665-1670

Lichter R (1982) Induction of haploid plants from isolated pollen of *Brassica napus*. Z Pflanzenphysiol 105: 229-237

Lichter R (1989) Efficient yield of embryoids by culture of isolated microspores of different *Brassicaceae* species. Plant Breed 103: 119-123

Mensuali-Sodi A, Panizza M, Serra G, Tognoni F (1993) Involvement of activated charcoal in the modulation of abiotic and biotic ethylene levels in tissue cultures. Sci Hort 54: 49-57

Pechan PM, Keller WA (1988) Identification of potentially embryogenic microspores in *Brassica napus*. Physiol Plant 74: 377-384

Stipic M, Campion B (1997) An improved protocol for androgenesis in cauliflowers (*Brassica oleracea* var *botrytis*). Plant Breed 116: 153-157

Takahata Y, Keller WA (1991) High frequency embryogenesis and plant regeneration in isolated microspore culture of *Brassica oleracea* L. Plant Sci 74: 235-242

Takahata Y, Takani Y, Kaisuma N (1993) Determination of microspore population to obtain high frequency embryogenesis in broccoli (*Brassica oleracea* L.). Plant Tiss Cult Lett 10: 49-53

Vergne P, Delvallee I, Dumas C (1987) Rapid assessment of microspore and pollen development stage in wheat and maize using DAPI and membrane permeabilization. Stain Technol 62: 299-304

Yang Q, Chauvin JE, Hervé Y (1992) A study of factors affecting anther culture of cauliflower (*Brassica oleracea* var *botrytis*). Plant Cell Tiss and Org Cult 28: 289-296

Subject index

Springer and the environment

At Springer we firmly believe that an international science publisher has a special obligation to the environment, and our corporate policies consistently reflect this conviction.

We also expect our business partners – paper mills, printers, packaging manufacturers, etc. – to commit themselves to using materials and production processes that do not harm the environment. The paper in this book is made from low- or no-chlorine pulp and is acid free, in conformance with international standards for paper permanency.

MIX
Papier aus verantwortungsvollen Quellen
Paper from responsible sources
FSC® C105338

If you have any concerns about our products,
you can contact us on
ProductSafety@springernature.com

In case Publisher is established outside the EU,
the EU authorized representative is:
Springer Nature Customer Service Center GmbH
Europaplatz 3, 69115 Heidelberg, Germany

Printed by Libri Plureos GmbH
in Hamburg, Germany